The Big Argument

Michael J. Westacott is a freelance editor living in Cairns, Australia.

John F. Ashton, PhD (epistemology, University of Newcastle), is a scientist and author. He is the editor of *The God Factor: 50 Scientists and Academics Explain Why They Believe in God* (HarperCollins, 2001) and *In Six Days: Why 50 Scientists Choose to Believe in Creation* (Strand, 2003).

The Big Argument

Twenty-four scholars explore why science, archaeology and philosophy haven't disproved God

Edited by Michael J. Westacott and John F. Ashton

STRAND PUBLISHING
Sydney

The Big Argument: Twenty-four scholars explore why science, archaeology and philosophy haven't disproved God
Copyright © 2005 Michael J. Westacott & John F. Ashton.
Copyright of the individual essays remains the property of the contributors.

First published 2005 by Strand Publishing.

Distributed in Australia by:
Family Reading Publications
B100 Ring Road
Ballarat Victoria 3352
Phone: (03) 5334 3244
Fax: (03) 5334 3299
Email: enquiries@familyreading.com.au
Web: www.familyreading.com.au

Quotations marked NIV are taken from *The Holy Bible, New International Version*®. Copyright © 1973, 1978, 1984 by the International Bible Society. Used by permission of Zondervan Publishing House. All rights reserved.
Quotations marked NKJV are taken from *The Holy Bible, New King James Version*. Copyright © 1982 by Thomas Nelson, Inc. Used by permission. All rights reserved.
Quotations marked NASB are taken from *The New American Standard Bible*. Copyright © 1960, 1962, 1963, 1968, 1971, 1972, 1973, 1975, 1977, 1995 by The Lockman Foundation. Used by permission. (www.Lockman.org).
Quotations marked NRSV are taken from *The New Revised Standard Version Bible*. © 1989, Division of Christian Education of the National Council of the Churches of Christ in the United States of America. Used by permission. All rights reserved.
Quotations marked ASV are taken from *The Holy Bible*, American Standard Version.
Quotations marked KJV are taken from the *Holy Bible*, King James Version.

This book is copyright. No part of this publication may be reproduced, stored in a retrieval system, or transmitted in any form or by any means—electronic, mechanical, photocopy, recording, or any other—except for brief quotations for printed reviews, without prior permission of the publisher.

National Library of Australia
Cataloguing-in-Publication data
The big argument : twenty-four scholars explore why science, archaeology and philosophy haven't disproved God.

　ISBN 1 876825 83 9.

　1. Apologetics. I. Westacott, Michael J. II. Ashton, John Frederick.

239

Edited by Owen Salter
Cover design by Joy Lankshear
Typeset by Midland Typesetters, Maryborough, Victoria
Printed by Griffin Press, Netley, SA

Contents

Introduction	vii

Part 1 God, Science and Philosophy

1 Has Science Disproved God? *Barry L. Whitney*	3
2 Fingerprints of the Divine Around Us *Danny R. Faulkner*	18
3 Where Did the Universe Come From? *Kenneth E. Himma*	34
4 Design by Information *Werner Gitt*	48
5 The Human Body: Evidence for Intelligent Design *Frank J. Sherwin*	75
6 Design in Nature: Evidence for a Creator *Ariel A. Roth*	99
7 The Scientific Case for Creation *George Javor*	123
8 A Question of Biology *David Catchpoole*	133
9 The Geological Evidence for Creation *Andrew A. Snelling*	150
10 Where Do Thoughts Come From? *Charles Taliaferro*	184

11	The Question of Moral Values *Steven B. Cowan*	195
12	The Problem of Evil *Jon Paulien*	212
13	Who Is God? What Is He Like? *Steven Thompson*	238
14	Will the Real God Please Stand Up? *Eric Svendsen*	245

Part 2 God, History and the Bible

15	Can the Bible Be Relied On? *Stephen Caesar*	269
16	Historical Evidence for the Biblical Flood *Jerry Bergman*	290
17	Archaeological Evidence for the Exodus *David K. Down*	306
18	The Historical Reliability of the Old Testament *Paul Ferguson*	326
19	Archaeology and the Reliability of the New Testament *John McRay*	352
20	What about the Bible's Scientific Reliability? *Timothy G. Standish*	368
21	Amazing Biblical Prophecies That Came to Pass *William Shea*	374
22	The Messianic Prophecies Fulfilled in Jesus *Arnold G. Fruchtenbaum*	400
23	The Evidence for Jesus' Resurrection *Michael R. Licona*	428

Conclusion

24	The Absurdity of Life Without God *Phil Fernandes*	450

The Contributors	469
Notes	473

Introduction

Imagine you are a brain surgeon in an operating theatre surrounded by some of the latest and most sophisticated medical equipment available. You are leading a medical team operating on a ten-month-old baby girl. Her brain has been growing faster than her skull, and the increasing pressure on the brain will eventually kill the child if she does not have this operation. You are about midway through the procedure—part of the skull has been removed—when your attention is distracted by the rapidly moving fingers of the anaesthetist adjusting his controls. Your eyes meet his as the words 'the patient's blood pressure is falling' are pronounced. The team pauses momentarily, then the anaesthetist urges, 'The patient is in cardiac arrest'.

With the help of your assistant you hurriedly pull the rolled-back scalp over the gaping wound and secure it with clips so the child can be rolled over for cardiopulmonary resuscitation (CPR). You and your highly trained and experienced team take turns. Five minutes of CPR pass but there has been nothing but a flat line on the EKG monitor. Ten minutes, then fifteen minutes of CPR and still no response. Twenty minutes and there has been nothing but a flat line on the EKG screen. At that moment someone asks you, 'Do you want to call it off?' You know the medical science. There has been no activity in the brain for twenty minutes. Were the child now to revive, she should be so severely brain damaged as to be beyond any hope of rehabilitation to live a meaningful life. Around you is the best life-saving equipment human minds have developed. Within arm's reach are human minds trained at some of the best universities in the world. What do you do now?

You have reached the limit of science, but you have not reached the limit of God. Do you know God?

This scenario actually happened to one of the world's leading brain surgeons, Dr Ben Carson, director of paediatric neurosurgery at Johns Hopkins University Hospital in Baltimore.[1] Beautiful little baby Shannon had been born to loving parents who had been trying for five years to have a baby. All went well at first, but later changes in her behaviour were diagnosed as a consequence of craniosynostosis and she needed a cranial expansion, a major operation. When Dr Carson was performing the surgery, events similar to those described above took place. The baby went into cardiac arrest and CPR was begun. After twenty minutes there was only a flat line on the EKG monitor and Dr Carson was asked if he wanted to stop the CPR—the baby was dying.

Dr Carson knew the science, but he had also been exposed to the message of the Bible in his youth and had come to know God. Instead of bowing to scientific knowledge he acknowledged God's power and decided to continue with the CPR while praying for God to intervene. The seconds ticked into minutes, and it was only after twenty-four minutes of CPR that there was a blip on the EKG screen. A few moments later Shannon began breathing.

Shannon was rushed to the paediatric intensive care unit, but when the anaesthesia wore off her pupils were still fixed and dilated, indicating no neurological response—a sign she was probably brain dead. Dr Carson again decided not to give up and continued to pray to God for the child to recover. It was just under twenty-four hours later that the baby's pupils began reacting, and Dr Carson reports that everyone in the intensive care unit team was amazed at the recovery.

But then the next day Shannon began having trouble breathing, and it was found that segments of her lungs had ruptured and were bleeding. The pulmonary specialist physician told Dr Carson that the child had less than a five per cent chance of surviving until morning. Again Dr Carson prayed for the child to be saved. Just after 3.00 a.m. he was phoned by intensive care unit staff to say that the last functioning segments of Shannon's lungs had just ruptured and the child would die in the next few minutes.

When Dr Carson later contacted the hospital to talk to the parents, however, he found that not only had the child not died but somehow her lungs had actually improved and she was rallying. In fact, over the next few days her lungs healed to such an extent that she was taken off the ventilator. When the rehabilitation specialists and the paediatric neurologist examined Shannon, however, they told her parents that their little daughter was so brain damaged that she was blind and deaf and would never be able to do enough for herself to justify any rehabilitation.

But these highly educated specialists turned out to be wrong for they had ignored the power of God. Prayers for Shannon continued, and three days later she could see and hear. Dr Carson reports that she recovered to become a completely normal child.

There was no medical explanation for Shannon's healing. To Dr Carson, God had brought the child back to life twice and then restored her sight and hearing, and these events were a powerful example of God doing today the same sorts of miracles the Bible describes.

Dr Carson's personal experiences of answers to prayer highlight the purpose of this book. Its aim is to bring together some of the powerful and corroboratory evidence that we now have available to us from archaeology, history, science, philosophy and theology which supports the validity and extreme importance of the Bible message. It is this message that brings us to a knowledge of God and why we are here.

Have you ever wondered how the universe, our planet and we ourselves came to be? Have you ever thought to yourself 'Is there a purpose to life?' or 'Is there a higher power—God—who actually cares about me?' Probably most people ask these sorts of questions but then get stuck on the follow-on puzzles, like 'Where did God come from?' or 'Why should there be anything at all?' Finding answers to these questions is not an easy task, and from my experience people usually put them into the 'too hard' or 'it doesn't really matter' baskets.

Many scientists and scholars believe that life formed by chance on our planet, which also formed by chance when matter and the universe came into existence as a result of some type of 'big bang' explosion of space and energy. If what they assert is correct then life has no meaning or purpose, and at best we should try to enjoy ourselves as much as we can before we die. On the other hand, there are also many scientists and other scholars who, like Ben Carson, believe there is a God and a purpose to our existence based on the teachings of the Bible, which records the experiences of men and women who spoke with God. If what these believers assert is correct, then the choices we make during our life do affect what happens after we die and there is much more to our existence than just this life on earth.

The purposeless evolutionary view is based on measurements which scientists have made on matter, energy, space and time and is then formulated on the basis of their own mental reasoning. Not surprisingly, other scientists reason from the same scientific facts and form a non-evolutionary view. The biblical view is based on the Bible record, which can be checked for accuracy against other historical records and archaeological findings. However, there are three additional claims that separate the Bible from other sources.

The first of these is the claim that the Bible gives an eyewitness account of creation and the time period up to the global flood. The eyewitness was the creator himself, God, who spoke to men and women of Bible times and revealed to them the past.

The second claim is that of prophecy. In the Bible we find

dozens of accounts where the future was described in detail well before it happened. The prophecies formed part of the message that the writers of the Bible received in visions or when God spoke to them. These prophecies were usually given to show the readers of the Bible that God had indeed given the message and that this message could be trusted, just as the events it foretold came to pass.

Some of these prophecies are amazing in terms of their detail and specificity. For example, one prophet named Isaiah, whose ministry in Jerusalem lasted from about 745 to 685 BC, prophesied that a future king of Jerusalem would be captured and the city plundered by the king of Babylon. He then went on to say that God would raise up a great conqueror with the specific name of Cyrus, who would come from the north and east and conquer Babylon, free the exiled Jews and authorise the rebuilding of Jerusalem and its temple.

Cyrus the Great did come from north and east of Babylon to conquer it in 539 BC. And he did release the Jews from their Babylonian exile and authorise the rebuilding of the destroyed temple in Jerusalem. Let's put this into perspective. Since there were about 146 years from the time of Isaiah's death to the fall of Babylon, his prophecy and specific naming of Cyrus was a century-and-a-half ahead of its time and more than a century before Cyrus was even born and named. The actions of Cyrus are well attested from a variety of ancient sources, confirming the accuracy of Isaiah's prophecy.

Isaiah was given many other prophecies including details of the Messiah's death on the cross which occurred more than 700 years after Isaiah's death. There are many other prophecies which demonstrate the unique divinity of the God of the Bible, who alone knows the future.

The third claim that separates the Bible not only from scientific theories but also from all other faiths is the assertion that the highly credible person Jesus Christ was actually God, who came for a short time to live among humanity in human form. No other book than the Bible asserts such a powerful claim—a claim which was confirmed in public before thousands of witnesses by the performance of miracles which were subsequently documented. In fact, it was the public claim by Jesus that he existed before Abraham, the founder of the Jewish race who lived two millennia earlier, that outraged the Jewish leaders to the point where they determined to have Jesus killed.

There is convincing evidence for all these important claims, which together make the Bible stand out from all other resources as the most important source of answers that we have for some of the most difficult and perplexing questions about life.

In the chapters that follow, scholars educated at universities around the world present evidence from archaeology, history, science, philosophy and theology that substantiates the Bible message and that provides answers to such questions as these: Has science disproved God? Where did the universe come from? What evidence is there for creation as opposed to evolution? Where do thoughts and moral values come from? Why is evil permitted? Who is God and what is he like? Can the Bible be relied on? What is the evidence from history and archaeology? What about the scientific reliability of the Bible? What is the evidence that the Bible contains accurate prophecies? What is the evidence that the resurrection of Jesus did occur?

These essays give a rare holistic insight into the accuracy, trustworthiness and importance to our daily lives of the Bible, which outlines the history and status of the relationship between God and humankind. They show that science has not disproved God, but rather, on the basis of the evidence we have available to us, that to live life without God is absurd.

John F. Ashton

Part 1

God, Science and Philosophy

1

Has Science Disproved God?

Barry L. Whitney

The vast majority of human beings have always believed in God. It is a testimony to the depth of this belief that it has persisted despite a new and aggressive challenge, one which has become increasingly prevalent in contemporary society. This challenge presents itself as science, but it is in fact a perversion of modern science more properly called 'scientism'.

By 'scientism' I refer to the ideology based on an antireligious bias which presumes, without proof, that human rationality and the empirical method of science are the sole means for determining what is true about reality. Ironically, there is no scientific proof for this assumption. Scientism fails to consider seriously whether there are other means for discovering truth, ignoring what theologians have argued for centuries and religious believers intuitively know: that besides empirical evidence (and rationality based largely on reflection on the empirical evidence), truth is revealed by God to our inner experience and accepted in faith. Human rationality and empirical verification, in themselves, are not the means for determining God's existence since, of course, God is not a physical object that can be observed and studied by science, or fully understood by human reason. In focusing solely on such rational empiricism, scientism has overreached the proper limits of science and misunderstood the nature of religious belief.

Despite the fact that scientism has an alarming number of adherents in the scientific and philosophical communities, it has not prevailed. Religious belief has withstood scientism's aggressive challenge not only by a convincing theological defence of belief in God, but by the contemporary revival of spirituality, especially its focus on seeking God's presence inwardly. This revival is a direct reaction to the meaninglessness and despair brought into society by decades of sceptical, anti-religious scientism.

It is important to note that science, unlike scientism, should not be a threat to religious belief. Science, to be sure, advocates a 'naturalistic' rather than a 'supernaturalistic' focus, and an empirical verification method for determining truths about the physical world and the universe. Yet the proper mandate of science is restricted to the investigation of the natural (physical, empirical) dimension of reality. It is this restriction that scientism has violated, replacing proper science with an illicit ideology that not only seeks to explain all things naturalistically, but assumes—without proof—that the spiritual realm is irrelevant, indeed non-existent. This unproven assumption is based on the mistaken belief that nothing exists unless it can be verified by the empirical scientific method. Such a belief is an invalid reductionism that reduces the explanation for all of reality to physicality. This 'physicalism' overextends the method and capabilities of science.

Had all scientists honoured the proper mandate of science, there would have been far less antagonism between science and religion, with each focusing on its respective field of study and contributing toward the production of an integrated view of reality that takes into consideration both physical and spiritual aspects.

Scientism's exclusive emphasis on human rationality and empirical verification is not the proper means by which to evaluate religious beliefs. There is, then, no justifiable reason to accept scientism's sceptical dismissal of religious beliefs as myth and superstition. Nor is there any good reason to accept scientism's claim that religious beliefs are irrational, meaningless and nonsensical. Such claims are themselves nonsense. Yet incredibly, many scientists and philosophers misunderstand this as they continue to engage in a scientism which ignores the fact that proper science is limited to a study of the physical world and that, as such, it has no competence to express negative opinions about the spiritual realm in the guise of scientific conclusions.

Overlooking important evidence

We may grant that God's existence cannot be verified as a physical object by the empirical method of science. God is a spiritual reality, not a physical object of scientific study. Yet scientism (and science) has overlooked important evidence for religious beliefs. There is not only an abundance of evidence for God, but much of this evidence is *indirectly* derived from rational reflection on the empirical data. The traditional arguments for God's existence—the so-called 'theistic proofs'—are prime examples. There is also more *direct* evidence for God in the inner experience of his revealed presence.

Theology, then, unlike science or scientism, accepts as sources of truth not only physical, scientifically verifiable evidence and the reasoning abilities of human beings, but also the spiritual truths revealed by God. These two sources are referred to, respectively, as 'natural' and 'revealed' theology (or as 'general' and 'specific revelation'). Natural theology refers to our natural or rational abilities to affirm God's existence by reason and empirical evidence. Revealed theology focuses on the more specific truths given by God in inner religious experience, and especially those experiences recorded in the Scriptures. While much about God can be known by general revelation, the evidence is inconclusive in itself; some things are revealed only in the certitudes of special revelation, truths which are accepted by believers in faith. From the perspective of faith, moreover, the evidence for God in general revelation is confirmed.

Science (unlike scientism) has understood the limitation of its method and the limits of human rationality. It is not, then, the empirical method of science but the inappropriate and overextended use of this method by scientism which has caused the antagonism between religion and science. Scientism has misunderstood the limits of human reason and sense experience. It is difficult not to infer that the basic motivation of scientism is not so much the seeking of the full truth about reality, but the promotion of an anti-religious bias which ignores and defames the spiritual dimension. The presumptuous claim that human rationality and the empirical method of science are not just *one* means for discovering truth but the *sole* means is an ideological, philosophical opinion which, in presenting itself as science, results in a pseudo (false) science.

Not all scientists have succumbed to this temptation to overextend the proper limits of science. Many have engaged in successful careers while maintaining belief in God. For example, in two recent books, *In Six Days* (2001) and *On the Seventh Day* (2002), more than eighty scientists were willing to have their testimonies and arguments for religious beliefs published.[1] They did so in response to sceptical ridicule by popular and influential scientists such as Stephen Jay Gould and Ernst Mayr who contend that no respectable scientist would believe in such things as the biblical account of creation by God. Examples of this sceptical, anti-religious scientism are unfortunately far too prevalent.

Philosopher A.J. Ayer, for example, a popular figure in philosophy classrooms, claimed that unless religious belief in God can be verified by the empirical method, we ought to reject such belief as nonsense. Ayer was aware that God's existence can neither be proved nor disproved by the empirical method, but his conclusion was not a respectful agnosticism that admits there is insufficient evidence to decide the question. Rather, he boasted that the empirical method demolishes the truth claims of religious experience on the grounds that God cannot be verified as the referent (source) of such experience. Ayer's view reveals not only the naivety of his scientism, but an anti-religious bias which displays an unconscionable lack of attention to longstanding theological argumentation for the validity of religious experience and beliefs.[2]

Carl Sagan was no less obvious in demonstrating his sceptical, anti-religious bias. His popular *Cosmos* book and TV series established him as a modern prophet of scientism. One of Sagan's infamous statements, 'The Cosmos is all that is or ever was or ever will be',[3] has become a cult-like mantra of scientism. This statement, of course, is not a scientific conclusion, but merely Sagan's sceptical opinion. A similar view is assumed, without proof, by too many scientists and philosophers, including popular figures such as Isaac Asimov, Richard Dawkins, W.O. Wilson, Stephen Jay Gould, Peter Atkins, Michael Martin, Paul Kurtz, Kai Nielsen, Anthony Flew, Francis Crick and Michael Newdow. The *Humanist Manifestos* (1933, 1973, 2000) also advocate a 'secular humanistic' ideology that rejects all religious beliefs as lacking the rationality of empirical verification. Again and again, in apparent ignorance (or defiance) of the limits of both the empirical test and human rationality, scientism extols these skills as the all-encompassing criteria for truth. Such a claim is nothing more than an ideological opinion, driven by sceptical bias.

I won't believe it unless I can see it . . .

Sagan presumes scientism in his infamous challenge that God should demonstrate his existence by displaying physical evidence which can be verified by science—for example, a burning cross in the sky. Such challenges, repeated by countless sceptics, reveal a disconcerting lack of awareness of the absurdity being proposed. They are little more than emotionalism which misrepresents the proper limits of science and the nature of God. God, of course, cannot be coerced or tempted by such challenges. Nor is God a physical object for scientific study, or comprehensible by the finite capacity of human reason.

Consider one further example, the pitiful taunting of American Christians by the first Russian cosmonaut. He sarcastically announced that he had not 'seen' God in outer space. His assumption—that since God is a physical object his existence must be verifiable by the scientific method—is sadly misinformed about the nature of God and the proper limits of empirical scientific proof. Further examples abound. It is staggering how prideful the presumption of scientism is in assigning to finite human minds the final authority for adjudicating all truths about reality.[4]

In sum, it is a sad irony that the test which supposedly determines all truths about reality, the verification test, is itself unverifiable. What confidence should reasonable people have in a test which, when applied to itself, condemns itself as meaningless? The empirical test is a naive, circular argument which engages in an illicit 'question-begging fallacy' by assuming what needs to be proven. Scientism answers the question 'How is truth about reality known?' not on the basis of a conclusion from convincing proof, but by simply assuming that 'truth is established only by the empirical test'. Nothing could be further from the truth.

The attitude of scientism, its 'I won't believe anything unless I can see it' dogma, is hardly a reliable guide to truth. As we all know, there is far more to life than meets the eye, more than can be verified by the senses or understood by human rationality. Indeed, it would be impossible to live our lives according to such means, since most of what we believe to be true is not empirically verifiable. We live and act in faith, not on the basis of conclusive, rationally verifiable proof, for most of what we do and believe. There are, in fact, a significant number of important beliefs which we accept intuitively as truths despite their not meeting the strict demands of the empirical test. These beliefs include not only religious beliefs but the main theories of quantum physics, aesthetic beliefs about what is beautiful and ugly, ethical beliefs about what is good and bad, historical statements about what occurred in the past, emotions, theoretical statements including the main theories and presuppositions of science (for example, that the universe is real, or that laws of cause and effect will continue to operate), and so on *ad infinitum*.

The perspective of faith

As noted above, belief in God is based *directly* on the experience of God, revealed to people's hearts (spirits) and accepted in faith. This belief is consistent with and confirmed by an abundance of *indirect* empirical evidence, demonstrating that our intuitive awareness of God's presence is not the irrational illusion or fantasy that scientism falsely claims it to be. The empirical evidence supports this inner belief in God, evidenced in the Scriptures (the Bible) and the 'theistic proofs' (some of which are discussed in the first part of this book). Other important evidences include the arguments for the authenticity of the Scriptures and their accounts of miracles, fulfilled prophecies and creation by God (discussed in the second part).

We must not overestimate the persuasive power of this evidence, since human rationality is a limited skill. But it would also be a mistake to underestimate the impressive nature of the empirical evidence. In itself, the evidence can neither create belief in God nor assure us of God's existence conclusively. Yet it can help to prepare the way for belief by refuting the obstacles presented by sceptical arguments against belief in God. More importantly, the empirical evidence, while inconclusive in itself, confirms the believer's faith in God, since the believer sees the evidence from the perspective of faith (see below).

Interestingly, the evidence for God is largely ignored or misrepresented by scientism, especially by those who seem engaged in an emotionally-charged crusade to promote an anti-religious bias—the stuff one finds in some sceptical magazines and on websites which seem oblivious both to the evidence and to proper methods of reasoned argumentation. I am aware, for example, of dozens of debates between theists and atheistic scientists or philosophers in which the latter display a pitiful lack of understanding of the Bible and theological scholarship, as well as an embarrassing lack of awareness of basic theological doctrines and the complex argumentation for their validity. This naivety, however, has not produced an appropriate humility, nor a toning down of the hostility which underlies the vendettas against belief in God.

Consider, for example, the complete lack of awareness of legitimate theological scholarship in the claim that 'the Romans wrote the Bible' and similar nonsense. Examples abound, such as accusations that belief in God is supposedly nothing more than a childish illusion, similar to belief in Santa Claus or the tooth fairy, crop circles, UFOs and ancient pagan gods like

Zeus. This all-too-typical kind of attack is a prime example of the fallacy of false association (also known as the fallacy of guilt by association). By associating belief in God with beliefs that are obviously fictional, the sceptic's contention amounts to little more than an emotional appeal, assuming that the reader will not notice how the 'argument' fails to take into account the basic fact at issue: that belief in God is grounded in formidable evidence while the fictional items associated with God are not.

Not enough evidence?

There is one common complaint from scientism's sceptics, however, which merits a serious response, a response which will bring us to the heart of the issue of belief in God and the essential role of faith. The complaint is that there is not enough evidence for God—that God ought to have made his existence more obvious, especially to those who are inclined toward scepticism or doubt. The well-known sceptic Bertrand Russell insisted on his deathbed that God had not provided 'enough evidence' for him to believe. Similar complaints have often been expressed. Why is God so hidden?

Any response to this question must concede that God is, indeed, hidden—that is, if we seek his full glory and complete answers to our theological questions. The Bible itself says that we see now only partially.[5] The empirical evidence, in itself, is inconclusive without the perspective of faith. Yet while faith assures us of God's existence, it does not present us with full intellectual understanding. The same can be said of the Scriptures, which, while enlightened from the perspective of faith by the Spirit of God, can be obscure in some passages to human understanding.[6] Likewise, the complex answers to important theological problems—the problem of evil and suffering, for example—can be obscure: God's apparent absence in times of trial and suffering is not fully comprehensible to human understanding.[7] Faith does not need, nor should we expect, complete intellectual understanding of such things. Faith reveals truths which are beyond empirical truth and beyond complete intellectual understanding. Faith is consistent with these truths and is sufficient in itself to assure us of God's existence and providential care. God is not hidden to those who believe in faith.

For those without faith, who seek God solely by empirical evidence and human rationality, evidence for God will remain inconclusive. Such means are far too limited in themselves to provide conclusive intellectual assurance. It is significant, of course, that millions upon millions of people *do* believe in God *without* conclusive empirical or rational proof. To account for such belief, social scientific explanations misleadingly attribute its origins to various cultural, psychological or biological factors. In all-too-typical scientific fashion—and this is the danger of science taking its limited understanding of reality as the complete explanation—such theories pretend to explain all facts, including spiritual realities, in naturalistic terms. As such, they ignore the long-standing theological explanation, namely, that faith is the gift of God whose presence is inwardly experienced by those who seek him with an open heart, unbiased by anti-religious scepticism.

Countless millions of people believe in God not on the basis of empirical evidence, but rather because they have accepted the gift of faith.[8] The empirical evidence for God may be sufficient for intellectual belief (as general revelation, and thus no one is 'without excuse'[9]); but it is not so overwhelming that it *coerces* belief. Coerced belief, like coerced love, would be meaningless. It is not our intellectual assent that God seeks, but a free response in faith to his revealed presence within our hearts. In faith, we accept and trust God's direct revelation to our hearts and as witnessed in Scripture (for faith arises in hearing the Word[10]). Likewise, through the eyes of faith, the empirical evidence confirms belief in God, interpreting the Scriptures and other evidence from the perspective of inner faith.

We are able to respond in faith to God's presence—revealed in our inner experience, in the Scriptures and in the empirical evidence of various kinds—because our hearts have let go of sceptical bias and rebellion. We have let go of our reliance on our own understanding, seeking God as the sole source of ultimate meaningfulness and purpose in our lives.[11] God provides the grace (as an unmerited gift) to make possible this change of heart and the free acceptance of the gift of faith. God's grace also makes it possible for us to desire the spiritual growth which progressively strengthens our faith as we become more trusting, loving and hopeful of God's blessings and trustworthiness. Our role is to accept these possibilities, granted to us by the mystery of grace.[12]

Faith in God is confirmed not only by the Bible (the record of God's historical revelations), but also by the similar faith in millions of people, which likewise is verified in its consistency with Scripture's revealed truths. Our inner experience is confirmed also by the spiritual transformations and growth which occur in those whose faith is genuine, producing the fruit of 'love, joy, peace, longsuffering, kindness, goodness, faithfulness, gentleness, self-control'.[13] Faith is an ongoing process of spiritual growth,[14] the development of a new awareness of ourselves as 'born anew'[15] and of the world seen far differently from the way it is seen through the eyes of scepticism.

For the believer, the darkness of meaninglessness, despair and the lack of purpose or direction is lighted by the reality of God's presence. Indeed, in the light of faith the empirical evidence of 'general revelation' is seen for what it is, as an awareness of God indirectly mediated through the senses and intellect. We experience God, as such, in both our hearts and our minds.[16] In the light of faith, empirical evidence of various kinds becomes a clear witness to how 'the heavens declare the glory of God'.[17]

Why scepticism rejects faith

But what of those who do not share this faith, this belief in God? Why does God remain hidden from them? One thing is certain: disbelief in God should not be based on conclusions drawn from human rationality and empirical evidence. These are limited and inappropriate means, in themselves, as we have seen.

We can agree with the great seventeenth-century scientist-theologian Blaise Pascal that the sceptic's rejection of God is not intellectual, but moral.[18] The answer to the question 'Why does God remain so hidden to those who remain sceptical about his existence?' lies ironically in the question itself. Simply put, the answer is that the sceptic has an anti-religious bias, a prideful inclination against the desire to believe in God. The Bible reminds us that, while God's light shines amid the darkness of our fallen nature, many love the darkness rather than the light.[19] It is this personal inclination toward rebellion and pride that keeps God hidden and hinders the work of grace in us.[20]

God has provided the grace needed to inspire all people to accept him freely in faith, and to appreciate the confirming evidence of his reality seen with the eyes of faith. Christian theology explains why all people do not respond in faith: we are born into a fallen, sinful and rebellious state, into a fallen nature filled with obstacles that hinder belief in God, making it difficult for many to accept God's gift of faith. Moral rebellion resists any acknowledgment of God's reality, a resistance which often is based on anger directed at God for some perceived wrong which caused great suffering, or on fear of accountability since to acknowledge God is also to acknowledge accountability to God. Prideful rebellion based in our own abilities and self-sufficiency leads to worship and service of the creature rather than the creator.[21]

The Scriptures promise that those who truly seek God with an open heart and mind will find God. God is the source of true knowledge and wisdom, a wisdom not found in human rationality.[22] God offers grace to all people that is sufficient for all to accept and to acknowledge his reality, but not irresistible. Grace makes possible the faith to believe, but does not do so coercively. The empirical evidence is sufficient for intellectual belief, but it too is not coercive.

Sceptical rebellion resists and distorts the awareness of God which otherwise would be far clearer in the empirical evidence and the Scriptures. Rebellious hearts and minds resist accepting faith in God,[23] grieving the Holy Spirit of God[24] who seeks the salvation of all people by faith.[25] Faith can be accepted and remain strong in us only if we are open to God's grace in trust and hope, and humbly admit and accept God's sovereignty and Christ's redemptive act, conceding that we ourselves cannot give ultimate meaning to our lives and cannot overcome the ills we suffer individually and collectively. By ourselves, we cannot achieve a lasting sense of meaningfulness, true happiness and fulfilment. Such are the fruits only of faith and trust in God. As long as faith is resisted by biased hearts and minds, there is estrangement and alienation from our true humanity, from one another and from the world, as well as from God.[26]

Without faith, empirical evidence and rationality are merely inconclusive human pursuits, sufficient for belief but not accepted by a biased, sceptical heart and mind. Only the inner revelation of God (in wordless intuition or, more specifically, in the Word of Scripture), freely accepted in faith, reveals the empirical evidence for what it is: evidence of the one true God who is powerful, eternal, loving and the creator of all things.

2

Fingerprints of the Divine Around Us

Danny R. Faulkner

Albert Einstein once noted that 'the most incomprehensible fact about the universe is that the universe is comprehensible'.[1] Indeed, there is no *a priori* reason why the world ought to be knowable, so this fact is startling to some people. Science would not be possible if the universe did not exhibit order and follow rules. For instance, science is based upon repeatability, the fact that we can repeat experiments and get the same results each time. However, if the outcomes of an experiment varied over time and location, then we could never replicate experimental results. There would be no point in determining some general rules describing behaviour of the physical world because those rules would constantly change.

Fortunately, the world does follow prescribed rules. Many ancient and Eastern cultures did not recognise this possibility. Instead, these cultures presupposed that the world was in constant flux. This philosophical basis did not permit these civilisations to anticipate the ordering of the world. If the world was in constant flux, then why would one expect it to follow rules? However, in the West, Christianity had a heavy influence. Of particular interest was the Protestant Reformation, which emphasised not only the worth and dignity of the individual no matter how lowly, but also the value of the natural world as God's creation. Since God created the world, it followed that he imprinted some of his character upon his creation. Since God is a God of order and operates by decrees, or laws, it seemed reasonable to the Western mind that God would have created an ordered world that follows consistent and knowable laws. In fact, the concept of physical law in science sprung from this assumption.

The belief that the creator God imposed order upon his creation fostered the development of science as we know it. Thus it is no accident that England in the seventeenth century was the cradle of modern science. For instance, Francis Bacon (1561–1626) is the father of the scientific method. A couple of generations later, the Royal Philosophical Society of London was packed with scientific giants such as Isaac Newton (1642–1727), Robert Boyle (1627–1691) and Robert Hooke (1635–1703). Christianity heavily influenced all the men involved in the birth of modern science. Most of these men assumed that the creator had endowed his creation with order. If we learned more about the natural world, we could learn more about God's nature. Among these men there was a general belief that the goal of their scientific work was 'to think God's thoughts after him'. Thus the study of science was a holy calling.

This foundation of modern science is unknown to most people today, and the majority of scientists now would consider such thinking unscientific. What happened to change the philosophy of science so drastically?

A complete discussion of this is beyond the purpose of this essay, so we will only briefly describe the causes here. The eighteenth-century Enlightenment made atheistic thinking respectable in the West. While only some of the Enlightenment writers were atheists, many others were theists who took a deistic approach to the world: they considered that while God may have created the world, he was not very relevant in its day-to-day operation. During the nineteenth century, atheism and deism led to the widespread acceptance of Darwinism, which caused scientists to think increasingly in terms of finding naturalistic explanations for everything, including origins. This philosophy reached fruition in the twentieth century when virtually any discussion of a creator became completely unwelcome in scientific studies.

Some contemporary writers, such as Phillip Johnson, have critiqued the modern approach to science. Johnson has used the term 'scientific materialism' to describe the belief that the physical realm is the only reality.[2] Accompanying this belief is the assumption, often tacit, that all explanations must be purely natural and physical. This attitude unfortunately rejects *a priori* any possibility of order imposed upon the universe by an external agent, opting instead for the assumption that order is just one of those things that the world happens to possess. If scientific materialism is true, then it follows that all truth, if it exists, can be learned only from science, because science is the methodology whereby we study the physical world.

I assume that the reader is scientifically literate. In the spirit of the original foundation of science, I ask the reader to consider the logical possibility that there may be something or Someone beyond the physical realm. I ask the reader to open his or her mind to the possibility that there is order and design in the universe, and to consider that this design suggests there is a Designer. Since God is spirit (non-physical), we cannot probe him scientifically. This does not mean God does not exist; it merely means the scientific method is limited in this matter. To dismiss this possibility is a logical fallacy.

In committing this fallacy, many conclude that since they see no physical evidence of the metaphysical, then the metaphysical must not exist. Of course, since science is limited to the physical world, it cannot probe any non-physical entity. This limitation of science and the knowledge that we derive from it ought to be obvious, but, alas, it is not. The late Carl Sagan exemplified this thinking when he said, 'The Cosmos is all that is or ever was or ever will be.'[3] Many people were impressed with this comment since it came from a brilliant scientist. However, it was a blatant metaphysical assertion masquerading as a scientific statement. Unless he was some sort of god, how could Sagan have known that the cosmos is all that exists, has existed or ever will exist? Because he had no knowledge or evidence that anything beyond the physical world exists, he merely asserted that nothing beyond the physical world exists.

The existence of the universe

The German mathematician and philosopher Gottfried Wilhelm Leibniz (1646–1716) was the first to frame the question of why something, rather than nothing, exists. Debate on this topic actually pre-dated Leibniz, and the debate continues even today. Before Albert Einstein published his general theory of relativity in 1916, most people thought of empty space as nothing. However, general relativity posits that space is something as well. Therefore, even if the universe were void of matter, something would still exist. For nothing truly to exist, not only must matter and energy not exist, but also space and time must not exist either. This is a very curious and profound notion. There is no physical reason why this universe—something—exists rather than nothing.

For scientific materialism based on purely physical, natural explanations for everything, this is not a satisfying prospect. Some scientists have proposed how the world could have come into existence through some physical process such as a quantum fluctuation. However, there are at least two problems with these sorts of explanations. One is that they are not verifiable by the scientific method, so they are hardly scientific answers. The other is that even if one of these explanations were true, it would not answer the question. The question is not *how* the universe came to be but *why* it came to be. Actually, while science can have great power in describing how things happen, it does a very poor job of explaining why.

There are two solutions to this dilemma. The first is to believe that there is no answer to the question of why. If there is no 'why' to the universe, then we are faced with the conclusion that the universe, and hence everything in it, has no meaning. To many people this is a bleak assessment. While many scientists find this explanation satisfying, again it is not a scientific answer, for it is merely the metaphysical assertion of scientific materialism. The second solution is to conclude that there is some Causative Agent beyond the universe that explains, not only how, but also why the universe exists. Many may object that this too is a metaphysical assertion, and they are correct. Therefore, we conclude that all answers to the question of why have a metaphysical assumption. We have a choice about which metaphysical assumptions we make, but we cannot escape making that choice.

Despite the confidence of some atheists to the contrary, it is reasonable to believe that God exists—no metaphysical assumption ought to be less valid than another, at least at the outset. Once we have made the reasonable assumption that a creator exists to explain why the world exists, it follows that we ought to include him in the question of how the world began. God also most likely ordained the current operation of the world—that is, the world bears an imprint of some design endowed by the creator. This brings us back to the philosophical basis of science previously discussed.

This design principle has gone by several names. *Teleology* was a common word for it in the past. Two notable authors in history wrote on teleology. One was the naturalist John Ray (1627–1705), a contemporary of Newton. A little more than a century later, William Paley (1743–1805) wrote on teleology as well. Very recently, teleology has been rechristened *intelligent design*. One of its prominent modern proponents is William Dembski.

Design in the universe

What sorts of arguments for design have people pursued? One is the inverse square law of gravity discovered by Newton. Newton and many others after him noted that if gravity followed any other dependence upon distance, then orbits would not be stable. It should be obvious that if the earth's orbit were unstable, then life on earth would not be possible.

How do those who reject design respond to this? They point out that the inverse square law of gravity is a consequence of the three spatial dimensions that our world has. The derivation of this takes some physics background, so we will not explore it here. But does this refute the design aspect of gravity? Not really. First, the derivation of the relationship between the inverse square law and dimensionality requires an assumption of a property of gravity that, while it appears to be true in this universe, is not *required* to be so. In other words, in a randomly selected universe, gravity may follow some other rules so that a three-dimensional universe does not inevitably lead to the inverse square law. Second, the fact that we live in a three-dimensional world ought not be taken for granted. The world has three dimensions, but are there any non-design reasons why it should? One can show that in a one or two-dimensional world the biochemistry of life is not possible, but could higher dimensionalities allow life? Perhaps rather than the inverse square law directly being evidence of design, it is the three dimensions of space requiring the inverse square law of gravity that are a design feature.

We can make some design inferences about the universe as a whole. The second law of thermodynamics states that the universe proceeds toward a state of increased disorder, or entropy. As the universe achieves the maximum entropic state, our ability to do work diminishes. Life relies upon the operation of work during complex biochemical reactions. Therefore, as energy to perform useful work becomes less available, the chemistry of life becomes impossible.

Yet a quick survey of the universe reveals that as it now exists, it is far from maximum entropy. This tells us a couple of things. First, the universe must have had a beginning at a finite time in the past. Second, the universe began with a lower state of entropy than it has now. To many people this suggests that the universe was wound up at the beginning as one might wind up a clock. This then suggests that it was the creator who wound up the universe, and so to some this constitutes evidence of God.

Modern cosmological arguments

The belief that the universe has a finite age is common today, but until recently it was not. The ancient Greeks generally thought that the universe was eternal, and this belief remained common in Western thought until well into the twentieth century. The primary appeal of the eternal universe among ancient writers was that an origin for the universe (and its required transcendent creator) was to them unthinkable. As Christianity became the basis for Western thought, the eternality of the universe ought to have been re-evaluated, but it was not. The eternality of the universe remained, as well as other ideas with pagan roots. As an historical example, Newton apparently thought that the universe was eternal but that the earth had a finite age.

This situation began to change in the 1920s. Einstein's general theory of relativity suggested that the universe is probably expanding or contracting. Astronomical observations revealed that most galaxies have a redshift, which suggests expansion. The Belgian priest and astrophysicist George Lemaitre (1894–1966) built upon this knowledge when he proposed his 'cosmic egg' from which the universe expanded. Lemaitre developed his model while fully accepting a non-eternal universe and that God created the cosmic egg. However, it is interesting to note that the idea of a cosmic egg is itself an ancient pagan idea. Lemaitre's model was modified and eventually became what we now call the big bang model.

The steady state theory was an alternative model to explain the expansion of the universe. The steady state model retained the eternal universe, without beginning or end. In 1965 Arno Penzias and Robert Wilson discovered cosmic background radiation, a prediction of the big bang model. Since the steady state model does not predict cosmic background radiation, most scientists rejected the steady state model. Therefore, since the mid-1960s, the big bang model has been the model of the universe generally accepted by most scientists.

Many contemporary Christian apologists and scientists believe the big bang model (though there are many exceptions, such as this author) and use the big bang as support for their Christian beliefs. As with Lemaitre, these Christians see the God of the Bible as the originator of the universe using the big bang. There is danger here: if we build our case with too much reliance on the big bang, then if this model is someday discarded (as scientific theories frequently are), what will happen to our theistic arguments?

One must realise that in a mechanistic big bang, the parameters of the universe, and probably even the form of physical laws and physical constants as well, are randomly generated. The universe does not care whether the values and conditions that exist in the universe are conducive to life. One therefore must ask what range of values and conditions the universe could have that would make life possible, and then assess the probability for the universe to exist within this range. The conclusion is that our universe is remarkably improbable and appears fine-tuned.

For instance, we have already discussed the low entropic state that the universe must have had initially. There is no reason why a big bang would have produced such an initial state. A big bang could have resulted in a universe created very near or at maximum entropy, but it did not. Another requirement would be the expansion rate of the universe. If the initial expansion rate was infinitesimally less than just right, the universe would have quickly re-contracted upon itself. If the expansion rate were slightly greater, the universe would have rapidly expanded to such low density that no galaxies, stars, planets or people would exist. Either way, we would not be here.

Another requirement for life to exist would be the clumpiness of matter in the early universe. The development of the observed structure of the universe would require that there be slightly more dense regions of the universe about which matter could gravitationally accumulate. If the matter in the early universe were perfectly uniform, there would be no gravitational clumping of matter, and there would be no galaxies, stars, planets or people. However, if the matter were too clumpy, then early in the universe nearly all the matter would form into many black holes, and again there would not be enough matter remaining to form galaxies, stars, planets and people. The range over how much matter in the early universe could clump and still result in humanity's existence is very narrow.

Many other improbable factors in the big bang cosmology seem to suggest that our existence is extremely improbable. Among these are the absolute values of the constants determining the electric, magnetic and gravitational forces, as well as their relative values. Space does not permit a full discussion here of these and other interesting 'coincidences' required in the big bang model. Suffice it to say that given it is the model of cosmic origin favoured by the vast majority of the (secular) scientific community, there are abundant improbable factors that make human life extremely unlikely. Yet here we are.

A fair evaluation of this situation ought to cause great reflection. Unfortunately, this is not the case. Many scientists who have investigated these facts have developed ideas to explain away the improbable. Let us examine a few of those.

Objections

In the early 1970s astronomers began using the term *the anthropic principle*, meaning that the universe possesses many properties that make the universe suitable for life. In its strongest form, this principle suggests that the universe exists or was designed to guarantee our existence. Of course, while this sort of idea was common among scientists three centuries ago, such thinking is most unwelcome today. How do scientists opposed to the strong anthropic principle respond to it?

The most common response is to appeal to the weak anthropic principle. The weak anthropic principle states that the universe exists in a state that allows life to exist somewhere in it. Contrast this with the strong anthropic principle: the strong anthropic principle *demands* that life exist while the weak anthropic principle merely *permits* life to exist. Barrow and Tipler wrote an exhaustive book on the anthropic principle.[4] After discussing the long history and current thinking on the matter, these authors conclude that the universe merely *appears* to be designed. In other words, while the universe does seem to be designed, this is merely the appearance of design and not actual design. If the universe did not exist as it does, then we would not be here marvelling how the universe looks. As improbable as the universe may appear, it happened, so its probability is one.

Of course, this answer is only satisfying to those who wish to disregard the evidence. If we applied such an approach to archaeology or physical anthropology, we would effectively destroy those fields. How can one ascertain that a particular stone was fashioned into a spear point? It is obvious that the stone bears evidence of having been reworked and shaped into a tool. That is, the stone obviously contains evidence of design that demands that there must have been a designer with some intelligence who crafted it. However, what is to prevent the critic from concluding that natural forces weathered the stone into its shape so that the stone only appears designed when in fact there was no designer involved? One could counter with a probability argument—that while occasionally a stone may weather in such a way as to produce what appears to be design, the sheer number of such shaped stones makes this nearly impossible. However, since we have information about only one universe, how could we begin to evaluate probabilities about the universe?

Scientists have offered other explanations. Some cosmologists have seriously suggested that there is not just one universe but many that together comprise a multi-verse. There are many variations on this theme, but the most common thread is that there are an infinite number of universes, and existing universes are constantly spawning new universes. In each universe, the laws of physics and other physical conditions may be different. Most of these universes will not be conducive to life, but a few will. It is no accident that we live in a universe that is conducive to life, or else we would not be here to make this observation. We could not exist in any of the far greater number of universes where conditions are hostile to life. Thus there is a selection effect—only in universes conducive to life will life exist. This entire line of reasoning is fascinating, but it is hardly testable. If an idea is not testable then it is not science. Therefore, the concept of a multi-verse is a philosophical conjecture masquerading as science. Sadly, its proponents generally are not open to the equally metaphysical possibility of a creator.

Another attempt to avoid theistic or design implications is a return to the eternal universe, such as the plasma universe theory. Proponents of this idea suggest that the universe is eternal and infinite, and that within the universe large, finite regions slowly and chaotically expand and contract. In their view, we live in a region that happens to be expanding at this time. Perhaps when regions contract, the usual flow of entropy reverses, so that during an ensuing expansion we observe the 'usual' movement toward increasing entropy. Another suggestion is that our universe is just one explosion in an infinite chain of big bangs, with each big bang being followed by expansion, contraction and subsequent re-expansion in the next big bang cycle. In between each cycle the laws of physics, including the amount of entropy, are reshuffled.

Of course, there is no way of scientifically testing any of these ideas. Furthermore, one must hypothesise processes, such as reversing entropy, that are contrary to the way our universe operates. If anything, this illustrates the desperate measures that many people take to avoid the consequences of design in the world.

There are also serious shortcomings in the big bang as a cosmological model. For example, in 1989 the COBE (COsmic Background Explorer) satellite was launched to measure slight temperature variations with position expected in the cosmic background radiation. After two years of data collection, it was found that there were no temperature variations within the predicted range—the cosmic background radiation appeared perfectly smooth. Only after very sophisticated statistical methods were applied to the data did researchers find temperature fluctuations, and those fluctuations were an order of magnitude below those predicted by the standard big bang model. Yet many cosmologists proclaimed this as a great triumph of the big bang model. How did they do that? They used the observed temperature fluctuations as an input to recalculate a new big bang model. In other words, data used to guide the construction of the model were used as evidence for the model. This is classic circular reasoning.

Circular reasoning is not new to the big bang model. The three most commonly quoted evidences for the big bang model are the existence of the cosmic background radiation, the expansion of the universe and the observed abundances of the lighter elements. The cosmic background radiation is a clear prediction of the big bang model, but not so the other two. The big bang model was devised to explain the observation that the universe is expanding, as was the steady state model. Today few people believe the steady state model, and consequently few would regard universal expansion as proof of that theory. As for the lighter elements, their abundances were measured and those values used as an input in the big bang model. It is not surprising that a theory devised to explain data is consistent with that data. Therefore, neither universal expansion nor the light element abundances are actually predictions of the big bang model.

Various astronomers have identified observations that appear to run counter to expectations within the big bang framework. Notable is the work of Halton Arp, who has found numerous instances where it is doubtful that observed redshift correctly gives distance via the Hubble relation.[5] If Arp is right, then we ought to question how many other redshifts correctly reveal distance. This could call into question whether universal expansion is real. Another difficulty is the work of William G. Tifft, who since the mid-1970s has accumulated data that show periodicities in redshifts.[6] His data show that redshifts appear to be quantised primarily around the value of 72 km/s with some secondary quantisation as well. If this effect is real, then the easiest interpretation is that we are near the centre of a finite universe. This possibility is unthinkable in the modern big bang cosmology.

Conclusion

In this brief chapter, we have explored the Christian basis for science, the retreat from that basis and the development of scientific materialism. We have surveyed only a few of the design inferences that one can make about the universe, even if we use what may be a tacitly atheistic theory.

Many scientifically literate people resist any hint of the metaphysical concept of a God. However, we found that the explanations devised to avoid God's existence are themselves metaphysical assertions. No one can smugly conclude that they make no metaphysical assumptions—we all do. Nor should atheists or agnostics think that their metaphysical assumption is superior to theism. The honest intellect must conclude that belief in a creator God is at least as reasonable as the atheistic, metaphysical alternatives.

3

Where Did the Universe Come From?

Kenneth E. Himma

The cosmological argument attempts to show that a personal God was the cause of the universe's existence and hence was its creator. Different versions of the argument rely on different causal principles: some versions hold that every material entity or event must have a cause; early Greek versions took the universe (and all else) to be eternal and held that every eternal entity must have a cause. In any event, all are grounded in the principle that the existence of certain kinds of thing can be explained only by postulating a cause of some kind. In the case of the universe, the cause is the all-perfect, personal God of classical theism.

The *kalam* argument described

Perhaps the most influential of the cosmological arguments is the *kalam* cosmological argument, originally formulated by Christian theologians to refute the classical Greek claim that matter was eternal and subsequently refined by medieval

Islamic and Jewish theologians.[1] The doctrine that all matter is eternal is inconsistent with the scriptures of all three major religious traditions, which teach that God is the creator of the universe.

Proponents use the *kalam* cosmological argument and its supporting reasons to show three claims with respect to the universe and its cause. The first claim is that the universe has a beginning and hence has not always existed. The second claim is that the beginning of the universe's existence is not an uncaused random event of some kind; the universe has a determinate cause. The third claim is that the cause of the universe's existence is a personal one, not an impersonal natural one. Thus the *kalam* cosmological argument and its reasoning explore three pairs of possibilities for the existence of the universe: (1) beginning versus no beginning, (2) caused versus uncaused, and (3) personal cause versus impersonal cause.

At the most general level, the structure of the argument is quite compact, consisting of two premises and a conclusion, and is fairly summarised as follows:

1. Any entity that has a beginning has a cause of its existence.
2. The universe has a beginning.
3. Therefore, the universe has a cause of its existence.

As is readily evident, the second premise coheres with the creationist commitments of classical theism and denies the Greek view that all matter is eternal. The first premise asserts a causal principle that explains the origins of things that come into existence in causal terms.

Defending the premises

In thinking about the *kalam* cosmological argument, we should keep in mind that the hallmark of a good argument is that it is adequate to convince a reasonable person with an open mind that its conclusion is true. Accordingly, it should ultimately rely on reasons that such a person is likely to accept as being true. In other words, the supporting claims of the argument should be considerably less controversial to its intended audience than its reasons; if the reasons are as controversial as the conclusion, then the argument simply cannot persuade someone who isn't already convinced of its conclusion.

The same is true of the *kalam* cosmological argument: to persuade open-minded unbelievers, it should ultimately rely only on claims that unbelievers would regard as plausible. This does not mean that the premises in the above summary of the argument must themselves be comparatively uncontroversial. But it does mean *either* that those premises should be uncontroversial or that they should rely ultimately on reasons that are uncontroversial.

In what follows, we shall attempt to determine whether the premises of the *kalam* cosmological argument satisfy this property.

Premise (1): Why think that everything with a beginning has a cause?

The first premise states: 'Any entity that has a beginning has a cause of its existence.' This is grounded in the intuition that something cannot come out of nothing. It needs little argument in its defence. The idea that any material entity could simply come into existence without any causal relations to some preceding event that involves another material entity seems, on its face, absurd. Indeed, as William Lane Craig has observed, the great sceptic David Hume took pains to point out that while he argued there is no proof for the principle that nothing could come into existence without a cause, he never intended to suggest that this principle was false.

Some theorists believe that certain results in quantum physics refute the causal principle, but the interpretation of these results remains controversial. On one interpretation, certain subatomic particles such as quarks and electrons come into and go out of existence without any cause; such entities have a beginning but come into existence, so to speak, randomly without having a cause of any kind. Many theorists, however, reject this interpretation, arguing that the results show only that these particles do not have a cause that we can *identify*. On this common interpretation, the existence of such particles has a cause that is *determinate* but not one that is *determinable*, and hence is consistent with premise (1) of the *kalam* cosmological argument.

In any event, it is one thing to assert the possibility that subatomic particles lack a cause; it is another thing to assert that the entire material universe (construed to include both space and time) could come from nothing at all. Regardless of their origins, subatomic particles like quarks come into existence in a material universe that contains many material entities that pre-exist quarks. In contrast, the existence of a universe that has a beginning but not a cause would have to come from absolutely nothing at all.

If the idea of uncaused quarks is not obviously absurd, the idea of a material universe that comes from nothing at all is. Accordingly, the first premise of the *kalam* cosmological argument seems to be on solid ground.

Premise (2): Why think the universe has a beginning?

It is true, of course, that the idea that the universe has a beginning has a strong theological foundation in the scriptures of each of the three major religious traditions. The idea that an all-perfect God created the universe logically implies that the universe has a beginning—namely, the point at which God brought it into existence.

But insofar as the *kalam* cosmological argument purports to show that God exists, it obviously cannot rely on the scriptures because the truth of the scriptures presupposes the existence of God. To assume the truth of the scriptures in arguing for God's existence is to assume the very thing that must be shown. No argument for God's existence can succeed in convincing a fair-minded unbeliever if its key premise straightforwardly assumes that God exists. For this reason, the proponent of the *kalam* cosmological argument must try to give a plausible reason for believing the universe has a beginning.

As it turns out, proponents have given two powerful arguments to think that the universe has a beginning. The first relies on the claim that 'actual infinites' cannot exist. The second relies on the claim that a process of successively adding members to a set can never result in an actually infinite set. These two arguments are taken up in order below.

The impossibility of actual infinites

To see what is meant by the notion of an actual infinite, it is helpful to consider an example of something that is merely a potential infinite. Suppose that an immortal being were going to spend the rest of eternity stringing letters together. The string constructed by that person would be *potentially* infinite in the sense that its length would approach, but never reach, infinity. In contrast, an actually infinite collection is one that doesn't just approach being infinite; it contains an infinite number of entities. An example of an actual infinity would be the entire set of natural numbers (if one could collect them in a set).

The first argument for the claim that the universe has a beginning is grounded in the claim that actual infinites cannot exist. Here is a schematic representation of the reasoning:

A. Actual infinites cannot exist.
B. If the universe does not have a beginning, then it would have existed for an infinite number of moments.
C. An infinite number of moments is an actual infinite.
D. Therefore, the universe has not existed for an infinite number of moments.
E. Therefore, the universe has a beginning.

The success of this argument crucially depends on the claim that actual infinites cannot exist; every other premise in the argument is obvious. So why think that premise (A) is true? The reason, as William Lane Craig has persuasively argued, is that if it were even possible for an actual infinite to exist, all sorts of absurdities we know cannot happen would be possible—and this is sufficient to show that actual infinites *cannot* exist.

Consider some of the properties of Hilbert's Hotel—a hotel that contains an actually infinite number of rooms.[2] Normally, if a hotel is full and a new guest arrives, there is no way to accommodate her without someone's checking out of the hotel. But this is not true of Hilbert's Hotel. If Hilbert's Hotel is full and a new guest arrives, we can accommodate her simply by asking each guest to move into the room with the room number that is obtained by taking her old room number and adding one to it. Thus the guest in room 1 moves into room 2; the guest in room 2 moves into room 3; the guest in room 3 moves into room 4; and so on. In this way, although we already had an infinite number of guests in an infinite number of rooms, we have now created a vacant room without anyone having to check out of the hotel!

But the situation is even worse than this. Despite being full, Hilbert's Hotel can accommodate an infinite number of new arrivals without anyone having to leave the hotel. Here's how. Simply ask each guest to move to the room that has the number obtained by doubling the room number of her old room. Thus the guest in room 1 moves to room 2; the guest in room 2 moves to room 4; the guest in room 3 moves to room 6; and so on. This manoeuvre opens up all the odd-numbered rooms. Since there is actually an infinite number of odd numbers, this is sufficient to open up enough rooms to accommodate an infinite number of new arrivals—still without anyone having to leave the hotel!

The absurdities abound. Depending on which guests leave, an infinite number of departures might leave either a finite number of rooms or an infinite number of rooms. If the hotel is full and the guests in the even-numbered rooms leave, then there will still be an infinite number of guests in the hotel—those who occupy the odd-numbered rooms. If the guests in the rooms with room numbers greater than 10 leave, there will be a finite number of guests left in the hotel—those who occupy rooms 1 to 10. But exactly the same number of guests departed in each example.

It is important not to overreach here. The proponent of the *kalam* cosmological argument would not deny that some infinites are not only possible but actually exist. Obviously, proponents are theists who believe that an infinite God exists. But beyond this, many people, theists and non-theists alike, believe that numbers are entities that really exist in the world—though they are obviously not physical objects that exist in space and time. Since numbers do not take up space, they exist, if at all, in a way that is timeless. If numbers exist in this 'abstract' sense, then there actually are an infinite number of them.

The argument for premise (A), then, should not be construed as denying the existence of an infinite God or the existence of infinite *abstract* objects like the set of all numbers. It purports to rule out only actual *physical* infinites—objects, unlike a personal God and an abstract object, that are constituted or comprised by material entities or events. In other words, the argument should be construed only as showing that there cannot be an infinite *collection* of things that exist in space and time. It is, of course, puzzling but true that the set of even numbers is a proper part of the set of all numbers, which has the same number of members. But the material universe simply cannot contain collections of physical objects that have such strange properties. Perhaps Hilbert's Hotel exists as an abstract object of some kind, but it simply cannot exist as a temporal physical object that takes up space in any material universe.

We know, then, that these kinds of entities might be thinkable in the minds of mathematicians, but there could never actually *be* an entity that has such ridiculous properties. Since it is the actually infinite character of Hilbert's Hotel that gives rise to such absurdities, the argument concludes that actual infinites simply cannot exist in space or time. Accordingly, an infinite number of moments cannot have elapsed.

An actually infinite set cannot be formed by successively adding members

The second argument for the claim that the universe has a beginning is grounded in the claim that an actual infinite can never be formed by a process of adding members, one after another, to a set. As philosophers put this point, the claim is that an actual infinite cannot be formed by 'successive addition'—successive addition being simply a process of adding members to a set one after another. Here is a schematic form of the argument:

a. Actual infinites cannot be formed by a process of successive addition.
b. The set of seconds the universe has existed is a set formed by a process of successive addition.
c. If the universe does not have a beginning, then the set of seconds the universe has existed is an actual infinite that has been formed by a process of successive addition.
d. Therefore, the universe has a beginning.

Premises (b) and (c) should be pretty obvious. The set of seconds gets bigger every second because its passage adds a new member to the set—clearly a process of successive addition; thus premise (b) is true. If the universe doesn't have a beginning, then it has actually existed for an infinite number of seconds, which means that the set of seconds formed by successive addition is actually infinite; thus premise (c) is true. It is for this reason that critics of the argument generally focus their attention on attempting to refute premise (a).

To see why premise (a) must also be true, suppose that some immortal person were right now to start making an ordered list of the natural numbers (0, 1, 2, 3, . . .) by writing down one number per second. This is an example of a process of successive addition. She continues to make the list bigger by adding one number to it every second. But notice that no matter how long she lists numbers, she can never succeed in listing an infinite number of them. At any point in time, only a finite number of seconds would have elapsed between that moment and the moment she began writing the numbers down. Thus her list would always be finite.

One might think the matter is different if the list doesn't have a beginning, but the idea that one can form an actual infinite by successive addition in this fashion is no less problematic. As William Lane Craig puts it, 'If one cannot count to infinity, how can one count down from infinity?' There is simply no reason to think that it is any easier, so to speak, to produce an infinite set by moving backwards than it is to produce an infinite set by moving forwards. A change of direction simply cannot make that kind of difference.

Once we see that the premises are true, the conclusion must also be true. The set of moments for which the universe has existed *must* be a finite set; and this means that it has existed for only a finite number of moments in time. But anything that has existed for only a finite number of moments must have a beginning. Thus the conclusion follows: the universe has a beginning.

Must the cause of the universe be personal?

Strictly speaking, the *kalam* cosmological argument does not, by itself, tell us much about the nature of the cause; indeed, the conclusion asserts only that the universe *has* a cause. For this reason, the non-theist might respond by agreeing that the universe has a cause, but then argue that the cause of the universe is an impersonal event: an explosion that occurs in a quantum vacuum, an event astronomers have dubbed the 'big bang'.

If the conclusion of the *kalam* cosmological argument itself does not exclude the possibility of such an impersonal cause, the arguments that justify the claim that the universe has a beginning also help us to see why the best explanation of why the universe exists includes the existence of a personal creator. One way or another, big bang theorists must tell some kind of story that explains the explosion that brought the universe into existence—and this requires describing the 'cause' of the event. Some big bang theorists have attempted to deflect this question by saying that the big bang could not have had a cause because causes have to come before their effects and there were no earlier moments in time. The big bang, on this view, brought both space *and* time into existence. But most big bang theorists realise that this is an inadequate response. In some sense of 'time', there had to have been a moment in time preceding the big bang.

These theorists speculate that our universe originated in a superspace—or multi-verse—that, in some sense, pre-existed our material universe. The event that caused the big bang occurred in this multi-verse and 'preceded' the big bang event in some temporal sense of the term. Indeed, some theorists go so far as to speculate that this universe is only one of many material universes that originated out of big bangs in the multi-verse; it would be odd, they reason, if such an event happened only once.

But once this move is made, the very difficulties that plagued the idea that the universe has always existed make trouble for the big bang theory. All of the same difficulties that appear to falsify the claim that the universe has no beginning also falsify the claim that the multi-verse has no beginning. The world, construed to include a multi-verse, cannot have existed eternally because this would imply the existence of actual infinites. There simply has to have been a first moment in which the world existed.

Indeed, any explanation of the world's existence in terms of purely impersonal 'causes' and 'effects' seems to imply the existence of an actual infinite—which we know is impossible. Impersonal causes and effects must be explained by a mechanism that occurs inside some system of 'time' that proceeds successively, one 'moment' after another. At some point, there must be some ultimate system of time that itself has a beginning that requires an explanation.

The only explanation of the existence of the universe that avoids these difficulties is that it was freely created by a personal, all-perfect God who exists outside of time and space. As William Lane Craig argues, the only eternal cause (one that is outside time) that could produce a temporal effect (one that is inside time) is a free act of some personal agent: 'a man sitting from eternity may will to stand up; hence, a temporal effect may arise from an eternally existing agent'. Although the *kalam* cosmological argument does not, strictly speaking, imply the existence of the God of the New Testament, it provides sufficient resources to justify thinking that the origin of the universe is a personal deity who freely brought it into existence as part of a divine plan.

Conclusions

While the *kalam* cosmological argument tells us that the universe has a personal cause in the form of an all-perfect deity, it does not purport to tell us anything about why that cause exists. In particular, it does not attempt to answer the questions of why God exists and why there is something rather than nothing. The answer to those questions must come from other arguments—such as the ontological argument that attempts to show that, as a logical matter, a perfect being must exist.

This calls attention to an important point about the arguments for God's existence, namely, that they are most powerful in tandem. There are a variety of other powerful arguments for the existence of God including the ontological argument, the design arguments, the fine-tuning argument, the argument from morality and the argument from mind. A number of these are discussed in this book. Taken individually, each of these arguments provides a strong and intellectually respectable reason that does not rely on scriptural reasons for believing a personal God exists. Taken together, they complement one another and provide a powerful cumulative case for the existence of an all-perfect personal creator that is not easily rebutted by the atheist. Accordingly, it should be kept in mind that the existence of God does not rest on just the cogent *kalam* cosmological argument.[3]

4

Design by Information

Werner Gitt

In our age of communication, *information* has become a fundamental entity.[1] However, there is no binding definition of information which is universally agreed upon by practitioners of engineering, information science, biology, linguistics or philosophy.

There have been repeated attempts to really come to grips with the concept of information. The most sweeping formulation was recently put forward by a philosopher: 'The entire universe is information.' Here we will set out in a new direction, by seeking a definition of information which will enable us to formulate laws of nature about it.[2]

Because information itself is a non-material thing,[3] this would be the first time that a law of nature has been formulated for such a mental entity. We will first establish a natural-law definition for information, then state the laws themselves, and finally draw eight comprehensive conclusions.

What is a law of nature?

If statements about our observable world can be consistently and repeatedly confirmed to be universally true, we refer to them as laws of nature. Laws of nature describe events, phenomena and occurrences which consistently and repeatedly take place. They are thus universally valid laws. They can be formulated for *material* entities in physics and chemistry (such as energy, momentum, electrical current, chemical reactions) and *non-material* entities (such as information, consciousness).

Due to their explanatory power, laws of nature enjoy the highest level of confidence in science. The following points about laws of nature are especially significant:

- Laws of nature know no exceptions. This sentence is perhaps the most important one for our purposes. If you are dealing with a real (not merely supposed) natural law, then it cannot be circumvented or brought down. A law of nature is thus universally valid, and unchanging. Its hallmark is its immutability. A law of nature is therefore in principle capable of falsification. One single contrary example would end its status as a natural law.
- Laws of nature are unchanging in time.
- Laws of nature can tell us whether a process being contemplated is even possible or not. This is a particularly important application of laws of nature.
- Laws of nature exist prior to, and independent of, their discovery and formulation. They can be identified through research and then precisely formulated. (Hypotheses, theories or models are fundamentally different from laws of

nature. They are invented by people, not merely formulated by them.) In the case of laws of nature for physical entities, it is often, though not always,[4] possible to find a mathematical formulation in addition to a verbal one. In the case of the laws for non-material entities, the present state of knowledge permits only verbal formulations. Nevertheless, these can be expressed just as strongly, and are just as binding, as all others.

- Laws of nature can always be successfully applied to unknown situations. Only thus was the journey to the moon, for example, possible.

What is information?

Information is not a property of matter
The American mathematician Norbert Wiener made the oft-cited statement: 'Information is information, neither matter nor energy.' With this he acknowledged a very significant thing: information is not a material entity.

Let me elucidate this important property of information with an example. Imagine a sandy stretch of beach. With my finger I write a number of sentences in the sand. The content of the information can be understood. Now I erase the information by smoothing out the sand. Then I write other sentences in the sand. In doing so I am using the same matter as before to produce this information. Despite this erasing and rewriting, this producing and destroying of varying amounts of information, the mass of the sand did not alter at any time. The information itself is thus *massless*. (A similar thought experiment involving the hard drive of a computer quickly leads to the same conclusion.)

Wiener, then, has told us what information is *not*; the question of what information really *is* will be answered in this chapter.

Because information is a non-material entity, its origin is likewise not explainable by material processes. What causes information to come into existence at all—what is the initiating factor? What causes us to write a letter, a postcard, a note of congratulations, a diary entry or a file note? The most important prerequisite for this is our own will, or that of the person who assigned the task to us. Information always depends upon the will of a sender, who issues the information. Information is not constant; it can be deliberately increased, and can be distorted or destroyed through, for example, disturbances in transmission.

In brief: information arises only through will (purpose).

Natural-law definition of information

Technical terms used in science, such as 'energy' or 'information', are sometimes also used in everyday language. However, if one wants to formulate laws of nature, then the entities to which they apply must be unambiguous and clear-cut. So one always needs to define such entities very precisely. In scientific usage, the meaning of a term is in most cases considerably more narrowly stated than (that is, is a subset of) its range of meaning in everyday usage. So a definition always does more than just assign a meaning; it also acts to contain or restrict that meaning. A good natural-law definition, then, is one which enables us to exclude all those domains (realms) in which laws of nature are not applicable. The more clearly one can establish the domain of a definition, the more precise (and furthermore certain) the conclusions one can draw from that definition.

Take, for example, the term *energy*. In everyday language we use this word in a wide range of meanings and situations. If someone does something with great diligence, persistence and focused intensity, we might say he 'applies his whole energy' to the task. But the same word is used in physics to refer to a natural law, the energy law. In such a context it becomes necessary to substantially narrow the range of meaning, which in everyday language is often very wide. Thus physics defines energy as the capacity to do work, which is *force × distance*.[5] An additional degree of precision is added by specifying that the force must be calculated in the direction of the distance. With this one has come to an unambiguous definition and has simultaneously left behind all other meanings in common usage.

The same must now be done for the concept of information. We have to say, very clearly, what information is in our natural-law sense. We need criteria in order to be able to unequivocally determine if an unknown system belongs to the domain of our definition or not. The following definition permits a secure allocation in all cases:

> Information is always present when, in an observable system, all of the following five hierarchical levels are present: statistics, syntax, semantics, pragmatics and apobetics.

If this applies to a given unknown system, then we can be

certain that we are within the appropriate domain of definition. Thus, for that system, all ten laws of nature about information (see later) apply.

The five levels of information

1. *Statistics.* In considering a book, a computer program or the genome of a human being, we can, for example, put the following questions: How many letters, numbers and words does the entire text consist of? How many individual letters of the alphabet (such as a, b, c . . . z for the Roman alphabet, or G, C, A and T for the DNA alphabet) are utilised? What is the frequency of occurrence of certain letters and words?

To answer such questions it is irrelevant whether the text contains anything meaningful, is pure nonsense or is just randomly ordered sequences of symbols or words. Such investigations do not concern themselves with the content; they involve purely statistical aspects. All of this belongs to the first and thus bottom level of information, the level of statistics. (This is the level on which Claude E. Shannon developed his well-known mathematical concept of information.[6])

2. *Syntax.* If we look at a text in any particular language, we see that only certain combinations of letters form permissible words of that particular language—that is, words agreed upon by pre-existing convention. All other conceivable combinations do not belong to that language's vocabulary. Syntax encompasses all of the structural characteristics of the way information is represented.

This second level involves only the symbol system itself (the code) and the rules by which symbols and chains of symbols are combined (grammar, vocabulary). This is independent of any particular interpretation of the code.

3. *Semantics.* Sequences of symbols and syntactic rules form the necessary preconditions for the representation of information. But the critical issue concerning information transmission is neither the particular code chosen nor the size, number or form of the letters—nor even the method of transmission. It is rather the semantics (from the Greek *sēmantikós*, significant meaning), that is, the message it contains—the proposition, the sense, the meaning.

4. *Pragmatics.* Information invites action. In this context it is irrelevant whether the receiver of information acts in the manner desired by the sender of the information, reacts in the opposite way or does nothing at all. Regardless of this, every transmission of information is associated with the intention, from the side of the sender, of generating a particular result or effect on the receiver. Even the shortest advertising slogan for a washing powder is intended to result in the receiver carrying out the *action* of purchasing this particular brand in preference to others.

We have thus reached a completely new level of information, which we call pragmatics (from the Greek *pragma*, action, doing). The sender is also involved in action to further his desired outcome (more sales/profit), such as designing the best message (semantics) and transmitting it as widely as possible in newspapers, TV, etc.

5. *Apobetics.* We have already recognised that, for any given information, the sender is pursuing a goal. We have now reached the last and highest level of information, namely

apobetics—the aspect of information concerned with the goal, the result itself. In linguistic analogy to the previous descriptions the author has here introduced the term *apobetics* (from the Greek *apobeinon*, result, consequence). The outcome on the receiver's side is predicated upon the goal demanded/desired by the sender—the plan or conception. The apobetics aspect of information is the most important because it concerns the question of the outcome required by the sender.

Laws of nature about information

In the following we will come to know the ten most important laws of nature about information (LNI).[7]

LNI-1: A purely material entity cannot generate a non-material entity. In our common experience, an apple tree bears apples, a pear tree pears and a thistle brings forth thistle seeds. Similarly horses give rise to foals, cows to calves and women to human babies. In the same way, we can observe that something which is itself solely material never creates anything non-material.

The universally observable finding of LNI-1 can now be couched in somewhat more specialised form by arriving at LNI-2.

LNI-2: Information is a non-material fundamental entity. The materialistic worldview has so widely infiltrated the natural sciences that it has become the ruling paradigm. However, it is an unjustified dogma. The reality in which we live is divisible into two fundamentally distinguishable realms: the material and the non-material world. Matter involves mass which is weighable in a gravitational field. In contrast, all non-material entities (for example, information, consciousness, intelligence, will) are massless and thus have zero weight. Information is always based on an idea; it is thus also massless and does not arise from physical or chemical processes. Information is also not correlated with matter in the same way as energy, momentum or electricity is.

LNI-3: Information is the non-material basis of all program-directed technological systems and all biological systems. There are countless systems which do not exhibit any intelligence of their own, but which can nevertheless transfer or store information or direct processes. Such systems can be found in the non-living realm (coupled computers, process direction in a chemical factory, fully-automated assembly lines, automatic car-washes, robots) as well as the living world (information-directed processes in cells, the dance of bees giving directions to the nectar source).

LNI-4: There can be no information without a code. Whenever information is stored, transmitted and processed, it requires a code system, utilising a symbol set that, after having been freely chosen, has been firmly defined.

LNI-5: Every code is the result of a freely-willed convention. The essential characteristic of a code symbol (character) is that it has at some point been freely (arbitrarily) defined. The character set which has come into being in this way represents, by definition, all permissible characters. They are designed so as to fulfil their intended purpose as well as possible (for example, Braille must be easily palpable; musical note symbols must be able to specify the duration and pitch of tones; chemical symbols must be able to specify all of the elements).

When we observe signals which have the appearance of

being code characters, such as signals from pulsars or the absorption lines of chemical elements in the solar spectrum, but which are in fact the result of the physical or chemical properties of the particular system, then this fundamental hallmark of information, the 'freely-willed (arbitrary) convention/agreement', is absent. Such signals are therefore not symbols in the sense of our definition.[8]

LNI-6: There can be no new information without an intelligent, purposeful sender. The process of generation of *new* information (in contrast to information which is only copied) always presupposes intelligence and free will.

LNI-7: Any given chain of information can be traced back to an intelligent source. In many cases, the originator (author) of the information is *not* or *no longer* observable. The mere fact that one can no longer observe the authors of historical documents would not lead anyone to conclude that they had no authors. Even in a library with thousands of volumes, no authors are generally visible. But no one would therefore claim that the books arose without any such intelligent source.

LNI-8: Allocating a meaning to a set of symbols is a mental process, which requires intelligence. Information arises through utilising a set of symbols (characters). The characters are placed adjacent to one another in such a way that units of information (words, sentences) are formed. This selection does not occur by chance, but always with the application of intelligence. Information always has a sender as its source. The sender has thought of it, and wants to transmit it to one or more receivers. As Figure 1 illustrates, all five levels of information are pertinent on both the sender's and the transmitter's side.

GOD, SCIENCE AND PHILOSOPHY

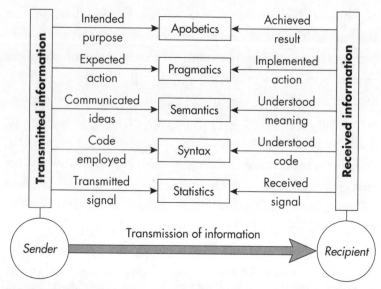

Figure 1: The Five Levels of Information

To fully characterise the concept of information requires the five aspects of statistics, syntax, semantics, pragmatics, apobetics. Information is represented (= formulated, transmitted, stored) as a language. From a stipulated alphabet, the individual symbols are assembled into words (code). From these words that have been assigned meanings, sentences are formed according to the firmly defined rules of grammar (syntax), which are the bearers of semantic information. Furthermore, the action intended/carried out (pragmatics) and the desired/achieved goal (apobetics) belong of necessity to the concept of information. All our observations confirm that each of the five levels is always pertinent for the sender as well as the receiver.

LNI-9: Information cannot originate in statistical processes. Promoters of evolutionary theory would score a massive breakthrough if they could show, in a real experiment, that information could arise from matter, left to itself, without the addition of intelligence. Despite the most intensive worldwide efforts, this has never been observed. To date, evolutionary theoreticians have only been able to offer computer simulations that do not correspond to reality.

LNI-10: The storage and transmission of information requires a material medium. Imagine a piece of information written on a blackboard. Now wipe the board with a duster. The information has vanished, even though all the particles of chalk are still present. The chalk in this case was the necessary material medium, but the information was represented by the *particular arrangement* of the particles. And this arrangement did not come about by chance; it had a mental origin.

The same information could have been stored/transmitted in Indian smoke signals, or in a computer's memory, through the arrangement of puffs of smoke or magnetised domains. One could even line up an array of, say, massive rocks into a Morse pattern. So clearly, the *amount* or *type* of matter upon which the information 'rides' is not the issue. Even though information requires a material substrate for storage/transmission, it is not a property of matter. In the same way, the information in living things 'rides on' the DNA molecule. But it is no more an inherent property of the physics and chemistry of DNA than the blackboard's message was an intrinsic property of chalk.

All these ten laws of nature about information have arisen from observations in our three-dimensional world, without an exception ever having been found. None of the laws rests upon philosophical assumptions or mere thought experiments, and to this point none of them has been able to be refuted or falsified by way of an observable process or experiment.

GOD, SCIENCE AND PHILOSOPHY

Eight comprehensive conclusions

Having firmly established the domain of definition, and familiarised ourselves with the laws of nature about information derived from experience, we can now zero in on effectively applying them. We arrive at eight very far-reaching conclusions that answer fundamental questions for us. All scientific thought and practice reaches a limit beyond which science is inherently unable to take us. This situation is no exception. But our questions involve matters beyond this limiting boundary, and so to successfully transcend it, we need a higher source of information. This higher source of information is the Bible.

We will proceed in the following sequential manner. First we will set out the (briefly formulated) conclusion itself. Then we will establish how we were able to reach this conclusion by applying the laws of nature about information. Lastly we will check the result against the Bible.

Conclusion 1: God exists.

> Because all forms of life contain a code (DNA, RNA), as well as all of the other levels of information, we are within the definition domain of information. We can therefore conclude that there must be an intelligent Sender. [Applying LNI-3, LNI-6, LNI-7]

Because there has never been a process in the material world, demonstrable through observation or experiment, in which information has arisen by itself, then that is also valid for all the information present in living things. Thus LNI-6 requires here, too, an intelligent author who 'wrote' the programs. Conclusion 1 is therefore also a refutation of atheism.

The top of Figure 2 outlines the realm which is, in principle, inaccessible to natural science: who is the message sender? To answer that the sender cannot exist because the methods of human science cannot perceive him is untenable according to the laws of information.

Conclusion 2: There is only one God, who is all-knowing and eternal.

> The information encoded in DNA far exceeds all our current technologies. Hence, no human being could possibly qualify as the Sender, who must therefore be sought outside our visible world. We can conclude that there is only one Sender, who must not only be exceptionally intelligent, but must also possess an infinitely large amount of information and intelligence—that is, he must be omniscient (all-knowing) and, beyond that, eternal. [Applying LNI-1, LNI-2, LNI-6]

According to LNI-6, at the beginning of every chain of information there is an intelligent sender. When one applies this to biological information, here too there must be an intelligent author of the information. In DNA molecules we find the highest density of information known to us.[9] Because of LNI-1, no conceivable processes in the material realm qualify as the source of this information. Humans, who can of course generate information (for example, letters, books), are also obviously excluded as the source of this biological information. This leaves only a Sender who operated outside our three-dimensional world.

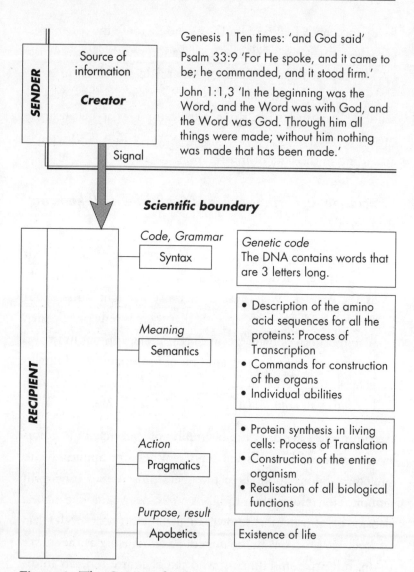

Figure 2: The Origin of Life

If one considers living things as unknown systems to be analysed with the help of natural laws, then one finds all five levels of the definition of information:

statistics (here left off for simplicity), syntax, semantics, pragmatics, apobetics. In accordance with the natural laws of information, the origin of any information requires a sender equipped with intelligence and will. The fact that the sender in this case is not observable is not in contradiction to these laws. In a huge library with thousands of volumes, the authors are also not visible, but no one would maintain that there was no author for all this information. If one penetrates beyond the boundaries set by the limits of natural science by consulting the Bible, the Sender reveals himself as the almighty creator.

After a lecture at a university about biological information and the necessary Sender, a young lady student said to me: 'I can tell where you were heading when you spoke of an intelligent Sender—you meant God. I can accept that as far as it goes; without a Sender, that is, without God, it wouldn't work. But who informed God so that he could program the DNA molecules?' Two explanations spring to mind:

Explanation A: Imagine that this God was considerably more intelligent than we were, but nevertheless limited. Let's assume furthermore that he had so much intelligence (and thus information) at his disposal that he was able to program all biological systems. The obvious question then certainly is: Who gave him this information, and who taught him? This would require a higher information-giver I_1—a 'super-god'—who knew more than God. If I_1 knew more than God, but was also limited, then he would in turn require an information-giver I_2—a 'super-super-god'. So this line of reasoning leads to an extension of this series—I_3, I_4 ... to $I_{infinity}$. One would require an infinite number of gods, such that in this long chain every $n^{th} + 1$ deity always knew more than the nth. Only once one reached the I_{infinity} 'super-super-super . . . god' could we say that such a god would be unlimited and all-knowing.

Explanation B: It is more simple and satisfying to assume only a single Sender—a prime mover, an ultimate creator God. But then one would need to also assume that such a God is *infinitely intelligent* and in command of *an infinite amount of information.* So he must be *all-knowing* (omniscient).

Which of these two explanations is correct? They are logically equivalent, but only one is realistic. In reality, there is no such thing as an actual infinite number of anything.[10] The number of atoms in the universe is unimaginably vast, but nevertheless finite, and thus in principle able to be counted. The total number of people, ants or grains of wheat that have ever existed is also vast, but finite. Although infinity is a useful mathematical abstraction, the fact is that in reality there can be no such thing as an infinite number of anything that can be reached by counting for long enough. Thus Explanation A fails the test of plausibility, leaving only explanation B. That means there is only *one* Sender. But this one Sender must therefore be *all-knowing*.

This deduction is a consequence of consistently applying the laws of nature about information. It has led us to the same conclusion as that which the Bible also teaches, that there is only one God:

I am the first and I am the last; apart from me there is no God.[11]

What does it mean that God, the author of biological information, the creator, is infinite? It means that for him there is no question that he cannot answer, that he knows all things. Not merely all things about the present and the past; even the future is not hidden from him. But if he knows all things—even beyond all restrictions of time—then he himself must be eternal. So through logical reasoning (without the Bible) we have found out why the Bible says that from contemplating the works of creation we can conclude the *eternal* power of God.[12] The fact that God is eternal is attested to by the Bible in many places.[13]

Conclusion 3: God is immensely powerful.

> Because the Sender ingeniously encoded the information in the DNA molecules, must have constructed the complex biomachinery which decodes the information and carries out all the processes of biosynthesis and was responsible for all the details of the original construction and abilities of all living things, we can conclude that the Sender wanted this all to be so and that he must be immensely powerful. [Applying LNI-3, LNI-7, LNI-9, LNI-10]

In Conclusion 2, we determined, on the basis of laws of nature about information, that the Sender (creator, God) must be all-knowing and eternal. Now we consider the question of the extent of his power. 'Power' encompasses all that which would be described under headings such as strength, creativity, capability and might. Power of this sort is absolutely necessary in order to have created all living things.

Because of his infinite knowledge, the Sender knows, for example, how DNA molecules can be programmed. But this knowledge is not sufficient to fashion such molecules in the first place.[14] Taking the step from mere knowledge to practical application requires the capacity to be able to build all the necessary biomachinery in the first place. Research enables us to observe these 'hardware systems'. But we do not see them come about other than through programming, and they themselves are required to transmit and carry out the programs. Thus they had to originally be constructed by the Sender. He had the task of creating the immense variety of all the basic biological types (created kinds), including the construction specifications for their biological machinery. There are no physico-chemical tendencies in raw matter for complex information-bearing molecules to form themselves. Without creative power, life would not have been possible.

We can't even begin to quantify the enormous degree of power required to create life on earth originally. But the Bible shows us the real extent of the Sender's power by presenting him as all-powerful—omnipotent, almighty:

> 'I am the Alpha and the Omega,' says the Lord God, 'who is, and who was, and who is to come, the Almighty'.
>
> For nothing is impossible with God.[15]

Conclusion 4: God is spirit.

> Because information is a non-material fundamental entity, it cannot originate from a material one. We can therefore conclude that the Sender must have a non-material component (spirit) to his nature. [Applying LNI-1, LNI-2]

Unaided matter has never been observed to generate information in the natural-law sense, that is, information with all five levels (statistics, syntax, semantics, pragmatics, apobetics). Information is a non-material entity and therefore requires for its origin a non-material source.

We have already reasoned our way to some characteristics of the Sender. Now we have a further one: he must be of a non-material nature, or at least must possess a non-material component to his nature. That is exactly what the Bible teaches: 'God is spirit, and his worshippers must worship in spirit and in truth.'[16]

Conclusion 5: There is no human being without a soul.

> Because people have the ability to create information, this cannot originate from our material portion (body). We can therefore conclude that each person must have a non-material component (spirit, soul). [Applying LNI-1, LNI-2]

Evolutionary (especially molecular) biology is locked into an exclusively materialistic paradigm. Reductionism, in which explanations are limited exclusively to the realm of the material, has been elevated to a fundamental principle. With the aid of the laws of information, materialism may be refuted as follows.

We all have the capacity to create new information. We can put our thoughts down in letters, essays and books, or carry on conversations and give lectures. In the process we are producing a non-material entity: information. (The fact that we need a material substrate to store and transfer information has no bearing on the nature of information itself.) From this we can draw a very important conclusion, namely, that besides our material body, we must have a non-material component. The philosophy of materialism can now be scientifically refuted with the help of the laws of nature about information.

The Bible, too, corroborates the conclusion that a person is not purely material when it says: 'May God himself, the God of peace, sanctify you through and through. May your whole spirit, soul and body be kept blameless at the coming of our Lord Jesus Christ.'[17] The body is the material component of a person, while spirit and soul are non-material.

Conclusion 6: The big bang is impossible.

Since information is a non-material entity, the assertion that the universe arose solely from matter and energy (scientific materialism) is demonstrably false. [Applying LNI-2]

It is widely asserted today that the universe owes its origin to a primeval explosion in which only matter and energy was available. Everything that we experience, observe and measure in our world is, according to this view, solely the result of these two physical entities, with nothing else added. Energy is clearly a material entity, since it is correlated with matter through Einstein's mass/energy equivalence relationship ($E = mc^2$).

Is the big bang theory just as refutable as a perpetual motion machine? Answer: Yes, with the help of the laws of nature about information. In our world we find an abundance of information, specifically in the cells of all living things, but also the enormous amount of information produced by people, which is today stored in the libraries of the world and on the World Wide Web. According to LNI-1, information is a non-material entity and therefore cannot possibly have arisen from unaided matter and energy. Thus the common big bang worldview is false.

The Bible, too, teaches that this world has not arisen from a process over billions of years, but through creation by an all-powerful God: 'For in six days the Lord made the heavens and the earth, the sea, and all that is in them, but he rested on the seventh day.'[18]

Conclusion 7: There is no evolution.

> Since (1) biological information, the fundamental component of all life, originates only from an intelligent sender, and (2) all theories of chemical and biological evolution require that information must have originated solely from matter and energy (no sender), we conclude that all theories or concepts of chemical and biological evolution (macroevolution) are false. [Applying LNI-1, LNI-2, LNI-6, LNI-8]

Judging by its worldwide following, evolution has become probably the most widespread teaching of our time. In accordance with its basic precepts, we see an ongoing attempt to explain all life on a purely physical/chemical plane (reductionism). The reductionists prefer to think of a seamless transition from the non-living to the living.[19]

With the help of the laws of information we can reach a comprehensive and fundamental conclusion: the idea of macroevolution—that is, the journey from chemicals to primordial cell to man—is false. Information is a fundamental and absolutely necessary factor for all living things. But all information—and living systems are not excluded—needs a non-material source. The evolutionary system, in the light of the laws of information, shows itself to be an 'intellectual perpetual motion machine'.

Now the question arises: Where do we find the Sender of the information on the DNA molecules? We don't observe him, so did this information somehow come about in a molecular biological fashion?

The answer is the same as that in the following cases:

- Consider the wealth of information preserved in Egypt in hieroglyphics. Not a single stone allows us to see any part of the sender. We only find these 'footprints' of his or her existence, chiselled into stone. But no one would claim that this information arose without a sender and without a mental concept.
- In the case of two connected computers, exchanging information and setting off certain processes, there is also no trace of a sender. However, all the information concerned also at some point arose from the thought processes of one (or more) programmer(s).[20]

The information in the DNA molecules is transferred to RNA molecules; this occurs in an analogous fashion to one computer transferring information to another. In the cell an exceptionally complex biomachinery is at work, which translates the programmed commands in an ingenious fashion. But we see nothing of the Sender. However, to ignore him would be a scientifically untenable reductionism.

We shouldn't be surprised to find that the programs devised by the sender of biological information are much more ingenious than all of our human programs. After all, we are here dealing (as already explained in Conclusion 2) with a Sender of infinite intelligence. The creator's program is so ingeniously conceived that it even permits a wide range of adaptations to new circumstances. In biology such processes are referred to as microevolution. However, they have nothing to do with an actual evolutionary process in the way this word is normally used, but are 'parameter optimisations' within the same kind.

In brief: the laws of information exclude a macroevolution of the sort envisaged by the general theory of evolution. By contrast, microevolutionary processes, with their frequently wide-ranging adaptive processes within a kind, are explicable with the help of ingenious programs instituted by the creator.

The Bible, too, emphasises repeatedly in the account of creation that all plants and animals were created after their kind. This is repeated nine times in the first chapter of the Bible, for example:

> And God said, 'Let the land produce living creatures according to their kinds: livestock, creatures that move along the ground, and wild animals, each according to its kind.' And it was so. God made the wild animals according to their kinds, the livestock according to their kinds, and all the creatures that move along the ground according to their kinds. And God saw that it was good.[21]

Conclusion 8: There is no life from matter.

Because the essential characteristic of life is a non-material entity, matter cannot have given rise to it. From this we conclude that there is no process in matter which leads from non-living chemicals to life. No purely material processes, whether on the earth or elsewhere in the universe, can give rise to life. [Applying LNI-1]

Proponents of evolutionary theory assert: 'Life is a purely material phenomenon that will arise whenever the right conditions are present.' However, the most universal and basic characteristic of life—information—is of a non-material nature. Thus we can apply LNI-1, which says: 'A purely material entity cannot generate a non-material entity.'

We repeatedly hear of the discovery of water somewhere in our planetary system (for example, Jupiter's moon Europa), or that carbon-containing substances have been demonstrated somewhere in our galaxy. These announcements are promptly followed by speculations that life could have developed there. This repeatedly reinforces the impression that, so long as the necessary chemical elements or molecules are present on some astronomical body, and certain astronomical/physical conditions are fulfilled, one can more or less count on life being there. But as we have shown, this is impossible. Even under the very best chemical conditions, accompanied by optimal physical conditions, there would still be no hope of life developing.

Since the phenomenon of life is ultimately something non-material, every kind of living thing requires a mind as its ultimate initiator. Australian scientists Batten, Ham, Sarfati and Wieland thus correctly write: 'Without intelligent, creative input, lifeless chemicals cannot form themselves into living things. The idea that they can is the theory of spontaneous generation, disproved by the great creationist founder of microbiology, Louis Pasteur.'[22] With this new type of approach, applying the laws of information, Conclusions 7 and 8 have both shown us that we can exclude the spontaneous origin of life in matter.

Summary

No one has ever observed water flowing uphill. Why are there no exceptions to this? Because there is a law of nature that universally forbids this process. Many plausible arguments have been raised against the teachings of atheism, materialism, evolution and the big bang worldview. But if it is possible to find laws of nature which contradict these ideas, then, since laws of nature have the strongest scientific credibility imaginable, we will have brought about their total downfall. We will have done so just as effectively as the way in which perpetual motion machines (those which supposedly run forever without any energy from outside) have been shown, using the laws of nature, to be impossible.

This is precisely what we have been able to show in this chapter. We have presented ten laws of nature about information. These give rise to comprehensive conclusions—about God, the origin of life and humanity, and the four abovementioned materialist ideas. With the help of laws of information we have been able to refute all of the following:

- the purely materialistic approach in the natural sciences
- all current notions of evolution (chemical, biological)

- materialism (for example, the idea that man is purely matter plus energy)
- the big bang as the cause of this universe
- atheism.

5

The Human Body: Evidence for Intelligent Design

Frank J. Sherwin

Evolutionary naturalists attempt to short-circuit the design argument for God's existence by saying that living things (such as the human body and its eleven systems) only *appear* designed—they aren't really. The biologist Stephen C. Meyer said in 1996, 'According to Darwin, living organisms only *appeared* to be designed by an intelligent creator; nature itself was the real creator.'[1] Darwinists attribute the exquisite detail in the living world to genetic mistakes (mutations) and natural selection that essentially take the place of a personal creator. As we shall see, however, these alleged mechanisms cannot explain the origin of the systems under discussion.

A popular tale that describes the position of the naturalist versus the supernaturalist (that is, their conflicting worldviews) concerns a theist and an atheist taking a walk through the woods. They suddenly come upon a beautiful crystal sphere, two feet in diameter, lying on the trail. Stopping, they examine the blemish-free structure. The theist asks, 'I wonder how this came to be here?' The atheist states that it's obvious somebody left it there: 'It didn't just appear.' Seizing the opportunity, the theist then asks, 'If the sphere was 20 feet in diameter, would someone still have to have left it in the woods?' 'Yes,' the atheist replies, 'logic dictates that if a small sphere had to be left here by someone, clearly a much larger one would also have to have the same means of origin.' 'What if,' the theist asks, 'the sphere was as large as this planet—would Someone have to have left it there?' The atheist realises the conundrum and quickly replies that *no one had to*—it would just be there.

The design argument is eminently logical. One naturally ascribes design (such as a crystal sphere) to a designer. Even Nobel Prize winner Francis Crick admitted as much when he wrote, 'Biologists must constantly keep in mind that what they see was not designed, but rather evolved.'[2]

Ice crystal formation is occasionally used by evolutionists as an example of sophistication that comes about randomly—design without a designer, as it were. But we cannot extrapolate this argument to organic life. Even a cursory look shows there is a significant difference between the design of a snowflake and a section of, say, the DNA molecule. That difference is *information*—stored data or knowledge. Life is all about complex information, ordered in such a way that it necessarily excludes a mechanistic or secular explanation.

Although ice crystals do follow the laws of physics and chemistry giving repeated order (that is, they are formed from a vapour in which the water molecules are oriented and moving according to random processes), *they do not convey any information*. An ice crystal is composed only of water, whereas something like an enzyme (a large protein molecule) is composed of hundreds of strategically placed amino acid molecules. It requires a code or blueprint to make a functional enzyme. Not so an ice crystal, which is simply a repeated order of identical water molecules.

Even with all the ingredients assembled in a kitchen, one would never expect a cake to come about through chance, time and natural processes. One must follow a recipe to get the delicious product (and even then the cake might be substandard!). Proteins such as enzymes are constructed by way of a very specific code (like a cake recipe). And a code requires a code-giver, just as a recipe requires some knowledgeable person to have written it.

Examples of design in humans

The human body is a marvellous testament to complexity and design. As a former college biology instructor who taught human anatomy and physiology to pre-medical students for many years, I can attest to this amazing detail. So much can be said about the tissues, organs and systems—and their interactions—that make up the created human body. Instances abound that point clearly to detailed design.

- The lungs that look like a pair of pink sponges in our chest, for example, contain about 600 million tiny air sacs called alveoli and have 750 woven miles of blood vessels. If the lungs were flattened out, they would cover a surface area of about one thousand square feet.

- Our bone is stronger than granite. A block of bone half the size of a computer mouse can support ten tons—four times the capacity of concrete. In an 'average' adult, the marrow of the flat bones (skull bones and ribs) also makes 2.5 million delicate red blood cells *per second*, as well as providing anchor attachments for our muscles.
- Our heart, composed of unique cardiac muscle, beats at least 2.8 billion times during the average lifespan—resting between beats. This means it pumps 600,000 tons of blood in the average lifetime through 60,000 miles of blood vessels. If skeletal muscle from the arm or leg tried to do what the heart does day out and day in, it would be useless within minutes. The heart is a wonder of engineering.
- We have two kidneys that contain 1.3 million amazing units of filtration called nephrons. With every beat of the heart one-third of the blood goes to the renal arteries, and thus through this pair of reddish-brown, bean-shaped organs. This means they filter about eight quarts of blood every hour. The kidneys are designed, at the level of the nephron, to return glucose, ions, water and other important substances to the blood while expelling two litres (three pints) of urine a day. Anatomists estimate our kidneys are about as complex as the brain. They are certainly not just a pair of blood filters, as many think.

There are many other illustrations of design in the human body. The following are some detailed examples.

Molecular machines

Perhaps one of the most fascinating discoveries in the field of cell biology was the revelation that our seventy-five trillion cells have tiny (submicroscopic) 'machines'. Two authors described these tiny machines this way:

> Most forms of movement in the living world are powered by tiny protein machines known as molecular motors. Among the best known are motors that use sophisticated intramolecular amplification mechanisms.[3]

The forces generated by these cell motors are described by scientists as 'gigantic'. These machines aid in various cellular processes including cell division, organelle transport, ciliary function and cargo transport along tiny protein filaments or tracks in the cytoplasm (the watery gel) of our cells:

> Myosin V [a member of a diverse protein family, the most popular of which is the myosin that makes up your muscles], a cellular motor protein, carries cargo within cells by moving along actin filaments. It takes 37-nanometre steps by placing one 'foot' over the other.[4]

These amazing motors are not made of metal, of course, but of protein molecules—the same protein found in the meat, fish and eggs that make up our diet. Research methods involving cell ultrastructure, which can only be seen with the aid of an electron microscope, have revealed organelles that could not even be imagined a mere three decades ago. Thanks to special electron micrographs, motors in the individual cell can actually be seen after using special techniques that create an artificial shadow.

Like so many areas of biology, research into molecular

motors is in its infancy, with the roles of many motors unknown. One question can be asked, however: If it takes the concerted efforts of designers and engineers to construct large motors for cars or power plants, what of these sub-microscopic protein motors that scientists themselves call 'ingenious'?

Another question is also justified. Could mistakes in the genetic code (DNA) called mutations produce such sophistication on such a small scale? No. Mutations may cause motor *defects* leading to severe flaws and possibly lethal diseases. Such ailments include degeneration of the retina (retinitis pigmentosa) and even anthrax susceptibility. But we would not expect 'the undirected process of mutation' to be creative, a fact recently acknowledged by evolutionary naturalists:

> Domain shuffling aside, it remains a mystery how the undirected process of mutation, combined with natural selection, has resulted in the creation of thousands of new proteins with extraordinarily diverse and well-optimized functions.[5]

Transcription and translation

People have an average of about seventy-five trillion cells. A majority of these have a membrane-bound nucleus that contains the amazing code of life, DNA. (Cells without a nucleus are the short-lived mature red blood cells.)

In order for cells to maintain their integrity and undergo life processes—such as breaking down food (carbohydrates) for energy—they must have a myriad of special molecules called proteins. These numerous organic compounds (there are about

100,000 different kinds) are the most important in the human body. Throughout a person's lifetime their cells are constantly undergoing a complex process called protein synthesis.

This process results in two basic protein molecules: *structural proteins* and *regulatory proteins*. The structural proteins are found, for example, in muscle, skin, internal organs and connective tissues, all of which make up about 20 per cent of our body weight. Regulatory proteins are huge, fascinating molecules called enzymes. These structures catalyse biochemical reactions without themselves being changed or used up in the process. (For decades the protein nature of enzymes was indisputable, but recent discoveries involving RNA have caused a re-evaluation of this thinking.) Other types of proteins include hormones such as insulin and antibodies (gamma globulins).

How do our cells make these important proteins? The process is as amazing as it is complex.

To begin with, the cells need a 'pool' of special compounds called amino acids that contain the important element nitrogen. The cells receive these amino acids from the foods we eat—mainly meat, fish and eggs. This is why, if you are a vegetarian, it is critical to understand what non-meat foods have all the essential amino acids.

Next, using these amino acids, the cells participate in a process known as 'transcription and translation'. Here's how it works.

As previously mentioned, our cells contain a nucleus that encloses DNA, the 'life code' or genetic blueprint. This code dictates to the cellular machinery what proteins to manufacture and when. DNA contains units of heredity called genes, and these specific sections of the DNA molecule are like recipes—code for a polypeptide (a chain of more that twelve amino acids) or a protein molecule. Cells in the pancreas, for example—called beta-cells—are designed to make the protein hormone insulin. The DNA within the cell's nucleus contains the recipe (gene) to make insulin.

Now since DNA does not leave the nucleus with the insulin production instructions, it must have a messenger to take the insulin message out to the cellular machinery designed to make the various proteins. The messenger is called messenger RNA (mRNA) and it *transcribes* (via enzymes) the insulin recipe gene from the exposed section of the DNA. The mRNA molecule—which is now the insulin recipe—leaves the nucleus and goes to a fascinating unit called the ribosome. This molecular machine is made of another kind of RNA (ribosomal RNA) and enzymes. The ribosome is designed to *translate* the mRNA by enzymatically assembling (synthesising) a corresponding sequence of amino acids to make insulin. Such protein synthesis is occurring right now at a fantastic rate in the cells of our bodies.

One must ask: how did these enzymatic processes of transcription and translation evolve? Was it by chance, time and natural processes—or was it by intelligent design? Keep in mind that it takes protein to make protein. In other words, in order to make an enzyme, one needs to have dozens of enzymes in place already. It is this organisation and order that non-evolutionary scientists recognise and that makes the biological sciences such a delight to study.

Homeostasis

Picture someone threatened by something or someone around them. They immediately prepare for action. Heartbeat and respiration increases and internal energy reserves are activated. This prepares the person for flight or confrontation. Their body even generates certain hormones involved with the 'flight or fight' syndrome.

These, of course, are familiar physiological reactions we've all experienced—another example of the body maintaining its internal equilibrium. This tendency of the body to resist change and to maintain itself in a constant state of stable equilibrium is called homeostasis, derived from two Greek words meaning 'remaining the same'. Achieving and maintaining this state of 'normalcy' is no easy feat for our body. So how is it that billions of people are able to go about their daily lives, not having to worry about their blood pressure or temperature dropping or shooting up to dangerous levels, or their heart suddenly racing for no apparent reason?

Homeostasis is often mediated by a principle called negative feedback regulation. This is a fancy way of saying that our bodies have been created to operate within narrow, predetermined limits. If something happens and one of our systems goes outside those limits, our bodies are designed to automatically return that system to specific tolerances.

One example is with body temperature. Here the body's regulatory systems operate much like a thermostat in a home. If it gets too cold and the temperature inside the house drops, the thermostat senses this and turns on the heating. If the house gets too hot, the air conditioner is activated. Our bodies are designed to respond in a similar way. Any deviation from a set point (for example, a body temperature of 37°C) involves activating the nervous system, various hormones and possibly conscious behavioural actions. For example, if we are standing motionless in a very cold area, the chances are we will begin to shiver uncontrollably. Shivering is a rapid and violent contraction of our skeletal muscles that produces great quantities of heat. We may also jump up and down, again producing muscle contraction generating heat. These acts, both conscious and unconscious, are intended to maintain stable internal conditions—homeostasis.

The brain

How do we see? This question seems basic. Hasn't sophisticated medical research all but answered it? Many people probably think scientists have clearly devised a step-by-step description of what happens when light (photons) enters our eye. But such is not the case, and this is because we still do not understand the main organ of sight: our brain. The brain continues to be an enigma—and a marvel of creation.

The central nervous system is composed of the brain and spinal cord. The weight of the average human brain is 1.3 kilograms (three pounds). Anatomists estimate that the brain is composed of 100 billion interconnected neurons (nerve cells). To give you an idea of this number, it's about the number of stars in the Milky Way galaxy. These neurons are linked to one another by 100 *trillion* synapses (tiny junctions between nerve cells, up to 50,000 in each neuron). The outer covering of the brain is the three millimetre-thick cerebral cortex, which makes up much of the brain's grey matter. The total surface area of this convoluted part of the brain is about 16 square feet and it involves almost everything that makes you human—from creative impulses to dreams and thoughts.

Our brain takes up a mere two per cent of our body, but it uses about 20 per cent of our energy. It is the brain that receives sensations and information from the world around us, resulting in various emotions and control for sleep, thirst, movement, hunger and every other important activity required for survival. In a blindingly rapid manner, electrochemical impulses pass from neuron to neuron at almost 180 miles per hour, according to one 1966 study. In this manner new information reaches the brain to be sorted, stored or ignored (if you consciously sensed *every* impulse from the environment you would rapidly go mad from overload).

A fascinating fact about the brain is that although it registers pain from receptors found throughout the body, it has not been created with pain receptors of its own. When a person has a headache, it is the sensory receptors 'firing' within swollen blood vessels (meningeal branches of the maxillary and occipital arteries) in and around the skull; it is not in actual neurons.

What are the chances that this three-pound wonder is a result of chance, time and genetic mistakes (mutations)? There are about a thousand regions in the brain designed to perform specific tasks. Zoology students are sometimes asked to 'trace the evolution of the brain'. Their answer is simply to draw and/or list the major areas and features of the brains of fish, amphibians, reptiles, birds and mammals—which are all vertebrates. But this is hardly brain evolution. It is merely recording major neural features of representative animals. The brain of a vertebrate animal, no matter how 'simple' the creature, is extremely complex, with no clue as to its supposed evolution.

The limbic system

Have you ever fainted, blushed or had 'butterflies in the stomach'? These are only some of the numerous gut reactions that accompany emotions due to activation of our limbic system. This amazing portion of our central nervous system is at the core of our brain and is involved with coordinating emotions (for example, behavioural drives and emotional states), processing memories and aiding the powers of reason.

The limbic system is made up of several components, including the reticular formation, the thalamus, and a bit of very sophisticated tissue called the hypothalamus. The later two are located above what is called the midbrain (mesencephalon).

Some scientists suggest that the limbic system appeared early in the evolution of the vertebrates, that the tissue is evolutionarily older than the cerebral cortex, and that the midbrain is less well-developed in people. But these musings are theoretical and have no scientific (observable and testable) basis.

The hypothalamus is quite amazing. It is located at the centre of the brain and is concerned primarily with regulating the physiological state of the body and maintaining homeostasis through the autonomic nervous system. The hypothalamus responds to thirst and hunger, regulates body temperature, blood composition, waking, sleeping and digestion, and is largely responsible for the physical effects of emotion such as fear and rage. For example, fear causes the hypothalamus to activate the sympathetic nervous system, directing the adrenal glands that sit atop our kidneys to secrete adrenaline (epinephrine). Nerve impulses from the hypothalamus also signal the pituitary gland to secrete hormones that control other parts of the body.

The reticular formation is a group of nerve fibres that carry information to parts of our brain including the hypothalamus and limbic system. These numerous nerve connections allow the reticular formation to modulate the central nervous system, and thereby control both the wake–sleep cycle and just how conscious we are while awake. Indeed, a portion of the reticular formation is the reticular activating system (RAS) in the brain stem that controls alertness and consciousness. Consider the RAS as a monitor that takes in all sensory information leading to the brain and controls our state of sleep and wakefulness. We focus our attention on events outside us based on how active the reticular formation is. To a large extent, sleep is the gradual shutting down of the reticular formation, while an alarm suddenly activates it. Destruction of fibres composing the reticular formation results in coma. Barbiturates and anaesthetics depress RAS pathways to the cerebral cortex.

What is the connection between the brain and the mind? The physical brain (grey matter) and what we call the mind (the seat of consciousness or the totality of mental activities) are *not* the same. Unfortunately, many biologists today, who might be termed 'behaviourists' or at best 'materialists', state that the mind does not exist because it is difficult to define and measure. Recently, however, those in the field of neurobiology have discredited much regarding behaviourism. Work continues with psychologists, neuroscientists and perhaps even theologians addressing the important question of the mind–brain relationship.

Portal systems

A portal system may be defined as an arrangement of two beds of tiny blood vessels in series in which the second bed of vessels receives venous blood from the first. Such a system provides an effective means of one-way communication by hormones and other materials such as food. The blood vessel that connects the two beds is called a portal vessel—named after its destination. The detail of these tiny vessels, and the manner in which they are organised to (for example) specifically take up hormones and deliver them, is evidence of intelligent design and organisation.

There are a number of very intricate portal systems in the human body. Here are just two examples.

Pituitary portal system

The master gland of our endocrine system is the pituitary. It is situated at the base of the brain, close to the hypothalamus. There are no nerve connections between the anterior pituitary and the hypothalamus, but there is a complex of capillary beds within the pituitary that forms a direct vascular (blood vessel) link.

The vessels between the hypothalamus and the anterior pituitary are designed to carry blood and hormones from one capillary network to another. Hypothalamus hormones can leave the bloodstream and affect cells of the anterior pituitary due to special design features in endothelial cells. These cells are unusually permeable, allowing large molecules to leave or enter the circulatory system. In this way, critical pituitary hormone secretion is regulated.

How could evolution possibly account for such fine-tuning?

Hepatic portal system

A delicious meal is enjoyed, and afterwards we are satiated—we have a feeling of fullness. Most of us don't think much about what happens to food after we eat it, but it certainly reveals intelligent design.

All food consumed goes to the liver, an extremely complex organ and the second heaviest in our body (the skin is the heaviest). The food we eat is mixed into a warm, yoghurt-like consistency called chyme that is then squirted into our small intestine via the enterogastric reflex. An efficient absorption process occurs whereby blood vessels surrounding the small intestine pick up these absorbed nutrients (fats, carbohydrates and so on) and convey them directly to the liver. We can say the liver has 'first shot' at all foods we eat before they enter the general circulation.

The hepatic portal system is a complex network of blood vessels (veins) that connects capillaries of the intestines, stomach, spleen and pancreas into the sinusoids—small, blood-filled spaces of the liver that are also capillaries. The vessel that connects these two beds is the hepatic portal vein. The blood in this vein is rich in foods such as sugars, amino acids and fats. What is amazing is that the liver 'knows' what to do with these products—excrete them if they are harmful, store some (such as sugars), and convert the rest into energy.

As you can imagine, food has the potential of having quite a bit of bacteria and other disease-causing material associated with it unless cleaned and/or cooked thoroughly. The liver is designed to give a certain amount of protection from these microbes and is an efficient cleansing and detoxifying organ, cleaning toxins from the blood and bacteria picked up from the intestines.

The immune system

Although we don't like to think of it, every twenty-four hours each person typically inhales from one to ten million disease-causing organisms (pathogens). With this kind of daily invasion of bacteria, toxins, parasites, venom and viruses, an excellent defence mechanism is a must. We call the ability of the body to resist infection and disease our immune system. Indeed, of all the systems of the body, nothing inspires wonder like wandering white blood cells, hungry macrophages, and untold millions of tiny proteins called antibodies.

How did such a detailed system arise? The following survey clearly supports the creation model.

Lymphoid tissue

The creator has designed the body with the thymus, spleen, tonsils and lymph nodes. These structures have lymph flowing through them that filters out bacteria and particles such as viruses. They are the site of critical lymphocyte production and programming.

About half of our immune system involves our thymus gland which is designed to aid in the maturation of the all-important T-cells (the T stands for thymus) that will be discussed in a moment.

Ovoid structures found in the axillary, neck and groin areas are called lymph nodes. They are penetrated by outgoing and incoming lymphatic vessels and are the site of lymphocyte development (the B-cells so important in antibody production).

The spleen is an amazing organ designed to serve as a blood reservoir. Its spaces contain blood, not lymph, and absorb and destroy worn white blood cells and platelets. Blunt trauma to the chest, such as in a car accident, frequently ruptures the spleen leading to a life-threatening condition of internal bleeding. Surgeons cannot fix a ruptured spleen and it is removed. The body can adapt to the loss of this organ with other organs such as the liver taking up the spleen's function.

White blood cells

White blood cells are called leukocytes ('white cells'), and two leukocytes are classified together: the monocytes and lymphocytes. Monocytes are the largest of the white cells and transform into macrophages at inflammation sites, ingesting organisms such as bacteria. Lymphocytes are cells that we perhaps have heard of in the news or remember from high school biology, specifically, the T cells (T-lymphocytes) and B-cells (B-lymphocytes). The T-cells interact directly with foreign invaders and are responsible for what is called the cell-mediated immune response, which is a specific immune response. As single cells go, the T-cell is an amazing work of creation. T-cells are able to detect and recognise invading micro-organisms and cancer cells. B-cells differentiate or specialise into antibody-producing plasma cells after they are activated by a foreign antigen. Such antibody production is part of the humoral immune response.

Seven out of ten circulating white blood cells are neutrophils, designed to engulf and digest bacteria and other foreign particles. They crawl about much like the single-celled paramecium you viewed under the microscope in biology class and release noxious chemicals identical to household bleach. In fact, one evolutionary naturalist called the neutrophil 'a very effective killing machine'. The majority of pus is comprised of dead neutrophils. Graphic as that description is, we can be thankful for this portentous property.

We have five classes of antibodies, special protein molecules designed to affix themselves, like a glove snugly fits on a hand, onto invaders such as bacteria or viruses. As previously mentioned, antibodies arise from plasma cells and are numerous and very effective. For example, a Y-shaped antibody is designed so that it can bind two invading molecules (antigens). The tight binding marks them for destruction by other portions of the immune system such as macrophages. The precise degree of cooperation between these immune subsystems is expected on the basis of creation.

Complement

One of the amazing subsystems of our immune system is what is called complement. The name comes from its 'complementing' other immune responses. Basically, complement proteins float in the blood stream and form pores or holes on the surface of invaders such as bacterial cells. When this happens the integrity of the bacterium is compromised and salts and fluids flow into it. Destruction of the invader soon follows when the bacterium expands and bursts.

Complement proteins are small and *very* specific—how could such a mechanism evolve by chance?

Blood clotting

If there was ever evidence for the creation model, it would have to be in the amazing process of converting blood from a liquid to a solid state known as blood clotting. Such a procedure ensures minimal loss of life-giving blood by closing the wound.

This is done through a complex cascade requiring twelve different clotting factors contained in the plasma (the cell-free, straw-coloured fluid of our blood). Clotting begins with special cellular fragments called platelets (thrombocytes). The platelets are tiny compared to the huge white blood cells from whence they come, but they are critical in the function of blood clotting. Platelets release several important enzymes designed to produce changes around the wound. Such changes reduce bleeding and notify large, wandering white blood cells. Evidence for intelligent design is seen in the release of a special chemical that platelets produce called serotonin. This amazing compound is a neuro-hormone, and among other feats it stimulates constriction of the blood vessels. An important protein complex suspended in our plasma is called fibrinogen. It is converted into insoluble fibrin by the action of an enzyme.

Fibrin contributes to the protein network in which blood cells become entangled—the blood clot.

Think of it, blood normally clots only where it is supposed to. When a clot forms, it doesn't spread throughout the body killing the individual, but stays at the site of the trauma. How did blind evolutionary processes accomplish this specific cascade of events without the creature or person bleeding to death?

Immune system evolution?

Creation scientists maintain that, like all other systems in the body, our immune system with its incredible detail could only have been designed. To their credit, evolutionary naturalists have admitted the enigma of immune system evolution:

> There is still much debate on how the vertebrate immune system evolved and even less consensus on its relationship to defense system in invertebrates.[6]

Evolutionist Charles Janeway states that the immune system 'is the best-characterized system in biology'. This is an amazing statement considering how much is not known about it and how evolutionary mechanisms do not support its evolution. After stating that a white blood cell called the macrophage 'perhaps' came about 'by an unknown evolutionary pathway', Janeway continues:

> One further mystery about macrophages is whether evolutionarily they are the source of dendritic cells and lymphocytes. The origin of lymphocytes . . . is a mystery in itself.[7]

The author uses words like 'perhaps', 'unknown' and 'mystery' in his evolutionary explanation of immune system features. But could it be that this 'best-characterized system in biology' never evolved at all?

Janeway comments that sophisticated means of host defence appear to have been in place by the time organisms diverged into plants and animals.[8] This observation says only that organisms have always had sophisticated means of host defence. Indeed, it supports the creation model that predicts systems were created complex from the start. We find more statements indirectly supporting the non-evolutionary model where Janeway observes that adaptive immunity appears abruptly in the cartilaginous fish.[9] Abrupt appearance is a hallmark of the creation model.

It is obvious from the previous description of the immune system that such an arrangement cannot come about by evolutionary mechanisms.

The Cori cycle

Have you ever had sore muscles after a stint of serious exercise the day before? Those tender muscles are due to lactic acid, a common end product of anaerobic ('without oxygen') respiration. Yet during vigorous exercise muscles that don't receive life-giving oxygen don't die due to lack of oxygen (which is used to break down glucose). There certainly would be no future in that for animals or people that were affected! In some way the body avoids muscle death, and decades ago scientists discovered how.

Muscles are able to break down blood sugar (glucose) for

energy in the *absence* of oxygen. This is called anaerobic respiration. Such exercise produces oxygen debt—a physiological state that occurs when muscle cells must respire without oxygen. The by-product of this respiration is lactic acid, which is toxic. The lactic acid is carried via the bloodstream to the body's detoxifying organ, the liver. In the liver the lactic acid is converted into glucose. These glucose molecules can then be returned to the active muscles or accumulate as a storage form of glucose called glycogen (we normally have two pounds of glycogen stored in our liver and skeletal muscles). The transport of lactic acid to the liver and new glucose molecules being shuttled back to the skeletal muscle is termed the Cori cycle.

Of course, such an anaerobic process is designed only to buy time. People cannot go on indefinitely producing quantities of lactic acid—just ask a distance runner who is not in shape! This is not a problem, however, because within two days the stores of glycogen normally found in the muscle are restored.

How did such an amazing process evolve? Enzymes play a crucial part in both aerobic and anaerobic respiration, and they are by their very nature extremely specific regarding the molecules they work on. How did they evolve—in response to massive muscle death of our primal ancestors? No, such a response would have come too late for the creature. The result would have been extinction, not evolution.

The Circle of Willis
The many blood vessels (veins and arteries) coursing through the human body are obviously quite complex. Cells that make

up tissues and organs need blood high in life-giving sugar (glucose) and oxygen. This is doubly true for organs such as the brain that can survive for only minutes without oxygen. With this in mind (so to speak), there is an amazing system of arteries connected in a unique manner on the floor of the cranial cavity below the hypothalamus, interconnecting the pairs of internal carotid arteries and vertebral arteries. This vascular structure is called the Circle of Willis and it supplies the brain with oxygen-rich blood.

What makes this group of vessels so remarkable is that it provides something called collateral flow—meaning that trauma or damage to any one of the three vessels will not seriously alter blood supply to the brain. The chances for an occlusion (blockage) of the circulation are thereby reduced.

These, then, are some of the astonishing features of the human body. There are many, many more. All indicate a profound degree of intelligent design in the amazing systems and functions of the body. It is impossible for such complexity of integrated design to arise by chance.

Charles Darwin addressed natural selection in his famous book *On the Origin of Species*, published in 1859. In the century that followed, secular scientists referred more and more to natural selection as being a creative force that could—and does—explain all of life and its complexities. But in 1989 a text was written by four evolutionary biologists who admitted that natural selection has no creative powers: 'Natural selection can act only on those biologic properties that already exist; it cannot create properties in order to meet adaptational needs.'[10]

This is what creation scientists have been stating ever since Darwin: natural selection certainly occurs in nature, but it cannot explain the *origin* of the species, systems or functions. Only the creation model can do that. We are the work of a wonderfully intelligent creator.

6

Design in Nature: Evidence for a Creator

Ariel A. Roth

How to make a butterfly

One of the great joys of childhood is playing with various kinds of creepy crawly things that are unfortunate enough to catch the attention of young investigative minds. Not all survive. However, from ants to turtles, these moving things are a superb source of entertainment. Caterpillars, varying in size from the tiny pacing measuring worm to big smooth or woolly ones, are easily captured and analysed. But if there is one adventure that evokes extreme wonder it is placing one of these worm-like caterpillars in a jar and noting that after a number of days it has transformed itself into a beautiful butterfly that can fly.

How could that happen? This is such a complete transformation, it borders on the unfathomable.

Also marvellous are the long and complicated developmental changes we note as so many living things grow from a single microscopic cell to mature organisms as varied as a jellyfish and a palm tree. We are learning that these changes are part of extremely complex processes in which all kinds of details—starting this or that part, neither too early nor too late, regulating the rate of growth, and stopping at the right time—all need attention.

The small larval caterpillar hatches from an egg. One of its important goals is to eat and grow. As it grows its hard body wall becomes too small, so it casts it off and forms a larger one; and this goes on several times until it reaches full size. This can represent considerable growth; the mature caterpillar of the carpenter moth weighs 72,000 times what it did at first. After the larval (caterpillar) stage of a butterfly or moth has reached full size, it then goes into what externally appears to be a quiet pupa stage. But inside, dramatic changes are taking place as the organism transforms itself into an adult that will be very different from what it looked like earlier.

Sometimes, to protect itself, the last caterpillar stage builds a cocoon around itself. In the case of the larva of the silkworm moth, this cocoon is built of silk thread. China became famous by making silk cloth from this thread. Amazingly, a single silkworm that is only eight centimetres long can spin out nearly a kilometre of silk thread in building its cocoon. The adult butterflies or moths that come from the pupa lay eggs and start the complex life cycle all over again. Occasionally, as in the case of the famous monarch butterfly, the adults migrate, sometimes for thousands of kilometres.

We are just beginning to learn some of the details of how caterpillars and other insect larvae change into very different adult stages. These larvae are programmed ahead of time to form adult stages. In the pupa stage, most of the tissues of the larva disintegrate and are used to build the new adult insect. The legs, skin and most of the gut of the larva are not at all related to those of the adult insect. In the larval stage there are over a dozen bodies of inactive cells called 'imaginal discs' located in various parts of the body; the adult will develop from these and a few other tissues of the larva.

We have discovered that a number of genes and hormones are involved in this process. Certain brain cells of the larva secrete what is called a 'brain hormone' that turns on glands in the thorax so they secrete the hormone ecdysome, which in turn stimulates certain genes into action that cause moulting and, at the right time, the growth of adult structures. On the other hand, a hormone called 'juvenile hormone' (which is actually five different hormones) prevents formation of the adult stage during the caterpillar stages, but then decreases in concentration so a butterfly or other adult can form. A number of other hormones have been found that are associated with the moulting process.[1]

All of this is only a minute representation of what is involved in changing a caterpillar to a butterfly.

Perceptive eyes

On one of the more memorable evenings of my life I was diving in the western Pacific Ocean off the coast of Enewetak Atoll, collecting samples of coral to determine how fast they grew during the night. It was getting dark and I needed more samples to get a good representative average.[2] I also was much concerned that some of the sharks in the vicinity might be hungry.

All of this changed into amazement as a shadow dashed over my head. Turning around I could see what looked like an octopus with its sinister eyes looking right at me. It quickly changed colour and darted in front of me. Two of its companions were scurrying from one location to another in an amazing frenzy of activity and colour changes. Unfortunately they soon disappeared into the darkness.

To my chagrin, I never could see them well enough to be absolutely certain they were not small squids or cuttlefishes, which can look very much like octopuses when rushing around in the dark. These three kinds of animals, which are all molluscs, are quite similar. They are especially well known for their soft bodies, many arms (tentacles) and very well developed eyes. Besides that, they are the most intelligent of the invertebrates (animals without backbones).

Stories about huge octopuses and squids are legendary. That these behemoths can grab small ships and haul them down to the bottom of the oceans is apocryphal. Nevertheless, reliable figures are impressive. The giant octopus of the north-western Pacific can reach five metres (16 feet) in length, and when its arms are spread out it has a span almost twice that. There are unconfirmed reports that double those figures.[3]

These are not animals a diver wants to ignore; apart from having a sharp beak, they have eight arms to tangle around a diver while he has only two to untangle with. They have a large tube-like funnel through which they can eject a strong stream of water that will jet them at great speeds. Furthermore, when necessary they can eject dark clouds of ink to hide behind and escape predators. If they lose an arm, they can regenerate one in a few months. Their soft bodies permit them to flow to safety through incredibly small holes, and they can easily change colour as they hide among the rocks. In their skin are special tiny sacks containing different coloured pigments. When the six muscles on the outside of each sack pull on them and spread them out like a plate, the pigment of the sack becomes much more visible. The muscles that control the pigment distribution are controlled from the brain, and by selecting which colour of sacks will be spread out the octopus changes its colour.

Octopuses can learn very fast, and the common ones, which are quite small, could make very interesting pets; however, they require a good supply of seawater to survive. They can easily escape and are both curious and destructive. One researcher reports that the expected survival time of a floating thermometer in a tank with an octopus is about twenty minutes.[4] It is easy to teach an octopus to recognise differences in size, brightness, orientation, shape and even the direction of polarised light (the direction in which light waves vibrate).[5] That is something human eyes cannot detect. On the other hand, as best we can tell, octopuses are colour-blind and cannot recognise the different colours they change into—yet those changes still protect them.

The eye of the octopus is remarkably similar in general structure to our eyes. However, the retina, which is the all-important light detecting layer that lines the inside of the eye, and the focusing mechanism are different. Eyes are very important to octopuses and even a small 250-gram laboratory octopus can have eyes as large as ours.[6] These eyes are incredibly complex organs. The reader is warned that the next few paragraphs are not easy reading; you may not remember or understand everything (no one understands everything about the eye), but persist and you will get a general picture of what we are dealing with.

The octopus eye has an iris, like the characteristic blue, green, brown or grey 'colour' part of your eye. This surrounds the black hole, the pupil, through which light enters the eye. In the octopus the pupil is usually rectangular and matches an elongated part of the retina that is especially sensitive to details. There are at least two sets of muscles in the iris that regulate the size of the pupil so as to control the amount of light getting inside the eye.[7] The octopus also tends to display wide-open pupils during courtship and when adopting a combative stance.

Behind the pupil is the all-important lens that focuses the incoming light so that a sharp image forms on the retina. The lens of the octopus is nearly spherical, which under ordinary circumstances would produce a blurred image, but this problem is solved by gradually changing the optical qualities of the lens. The refractive index continuously decreases towards the edge, causing the light rays to bend less and focus on the retina. A specialist on the visual process in animals has commented, 'This is really quite a sophisticated optical invention not yet duplicated by optical engineers.'[8] The octopus focuses on close objects by moving the lens out and on distant objects by moving it in. It appears that several sets of muscles around the edge of the lens help to push it in, while it is pushed out when a multitude of small muscles in the outside wall of the eye contract.[9]

The retina itself is incredibly complex. It contains twenty million light sensitive cells, and each of these cells—at least as we see in the squid eye, which has very similar cells—contains several sets of 200,000 to 700,000 microscopic tubes (microvilli). These tubes lie at right angles to each other as they face the incoming light, and this arrangement may contribute to the sensitivity of the octopus eye to polarised light. In these tubes lies a light sensitive pigment called rhodopsin. This special protein molecule bends when light strikes it and this initiates an avalanche of biochemical changes that electrically depolarises the charged light sensitive cell, thus starting an impulse to the octopus brain indicating that light has been seen by that cell. There is another kind of molecule related to rhodopsin that is believed to help in restoring the used rhodopsin to its original state so it will again be sensitive to light.

The light sensitive cells, which are very long, provide another mechanism beside pupil size to help the eye adjust to bright or dim light. These cells contain dark pigment granules that absorb light and migrate out towards the light under bright conditions, and get out of the way in the base of the cell when light is dim.

You don't have a pigment migration mechanism to regulate the light intensity in your retina; on the other hand, like the octopus you have six muscles on the outside of each of your eyes that move it around in various directions. In the octopus we find some three thousand nerve fibres, each about one thousandth of a millimetre in diameter, that conduct impulses from the brain to those six muscles so as to carefully control the movement of the eyes. Octopus eyes can follow moving objects, and when moving rapidly they jump quickly, like your eyes do, from one picture to another (nystagmus) so as to reduce blurring. Next to the brain of the octopus are two very complicated spherical organs (statocysts) that can sense both orientation and motion.[10] Without these the octopus is unable to keep its eyes in their normal horizontal position or to discriminate between horizontal and vertical objects.

Bundles of nerve fibres carry impulses from the retina to the two optic lobes of the brain,[11] which are by far the largest part of the octopus brain. Some nerve fibres also carry impulses in the opposite direction, from the brain to the eye, controlling various parts inside the eye such as the lens and iris. In the optic lobes the impulses from the retina are processed by several kinds of nerve cells, and several kinds of chemical transmitters serve to conduct impulses from one nerve cell to another. The visual memory of the octopus also seems to be located in the huge optic lobes. These are very complex structures harbouring some 130 million nerve cells. The rest of the brain has some forty million cells, a small number compared to what is found in the eight very versatile arms that the octopus uses for exploration, attachment and capture. There one finds an enormous 350 million nerve cells![12]

Many kinds of animals have eyes, but they vary greatly. Earthworms have many simple light-sensitive spots that are especially numerous at both ends; some marine worms have as many as 11,000 light sensitive 'eyes'; and a one-celled protozoa can have just one tiny light sensitive eyespot or none at all. More advanced animals like flies and eagles have eyes that generate comprehensive images, but the eyes in these two kinds of organisms are extremely different in structure and how they work.

Closely related to the octopuses are the squids, which have a more elongated body covered by a thick muscular mantle. As with the octopus, squids use a jet of water from a funnel for transport. When the muscular mantle contracts rapidly, a very strong jet can engender extremely rapid motion that can help in escaping predators such as the sperm whale. Some squids are gigantic, with a mantle over five metres long and a total length, including the arms, of 21 metres (65 feet). These are the largest living invertebrates known. These behemoths, which live in deeper ocean waters where light is very dim, need large eyes to collect as much light as possible. The eyes of one of these giants that washed ashore in New Zealand measured 40 centimetres (16 inches) in diameter; and that is significantly larger than an ordinary 30-centimetre world globe! These are the largest eyes we know of; it is estimated that they may have a billion (1000 million; 10^9) light-sensitive cells. And that's not all: just the DNA in each cell of a squid is estimated to have 2,700,000,000 bases.[13]

Some ordinary-sized squids, in the one-metre range, have turned out to be marvellous animals for the study of nerve cells.[14] They have a few huge nerve fibres close to one millimetre in diameter, and that is one thousand times the diameter of the nerve fibre of an ordinary nerve cell. The nerve impulses travel much faster in these large fibres. Their purpose is to carry impulses rapidly to the farthermost muscles of the mantle so that the whole mantle will contract at exactly the same time, resulting in a powerful jet from the funnel.

Much of the knowledge that science has obtained about nerve cells during the last half century has come from the study of these giant nerve cells of the squid. Nerve cells, whether in squids or other animals, can be over a metre long, with the nucleus of the cell near one end and a long nerve fibre transmitting impulses to the other end. An interesting finding is that there are special protein molecules in these cells that carry needed cargo packages back and forth so that the far end of the fibre is provided with special molecules manufactured at the opposite end in the region of the nucleus.[15] Each kind of molecular cargo has its own specific protein-transporting molecule that moves it along minute tubules in the fibre of the nerve cell.[16] While nerve impulses travel at a rate of several metres per second along a nerve fibre, the cargo transporting system, which does not have to move rapidly, takes several days to move molecules one metre. It turns out that in some 'simple' invertebrate animals, not only are we dealing with all kinds of relationships between millions of nerve cells, but inside those nerve cells a multitude of things are happening to keep them functioning properly.

Also very remarkable are the eyes of the now extinct fossil trilobites. These animals, which were remotely similar to a horseshoe crab, had compound eyes like those of a butterfly, with many light sensitive tubes, each pointing in a slightly different direction. Each tube had a lens made of the mineral calcite. Calcite has a complicated but orderly arrangement of atoms, and the resulting crystal bends light at different angles according to how it is oriented. In trilobites the mineral lens of each tube was oriented in the right direction so as to give the proper light input. Furthermore, the lens was shaped in a complex way so as to relate to another light medium and eliminate the fuzziness created by the problem of spherical aberration. Such sophistication in design is comparable to our modern optical systems.[17]

It is hard to think that these highly defined instruments that produce precise images just happen to come together by random changes. It seems more reasonable to think that some very knowledgeable designer was involved.

Seeing without eyes

Two centuries ago the famous Italian scientist Lazzaro Spallanzani noted that when he brought owls into his study room, they would fly when light was very dim but refuse to fly in total darkness. When he repeated the test with bats, they flew confidently all around in total darkness, avoiding wires that had bells attached so Spallanzani could tell if they had been touched. When he plugged up the bats' ears, they ran into all kinds of things. Why? For a century-and-a-half, these facts presented a baffling mystery and added more intrigue to the bat's sinister reputation.

The mystery of how bats navigate in the dark was not solved until an instrument was developed that could detect sounds much higher than the highest notes we can hear, which is at best around 20,000 cycles per second. When bats were brought in around this new instrument, it was discovered that what were thought to be nice, quiet animals were really extremely noisy, emitting all kinds of loud sounds in the 20,000 to 150,000 cycles per second range. Bats and a few other animals (including some shrews, dolphins and whales) use what we call 'echolocation' to see in the dark. They send out a sound from their throats and listen for its echo as it bounces back from an object.

We have discovered that bats perform extremely fast and precise analyses of the returned echo, and it works very well. A bat can easily collect 500 insects an hour in the dark, and that averages out to one every seven seconds. Furthermore, they can avoid running into very small objects; some bats can avoid a wire as thin as a human hair. Bats readily follow a small stone tossed into the air, but they quickly learn to distinguish between stones and insects.

More trivia tells us how these animals perform their astounding feats. The numbers are impressive. When searching in the dark a bat will emit around ten pulses of sound per second, but when it approaches an insect the rate goes up to twenty times faster as it analyses the sound that is bouncing back. It can vary the sound from a constant pitch note to a variable one giving different information. Depending on how long the sound takes to come back, it can tell how far an object is. It can tell its direction, probably as you do, by analysing the quality and timing of the sound coming to its two ears.

Furthermore, bats can easily direct their ear cones in various directions. You no doubt have noticed that the sound of a horn or the siren of an emergency vehicle approaching you is higher pitched than when it is leaving you (the Doppler effect). Remarkably, we have discovered that some bats are able to compensate for this kind of shift, caused by their own motion, by changing the pitch of the sound they emit, and thus bring the returning sound to a preferred processing frequency. In performing this astounding feat they are able to identify changes in sound frequency that are less than one part in a thousand.[18]

The ear of the bat is basically like that of other mammals, but it can detect sound waves vibrating many times faster.[19] At these high frequencies the movement of the eardrum is estimated to be less than one millionth of a millimetre. These very fast microscopic changes are transferred to the inner ear by way of three tiny bones commonly known as the hammer, anvil and stirrup. These bones serve as a lever system that performs the extremely important function of efficiently transferring the sound waves that have travelled in air to smaller but stronger waves in the fluid of the inner ear. Two muscles are attached to these bones and are believed to help compensate for very loud sounds. The sounds are transferred to nerve impulses in the cochlea, which is shaped a little like the shell of a snail. In the bat it is only about three millimetres in diameter and has been called a 'micromechanical wonder'.[20] There, up to 55,000 nerve cells are attached to special hair cells sensitive to vibrations in a membrane that responds in different places to different frequencies of sound. These nerve cells send impulses to the brain according to what is heard, while other nerve cells in the brain send impulses to inhibit neighbouring hair cells, thus focusing the pitch of the sound.

There are at least half a dozen nerve centres in the brain of the bat that function in processing and analysing the sounds coming in from both ears. We find here, as in most biological systems, a multitude of interdependent parts that cannot function without others. This all reflects complexity that seems much more to originate from purposeful design than from random undirected changes.

Interdependent parts

My friend was exhausted. He had worked late into the night and then had to drive a long distance back to the college he was attending. Rest, classes and homework awaited his attention. However, it would be a long time before he would ever get back to college. He was driving along a sparsely travelled country road when weariness overcame him, and his unguided car plunged into the waters of a nearby stream. He survived the ordeal, but his injuries were severe. With the nerves in the lower part of his spinal cord severed he no longer had control of his legs, and he was destined to a wheelchair for the rest of his life.

Healing, if you can call it that, took a long time. Fortunately this friend of mine was no ordinary person. He was not the type who would allow serious problems to transform him into a burden on society. He was going to help others in spite of all the obstacles he faced. He finished college. His engaging personality and dedication to his God helped him as he successfully served as a teacher, editor, chaplain and pastor. His friendliness and understanding warmth helped many.

However, all was not going well. With the nerves to his legs cut, his useless limbs were a constant source of problems. Their tendency towards deterioration became so pronounced that five years after the accident he had them amputated.

The problem my friend had with his legs demonstrates how various parts of living organisms are dependent on other parts.[21] A simple example illustrates this further. If we have a muscle moving a bone in a leg, that muscle will not work unless there is a nerve going to that muscle to activate it. But neither the muscle nor the nerve will work unless there is a system in the brain (or central nervous system) to control the activity of the muscle. The controlling mechanism in the brain sends impulses by way of the nerve to cause the muscle to contract and move the bone. The three parts—the muscle, the nerve and the controlling mechanism—are examples of *interdependent parts;* they need each other in order to function.

In these systems, nothing works unless everything works. Some scientists characterise such systems as having 'irreducible complexity'.[22] The word complexity refers to systems whose various parts are functionally related to each other.

Systems with interdependent parts are abundant in all living things, and are usually much more complex than the simple example of muscles, nerves and control mechanism described above. In the cells of the squid there are an estimated 35,000 different kinds of enzymes.[23] Most of these enzymes function in governing chemical changes related to other chemical changes performed by other enzymes, and as such they represent a vast, complex array of interdependent parts.

The randomness of evolutionary changes

If twenty children are set free in a toyshop, something is bound to happen. Assuredly, the neatly ordered stock of toys will become less organised, and the longer the children are revelling in the store, the more mixed up the stock will become. Active things naturally tend to get mixed up. Molecules of perfume diffuse out of an open bottle into the air; they do not collect from the air to become concentrated in a bottle. Pollution poured into the sea tends to become well mixed up and diluted in the world's oceans. Activity tends to mix things up more and more.

The tendency towards things becoming mixed up in nature runs counter to evolution. Evolution proposes that randomly distributed molecules formed 'simple' life forms that, although small, were actually highly organised. In this case 'small' cannot be equated with 'simple'. Evolution is then further assumed to have formed much more complicated organisms with specialised tissues and organs that include flowers, eyes and wings. Some evolutionists suggest that the occasional self-organisation of simple matter, such as seen in the formation of a salt crystal or the rare wave pattern that sometimes forms when chemicals migrate through solid matter, might be a model for the self-organisation of matter into living organisms. But there is a vast chasm between simple crystals and complex living systems. The development of interdependent functional complexity runs counter to the general tendency in nature toward unorganised mixing. This is a major problem of the theory of evolution.

Evolution usually places major emphasis on the occasional unnatural random change in an organism's heredity mechanism (DNA). Such changes, called mutations, combined with natural selection, are generally considered to be the basis for evolutionary advancement. However, such random events would usually tend to mix things up, not organise them. Neither random mutations nor natural selection have any foresight to plan ahead so as to guide the evolutionary process in the gradual development of systems with interdependent parts. And it is unrealistic to think that a number of 'just right' mutations just happen to occur all at once to produce working systems.

Furthermore, mutations are almost always detrimental to the complex systems of living organisms. An estimate of only one favourable mutation out of a thousand is being generous to evolution. In dealing with complex systems with many interdependent parts, just a small mutation can cause a whole system to stop working. It is somewhat like severing the nerves going to my friend's legs; it spoiled his whole legs. Likewise it is much easier to ruin a watch so that it no longer keeps time than to make one. Few would argue that there isn't a tendency towards randomness in nature and that evolution needs to explain the opposite.

Natural selection: a problem for evolution

Last century, Charles Darwin developed the concept of natural selection, which is the most accepted model for an evolutionary mechanism. He observed that there are small variations among living organisms. There is also overproduction of offspring that results in shortages of food and space, and this results in competition for survival. Only the fittest (the best ones) of new varieties of organisms would survive the competition, and they in turn would produce similarly fit offspring. Thus the fittest, which are considered more advanced, survive through this process. It is called 'natural selection', or 'survival of the fittest'.

Despite the fact that random mutations tend to be

overwhelmingly harmful, this mechanism is often used to suggest evolutionary advancement. While it appears that in nature natural selection does function as a means of eliminating weak aberrant organisms, it faces a major problem when it comes to the evolution of systems with interdependent parts—and almost everything that is alive is full of interdependent parts.

That my friend had his legs amputated illustrates a basic problem faced by Darwin's natural selection model. Useless structures can be cumbersome impediments, and organisms can usually get along better without them. They require extra energy both to produce and to maintain. The problem facing evolution is that many parts of gradually evolving organs or systems would be useless impediments until all the necessary interdependent parts had evolved. Until that time, organisms would get along better without these extra useless parts, and natural selection would tend to eliminate them, thus interfering with the evolution of any complex systems. Only after *all* necessary interdependent parts are present can these systems work and be useful so as to provide any reason for survival through the natural selection process.

Evolutionists have addressed this problem. Often it is suggested that the intermediate stages in the development of a complex structure are useful and have survival value. While it can at times be difficult to state that what appears to be an isolated developing organ or a protein is completely useless, there are so many instances where this does seem to be the case that justification for the argument rests with the evolutionists. For instance, we find hundreds of muscles, nerves and control centres for muscles in even just one small advanced animal like a mouse. It would appear that almost all, if not all, of these muscles would be useless without a specific nerve to stimulate each one. Furthermore those nerves would be useless unless there was a special control centre sending the proper kind of impulse at the proper time to make the muscle work as it should. There would be no survival value for any of these parts without the others.

A second evolutionary explanation is that some structures have changed function. One of the outstanding examples is the claimed scenario that the gill-bar structure of fishes that don't have jaws evolved into the jaws of fishes that do, and part of those jaws later evolved into the hammer and anvil bones of the ear of more advanced animals.[24] Evolutionists have amassed impressive diagrams illustrating how this supposedly happened, but any suggestion of authentication remains only speculative. It is hard to imagine that all of a sudden, all kinds of the right mutations happened at once to produce all these complicated changes to provide survival value. Alternatively, it is also hard to imagine how you could gradually add two more bones to the ear's sound transferring system without awkward intermediate stages that would not provide survival value. According to the survival of the fittest principle, unless each change provides an advantage over the previous stage, natural selection cannot preserve it.

Since so many of the systems in nature have interdependent parts that don't work until all the necessary parts are present, it is not surprising that many examples can be suggested where more than one change is necessary in order to provide any survival value. Many of these involve very complex changes, and too many coincidences are required to favour evolutionary plausibility. Examples related to our earlier discussion include:

1. In the caterpillar, the brain hormone that stimulates the hormone from the thorax to turn on the genes affecting moulting represents three interdependent components.

2. The eye is useless without a brain to interpret what is seen; you need both to provide survival value. For instance, the ability of the eye to distinguish colours is useless in the evolutionary process unless you have comparable advancements in the brain to interpret colours.

3. Why would the muscles that expand and those that contract the pupil of the octopus evolve unless there were nerves for the muscles, and a system in the brain responding to the amount of light so as to exert proper control over pupil size?

4. Why would a system to control the focus of the lens of the octopus eye evolve until you had a system to move the lens? You also need a system to detect the sharpness of the focus.

5. Why would the muscles on the outside of the eye that move it around evolve unless you had a complicated system in the brain to direct this?

6. Why would a bat evolve the ability to analyse the Doppler shift if it did not have a system to rapidly adjust its tone to compensate for this?

7. Why would the muscles that move the minute bones (hammer and stirrup) in the ear evolve until you had the nerve cells, including the right nerve cell connections to control their action? At least four different nerve cells are involved in this system,[25] and they need to be connected the right way.

We could easily multiply this list by a thousand. While it may be possible to postulate some evolutionary reason for some of these parts, the sheer number of interdependent parts suggests that we look elsewhere than survival of the fittest for an answer. For two centuries, evolutionists have been looking for a satisfactory mechanism to try and explain the spontaneous origin of complex systems with interdependent parts. Their persistence is commendable, but unfortunately no satisfactory answer has come forth.[26]

Furthermore, if evolution is a real ongoing process, as evolutionists hold, we should expect to see many examples of new developing organs or systems (legs, eyes, livers, new kinds of organs and so forth) trying to evolve in those organisms that do not have them. Yet as we look at well over a million different species that have been identified over the surface of the earth, we do not seem to see any. This is a major indictment of the widely accepted evolutionary concept. In a broader context the question is this: How can mostly detrimental random mutations, which have no foresight, gradually produce complex biological systems that have no survival value until all interdependent parts are present, so that they can work together and have survival value?

Those who believe in God are sometimes accused of believing in miracles, but you really have to believe in miracles if you think all these problems can be overcome by chance. The data of nature is better explained by the concept of a very knowledgeable intelligence who designed all these complexities. And that is where God comes into the picture.

Doesn't God need a designer?

This is an old question. And if you think that God needs a designer, then you can wonder who designed God's designer, and then who designed the designer of God's designer, and so on *ad infinitum*. This question does not lead to a very satisfactory answer. It is another form of the commonly asked question 'Where did God come from?' Sometimes the implication is that if you don't know where God came from, your information is sketchy and you need not imply that there is a designer God at all. Of course, one can easily equalise the query and ask, 'How did the *universe* come to be?'

However, there is another aspect to this question. Just because I do not know how God came to be does not mean he does not exist, any more than if I don't know how the universe came to be it does not exist. Our questions about ultimate origins are very real, but we have few good answers. The question 'Why is there anything instead of nothing?' baffles us because we all know there is something, but we can't easily answer how or why that something came about.

At present, many scientific interpretations suggest that what happened before or during the first few moments of the postulated 'big bang' is not known. The Bible of the Christian tradition indicates that God has always been and always will be,[27] but we have trouble with that. We especially see a conflict between this idea of God and our ordinary linear concept of time that runs from past to future. However, relativity experiments indicate that time can change, and some scholars suggest a broader concept of time (for example, time not being just linear but having other dimensions such as width). There is a lot we don't know.

Design in Nature

The question of whether God needs a designer is interesting, but it is a very different question from the one of whether there *is* a God or designer. I am certainly willing to accept the existence of many things even though I don't know how they came about. If a crocodile is chasing me, I am willing to concede its existence long before I know how, why or where it came from. Likewise in nature, we can see the evidence of a designer even though we may not know how, why or where the designer came from.

Some also ask, 'But where is this designer God?' We may not see him, but the evidence in nature of his existence is so abundant that it is difficult to deny. We don't have to see God in order to believe he exists. If I examine the remains of a house after it has been gutted by fire, noting the charred timbers, burned roof and melted belongings, it is obvious that there was a fire even though I don't see any flames. Likewise, if in a clearing in a forest I find a well-manicured garden with no weeds and well-ordered rows of flowers, I may not see the gardener there, but I am sure he exists. Although we cannot see God, the evidence of his activity is overwhelming.

For centuries humanity has debated the question of the existence of God. Recent advances in our study of nature are revealing such complex intricacies that it is becoming more and more difficult to think that all these wonderful things came about just by accident. Nature tends toward disorganisation, not organisation. The organisation we see is so complex that an intelligent organiser must be involved. From the minute complex single cell to man, we find a vast array of all kinds of biological systems that would not work until all essential interdependent parts are present. These systems pose a serious challenge to the commonly accepted concept of evolutionary advancement by natural selection because these systems do not provide any survival value until all essential parts are present so the systems can work.

It is time for scientists to seriously consider the possibility that there is a very knowledgeable Designer behind all the complex things we are finding.

7

The Scientific Case for Creation

George Javor

The headline caught my eye: SCIENTIST: LIFE RECIPE EASY. Written by Paul Recer of the Associated Press on October 15, 1998, the accompanying article quoted Bruce M. Jakosky, a University of Colorado planetary scientist, speaking at a meeting of the American Astronomical Society. His recipe for life was: 'Take some chemicals, stir in a source of energy and mix with mild temperatures for a few hundred million years.'

Most biological scientists who actually work with living matter, and who are aware of the incredibly complex structures and processes required for life, would quarrel with the adjective 'easy'. But otherwise they would concur. As far as the scientific establishment is concerned, life on earth evolved spontaneously billions of years ago. This is also what textbooks teach to new generations of students: 'The Earth was formed from a cloud of condensing cosmic dust and gas about 4.5 billion years ago. Life arose soon thereafter.'[1]

The concepts of evolution come to us from ancient Greek

and Roman philosophers and are foreign to the Judaeo-Christian worldview. Nevertheless, among current adherents of evolution are atheists, Jews and Christians.

A straight reading of the first two chapters of the biblical book of Genesis informs the reader that God created earth and all living matter on it in six days. This foundational concept is an unshakeable pillar on which subsequent biblical writers build their histories. But the modern secular mind is not prepared to deal with the possibility that the world came about rapidly by the decisive action of an almost unimaginable power.

Differences in beliefs about origins reflect a diversity of attitudes toward reality in general. For atheists, the universe is governed by impersonal forces of attraction and repulsion. There exists neither supernatural planner nor supernatural implementer. Rather, forces of nature, chance and the passage of time bring about worlds such as ours, provided all necessary conditions for life exist.

In contrast, creationists attribute everything in the universe to an original cause, the will and ability of the creator as described in the Bible. Their world is filled with evidences of superb design and grand purposes.

Between these two diametrically opposite positions stand the theistic evolutionists, who hold onto the concept of God as the first cause but dismiss a literal reading of the Bible. Though unable to marshal theoretical support from either the Bible or the writings of evolutionary theoreticians, the adherents of this position hedge their bets by keeping one foot in both camps.

There are also those who refuse to be drawn into the

controversy. Some take the agnostic stance, declaring it a waste of time to fret over such problems since obviously they cannot be solved. Others maintain that the question of where and how everything started is a matter of personal preference, like one's favourite food or colour—certainly not worth getting all worked up over. In fact, they do not care one way or another. 'The main point is, we're here. Let's concentrate on the here and now!'

But giving up on origins comes at great expense. It means giving up on the meaning of who we really are. It reduces humans to the level of organisms that are incapable of philosophy. Such beings are driven by instincts rather than by principles.

The modern theory of the chemical evolution of life, the process which brings living matter into existence spontaneously from non-living precursors, is not a novel concept. Aristotle (384–322 BC), in his *Historia animalium*, describes spontaneous generation of organisms as well as generation from 'seed'. William Harvey (1578–1657) wrote in his *De generatione animalium*: 'Animals have one thing in common with plants, that some arise from seed and others spontaneously.' Jan Baptista van Helmont (1580–1644), an alchemist, gave the following recipe to generate mice: 'A dirty shirt is stuffed into the mouth of a vessel containing wheat; within a few days . . . the ferment produced by the shirt, modified by the smell of the grain, transforms the wheat itself, encased in its husk, into mice.'

It took two more centuries—and the work of a number of scientists, among them Francesco Redi (1626?–1699?), Lazaro

Spallanzani (1729–1799), Theodor Schwann (1810–1882) and Louis Pasteur (1822–1895)—to convince the scientific establishment that spontaneous generation of life could not happen in the world that now exists or that existed in historic times. Modern evolutionists therefore propose that conditions must have been very different on a primordial, prehistoric world, permitting 'abiogenesis' (formation of life from non-living matter).

Besides life evolving spontaneously on earth, another logical alternative to creation of life here is 'panspermia', the concept that life arrived from somewhere else in the universe. But the search for life on Mars and elsewhere in the solar system has turned up nothing. Understandably, panspermia has fallen out of favour in recent years.

Thus we are left with two alternatives. Either life arose spontaneously or it was created. At first blush, this may seem as far as this topic can be taken. After all, there were no eyewitnesses present at the birth of life. The ancient reports of this event recorded in the Bible are stories preserved through many generations from a dim past. Their authenticity cannot be verified by independent means.

However, if one of the two alternatives can be eliminated decisively then, by default, a clear answer presents itself. The remainder of this chapter is devoted to showing that *life, as we know it, cannot evolve spontaneously under any condition.*

Life is the manifestation of very unusual types of matter. Living matter is able to incorporate certain energy rich substances or even pure energy (light) and use such energies for growth, reproduction and movement. It is capable of sensing its environment and, to some extent, adapting to changes.

The smallest living entities are cells. Some organisms, such as the tiny bacteria, are single cells. Larger and more complex life forms—plants, animals and humans—are composed of many cells. The lives of these organisms depend on the composite interactions of their organs, and these organs in turn depend on the complex interactions of their cells.

The millions of different life forms in existence today arose from their parents. One of the important properties of living matter is that it can reproduce itself. Life only comes from life. Yet the evolutionary paradigm suggests that on a primordial earth, a simple cell, a 'progenote', arose spontaneously and all other forms of life descended from this ancestral cell.

Abiogenesis, both spontaneous and directed, is merely a concept. It has not been observed or demonstrated in the laboratory. Moreover, no one has come up with a convincing scenario of how it could happen, even on paper. Instead, textbooks, monographs and articles offer the reader a mixture of scant laboratory results, vague speculations, hand waving and redefinitions of 'life' to some convenient but unconvincing concept.

Any theory of spontaneous abiogenesis has to traverse the long theoretical road from a barren, sterile, ocean-covered earth, surrounded by an oxygen-less atmosphere, to the formation of at least one live cell.

Chemical analysis of many different types of live cells reveal that they are composed of approximately 70 per cent water, 29 per cent complex organic (carbon-containing) biopolymers (proteins, nucleic acids, complex sugars and fatty substances), and one per cent simpler organic substances plus minerals. In living matter, all of these chemicals are in a dynamic flux, that is, they are continually synthesised and degraded, but the cell's overall chemical composition remains constant.

Life is based on chemical transformations in cells. In a typical bacterial cell many hundreds of such events are occurring simultaneously. Each individual chemical transformation results in one or more new substances. There is nothing magical about these chemical conversions. For example, when glucose is taken up by the bacterium *Escherichia coli*, it combines with phosphate to form a new compound, glucose-6-phosphate. This and every other chemical change is catalysed (promoted) by proteins called enzymes. Enzymes bind specific substances (substrates) and convert them to new compounds (products).

Individual chemical reactions can be duplicated in test tubes. In this example, glucose, ATP (adenosine triphosphate, the source of phosphate) and the enzyme hexokinase would be suspended in a neutral (pH=7) buffer, and the solution incubated at 37ºC. In a few minutes most of the glucose would be converted to the product with a corresponding decrease in ATP concentration and the reaction would be over. This is 'equilibrium', the end of the reaction, where the ratio of product to substrate is constant. When equilibrium is reached, no further observable chemical change occurs.

Since life depends on chemical changes, if all reactions in a cell reach equilibrium, the cell perishes. But in live cells, none of the hundreds of chemical transformations are at equilibrium.

This is most remarkable in light of the fact that every reaction is pushed toward equilibrium by its catalyst.

When glucose is phosphorylated in *E. coli*, the product of the reaction, glucose-6-phosphate, does not remain in the cell. Instead, it is instantly transformed by a second enzyme, an 'isomerase', to a new substance, fructose-6-phosphate. As soon as it is made, this compound is changed to yet another substance. In fact, the original glucose is transmuted step-by-step into ten different kinds of metabolites in a biochemical chain reaction called glycolysis. In this process, some of the original energy from glucose is released and utilised.

Equilibrium is not reached in any of the chemical reactions of glycolysis because these processes are interconnected. The product of one reaction is the starting substance of the next. Like a juggler who has to keep all of the objects flying through the air, the cell needs to keep all of its reactions away from equilibrium. What makes this possible is the existence of interconnected networks of chemical pathways in the cell. In addition, elaborate regulatory processes ensure that the outputs of all biochemical pathways are neither more nor less what is needed by the cell. There are no loose ends; every atom is accounted for and utilised.

The formidability of chemical dynamics within a live cell may be appreciated by considering what happens to an *E. coli* cell when it is treated with a few drops of organic solvent, such as toluene. The cell is killed as the solvent creates holes in the cytoplasmic membrane. With the destruction of its membrane's integrity, the bacterium loses the ability to generate ATP, the chemical currency of energy. Without ATP, energy-requiring reactions come to a halt, and generalised equilibrium soon sets in.

The remarkable thing about this dead bacterium is that it still possesses all of its proteins, nucleic acids and other important components, all in their proper locations within the cell. The only missing component is the steady state non-equilibrium chemical dynamics.

Chemical evolutionary scenarios outline possible schemes whereby the requisite proteins and nucleic acids could come into existence spontaneously. But the example of the dead *E. coli* cell shows that the presence of these important biopolymers, all in their correct intracellular configurations, is still not sufficient to have a live cell.

When the first formal proposals of abiogenesis were made in the 1920s by Aleksandr Oparin and J.B.S. Haldane, modern biochemistry was in its embryonic state. At that time it was just being established that enzymes were proteins by J. Sumner, who crystallised urease. The discoveries of the urea and citric acid cycles occurred five to ten years later, the chemical identity of genetic material was still fifteen to twenty years in the future, the first protein's amino acid sequence came thirty years later, and Dolly the sheep was cloned only in the 1990s.

Proponents of chemical evolution in the third millennium are choosing to stick with a simplistic eighty-year-old model, against a background of a tidal wave of evidence that abundantly shows the mind-boggling chemical complexity undergirding life. Their task is to show how the following processes could have taken place on a hypothetical primordial earth:

1. formation of bio-monomers
2. formation of biopolymers
3. formation of connected metabolic pathways
4. formation of a live cell, where the chemical transformations are in steady state non-equilibrium.

Over the past fifty years dedicated biochemists have striven valiantly to demonstrate the feasibility of these processes. The impetus came from the classic experiments of the graduate student S. Miller working with Dr H. Urey. They demonstrated that a mixture of methane, ammonia, hydrogen and water vapours, exposed to electric sparks for an extended period, formed several amino acids and other biologically relevant substances.

Modifications of this experiment produced most of the twenty different amino acids needed for protein synthesis, as well as the nucleobases of nucleic acids. Thus the feasibility of the first step of abiogenesis had been demonstrated.

But then the laboratory efforts ran into the brick wall of reality. To this day, it has not been possible to perform step 2, linking amino acids into proteins, or nucleobases to nucleotides and nucleotides to nucleic acids under prebiotic conditions that could be reasonably expected on an early earth. The experimental proof of chemical evolution got bogged down on step 2.

If proteins and nucleic acids could be produced under primitive conditions, the sequences of amino acids and nucleotides would be random. Biological information resides in the precise sequences of amino acids of proteins and nucleotides of nucleic acids, just as the meaning of words depends on correct spelling. What is missing is an 'editor' who would preserve biologically useful polymers and discard the rest. The 'editor' would need to 'know' which biochemical processes undergird life.

Finally if, with the help of an 'editor' and a functioning biopolymer synthesising system, cell-like structures were to emerge, such constructs would be equivalent to the dead bacterial cell, because in the process of assembly, the enzymes would create chemical equilibrium among the components.

Our inability to bring back to life intact but dead *E. coli* cells in the laboratory shows the impossibility of spontaneous generation of life. We actually know what needs to be done—changing from equilibrium to non-equilibrium the hundreds of interconnected biochemical pathways. We are just not able to make this happen with our current technologies.

Max Delbruck, the Nobel Prize-winning biologist and a devoted evolutionist, wrote in his book *Mind from Matter?* that he decided not to read 'this literature on prebiotic evolution until someone comes up with a recipe that says "do this and do that and in three months things will crawl in there".'[2]

Far from being 'easy', the evolutionary recipe for life belongs to the same category as the recipe for the perpetual motion machine. It is science fiction at its worst.

A Question of Biology

David Catchpoole

What exactly *is* 'natural selection'? It is often referred to as 'survival of the fittest', meaning that only the strongest organisms survive to pass their genes on to the next generation. This is really a straightforward and commonsense insight from what we observe in the world around us—something anyone thinking on this subject would know to be intrinsically true. Given that in popular parlance the terms 'Darwinism', 'natural selection' and 'evolution' often seem to be used interchangeably, it is not surprising that, in many people's minds, evidence of natural selection is taken as being evidence of evolution.

The confusion is understandable when one considers that most presentations of evolution to the public do not clearly point out that natural selection is *not* evolution since by itself it cannot make new organisms. Natural selection can only eliminate weaker organisms (and thereby their genes) from the population. A group of creatures might become better adapted to a colder environment, for example, by natural selection weeding out those which don't carry the genetic information to make thick fur. But that doesn't explain the origin of the information to produce thick fur.

So by what mechanism could evolution have possibly added all the extra information to progressively transform the alleged first single-celled organism into trees, birds and people? Evolutionists propose that 'mutations' are the key to explaining the new information that evolutionary theory requires. In one sense, mutations fit the evolutionary theory's requirement that when an organism reproduces itself, the copying process must not be completely accurate. But mutations are the *wrong type of change*—they destroy or corrupt genetic information, not create it. This is because mutations are accidental mistakes which occur when the genetic information (the coded set of instructions on the DNA which specifies the blueprint for construction and operation of organisms) is copied from one generation to the next.

Closer examination of alleged 'examples of evolution' shows that the changes we observe in living things are not the types of changes needed to turn microbes into man.

Peppered moths

High school and university courses in biology have for many years taught this supposed 'classic example' of evolution in action. The textbook story was that when the Industrial Revolution blackened the trunks of trees where dark and light peppered moths (*Biston betularia*) rested during the daytime, the light-coloured moths in the population became visible to bird predators, resulting in a preponderance of the dark-coloured forms.

But as it stands, that account is an example of natural selection, not evolution. Famous evolutionary biologist L. Harrison Matthews, writing in the Foreword to the 1971 edition of Darwin's *Origin of Species*, said the peppered moths are a beautiful demonstration of natural selection 'but they do not show evolution in progress'. Dark and light moths existed before and after the Industrial Revolution.

So the kinds of evolutionary changes capable of turning microbes into moths were not observed. What is more, recent closer scrutiny by evolutionists has exposed that the peppered moth story is not as originally portrayed. For example, it has now been shown that moths don't rest on the tree trunks by day. (Apparently, photos and film of the moths resting on tree trunks were fraudulently staged, with some even being *glued* to the tree.) The story is further eroded because the resurgence of the light-coloured form occurred while tree trunks were still blackened.[1]

When evolutionist Jerry Coyne heard of the numerous flaws in the peppered moth story, he wrote in the journal *Nature* that the peppered moth should be discarded as 'a well-understood example of natural selection in action'.[2]

Cave fish with sightless eyes

The non-functional eyes of cave-dwelling fish are often claimed as evidence for evolution having occurred from sighted fish. But let's have a closer look.

It is indeed correct to say that cave-dwelling fish with non-functional (or malformed, or absent) eyes are descended from fully-sighted ancestors. For example, the cave-dwelling (eyeless) and surface river-dwelling (eyed) forms of the Central American banded tetra fish (*Astyanax fasciatus*) demonstrate this well. (When the two forms interbreed, viable young are produced, indicating they are the same 'kind'.[3,4])

But this is *not* evolution in the overall upward information-adding sense that evolutionists say led to our existence. Rather, the loss of eye function is the result of a downhill mutational change, a *corruption* or *loss* of the genetic information coding for eye manufacture. For fish living in light, such mutations are terribly disadvantageous, but for fish in underground environments, sightlessness does not give any disadvantage at all. In fact, a complete lack of eyes is an advantage when bumping into cave walls in the darkness, as eyes are vulnerable to injury and subsequent infection, possibly leading to death.[5]

In a totally dark cave, then, eyes are not an asset but a handicap. Thus, via a straightforward process of natural selection, eyeless fish will be more likely to survive and reproduce, passing their genes on to their offspring. After just a few generations it would not be surprising to find only sightless fish in that environment, as natural selection eliminated the genes for functional eyes. But this legitimate example of natural selection does not support the idea of evolution, as sightlessness is the result of a *loss* or *corruption* of the genetic information needed to produce eyes. And how could a corruptive or degenerative process ever have generated seeing eyes in the first place?

In contrast, the example of blind cave fish fits the biblical account of creation. The fact that surface (eyed) and cave-dwelling (sightless) fish can interbreed indicates they are descendants of the same original kind created by God to reproduce 'after their kind'.[6] Originally, all the various kinds of creatures were created perfect.[7] But the corruption of DNA evident in organisms today fits with God having cursed the whole creation when Adam sinned, with the result that everything is now 'in bondage to decay'.[8]

A tragic world of mutations

This deterioration of a once-perfect creation, as the Bible describes, is essential for a correct understanding of why the world is the way it is. With the copying of DNA from each generation to the next, the observable accumulation of information-losing mutations across species right around the world unarguably fits the biblical account.

The mutant 'feather-duster' budgerigar, for example, suffers from having curly feathers which never stop growing, so it can't ever fly.[9] The mutation probably damages the control gene responsible for turning off feather growth. With precious nutrients and energy thus diverted away from other parts of the body, most of these mutant birds die by the time they are eight months old (compared with fourteen years for a normal budgerigar).

Similar (but less debilitating) examples of control gene mutations affecting hair growth can be seen in Australian merino sheep and angora goats—an economic benefit to commercial wool producers and, with regular shearing, non-fatal to the animal. Some bacteria have a mutation disabling the control gene that regulates manufacture of a chemical that destroys penicillin, resulting in unrestrained production, thus rendering them immune to penicillin.

These examples show that while some mutations lead to early death, others are more innocuous changes that do not seriously impede the health of an organism. Sometimes a mutation can even confer an advantage in a particular environment, as in the penicillin-resistant bacteria, the eyeless cave fish, and wingless beetles on windy islands (as wingless beetles can't fly they can't be blown out to sea).[10] But in virtually every instance, rather than adding new information, mutations destroy genetic information or corrupt the way it can be expressed—the control genes are damaged and no longer able to perform their original function.[11]

The impact of a mutation is not always an obvious physical defect like curly feathers or eyelessness. Sometimes it is subtler. Commonly the loss of gene function predisposes an organism to disease. In humans, a staggering 9500 specific DNA defects have been identified that cause, or predispose towards, over 900 different human diseases, with a further 3400 mutations identified but not yet confirmed.[12]

Unfortunately for the theory of evolution, no mutation has yet been shown to add the sort of new information to a creature's DNA that the theory requires. When biophysicist Lee Spetner examined mutational changes that evolutionists have claimed have been increases in information, in every case studied there was a loss of specificity (that is, a loss of information).[13]

But what if, one day, such a mutation were found? Even *if* this happened, evolutionists would still need to find hundreds more examples before mutation and selection could be considered truly adequate to explain microbes-to-man evolution.

Antibiotic and pesticide resistance

Pathogens are often said to have 'evolved resistance' to antibiotics. Similarly, the build-up of resistance to insecticides by insects is often thought of as being 'proof' of evolution. Upon examination, we observe that resistance can come about in a variety of ways, but in every case there has been no increase in genetic information—any changes are downhill, or at best horizontal.

What often happens is that some bacteria/insects in a population already had the genes for resistance to the antibiotics/insecticide. (In fact, some bacteria obtained by thawing sources frozen before the development of antibiotics have been shown to be already resistant to subsequently developed synthetic antibiotics!) When antibiotics/insecticides are introduced to a population, those lacking resistance are killed, and any genetic information they carry is eliminated. So the population now carries less information, but they are all resistant. Thus natural selection and adaptation involve *loss* of genetic information.

Sometimes antibiotic resistance is the result of a mutation, but all known cases have involved information loss, as in the penicillin-resistant bacteria mentioned earlier. The so-called 'supergerms' are really weaker than their normal cousins, unable to compete with them in a normal environment.[14]

In other cases, antibiotic resistance is acquired when pieces of genetic material are transferred between bacteria, even between different species. But this does not represent evolution of new genes because the genetic information *already existed*.

Darwin's finches

Often portrayed as an example of 'contemporary evolution', the adaptation of finches to a diverse range of habitats on the Galapagos Islands is actually good evidence for the biblical account of history, as we shall see.

Thirteen species of finches on the islands visited by Charles Darwin have been identified, with Darwin concluding that they had all descended from finches that had earlier migrated from the mainland.[15] The key differences between species relate to beak sizes and diet, and it has been estimated that a new finch species could arise in only 200 years.[16]

However, the 'evolution' of new species on the Galapagos is not molecules-to-man evolution. Rather, each new 'species' is carrying a subset of the original genetic information carried by the original mainland finch immigrants. Note that the finches are still finches. One species has strong and wide beaks for cracking nuts; another has longer beaks useful for extracting food from cracks in logs; but both are descended from a finch population carrying the genetic information for a wide variety of beak sizes. And the rapid speciation fits with the biblical account of all birds and air-breathing land animals today being descended from male/female pairs coming off Noah's Ark around 4500 years ago. (Noah didn't need to carry thirteen species of Galapagos finches—he only needed the single 'kind' from which all finches today are descended.) In contrast, rapid speciation is often a surprise to evolutionists with their framework of millions-of-years thinking.[17]

We should also note that there is often a price associated with specialisation, or adaptation—namely, the *permanent* loss of information in a particular group of organisms. This reflection of the built-in limits to variation often surprises evolutionists, who are used to thinking in terms of evolution being a creative process, virtually without limits. For example, evolutionists were recently astonished to find that a highly specialised rainforest fly was totally unable to adapt to drier conditions.[18]

The biological evidence does not support the scenario that microbes turned into today's array of life forms, adding information by mutations and selection. Instead, the decrease in genetic information from mutations and selection/adaptation/speciation (and extinction) is consistent with the biblical framework of originally created gene pools with immense in-built variety having since been depleted.

The facts of biology fit the Bible's description of history: organisms created perfectly to reproduce 'after their kind', then subjected to the curse because of the sin of the first man, Adam, but carrying sufficient genetic variation to successfully repopulate diverse habitats throughout the earth after the global devastation of the Flood.

Are similarities in living things evidence of common ancestry?

According to standard evolutionary theory, the immense biological diversity in the world today evolved from some kind of single-celled organism which happened to arise from non-living matter. On that basis, evolutionists emphasise that the physical and genetic similarities between humans and other living organisms is evidence of common ancestry.

All birds, for example, have feathers, lay eggs and have a specialised lung, and therefore they must have had a common ancestor with these features—or so the evolutionary reasoning goes. Similarly, the oft-mooted idea that humans and chimpanzees have some 96–98 per cent similarity in their DNA is often claimed to prove that human beings evolved from apes.

However, the 'common ancestor' argument is not a direct finding but an interpretation—an interpretation made from an evolutionary framework. From a biblical framework of thinking, creationists would say that the similarities exhibited by the various kinds of biological organisms are not due to common ancestry but indicate a *common designer*.[19] The similarities among birds reflect their having been created with a common basic plan.

We can see this in examples from industry. The early Porsche and the VW Beetle, for example, though very different cars, actually shared many similarities (such as rear aircooled, horizontally opposed engines). Why? Because they shared the same designer. So we shouldn't be surprised if a designer for life used the same biochemistry and structures in many different creatures. Consider this: if similarities were not evident across life forms, we might be inclined to believe in many creators, not just one. In the Bible, we read that God's very nature is revealed to us in what he has created.[20] So the unity of creation is testimony to the One True God who made it all.[21]

It would seem, too, that God created things in such a way that the patterns of similarity we see actually defy any 'naturalistic' (that is, everything-made-itself) explanation of our origins, as the examples below will show.[22]

The more similar creatures are, according to evolutionary reasoning, the more closely they should be related (that is, share a recent common ancestor). Evolutionists use this argument to link humans with frogs. They say that, because frogs and humans share a five-finger/toe per hand/foot pattern, and limbs with two bones attached to the hand/foot joined to a single other major limb bone, then we must share common ancestry.

But this argument breaks down when one looks at how they develop. In the womb, the human embryo develops a bony plate, then material between the digits dissolves. But in frogs, the digits grow outwards from buds. These starkly different growth processes argue against the evolutionary 'common ancestry' explanation for the similarity.[23]

And if evolutionists use similarities between organisms as evidence of recent common ancestry, how then do they explain similarities between organisms that evolutionists *don't* believe are closely related? Haemoglobin is a complex molecule that carries oxygen in our blood and results in its red colour. This molecule is found in vertebrates—so, if similarities indicate common ancestry, then all vertebrates must be closely related. However, haemoglobin is also found in some earthworms, starfish, crustaceans and molluscs—but evolutionists don't say we are closely related to earthworms! Also, the haemoglobin of crocodiles has a greater similarity to that of chickens (17.5 per cent) than to that of their fellow reptiles the vipers (5.6 per cent). Another puzzling anomaly for evolutionists is that camels and nurse sharks both have the same distinctively structured antigen receptor protein, but this cannot be explained by a common ancestor of sharks and camels!

As mentioned above, the idea that humans and chimpanzees have around 96–98 per cent similarity in their DNA is often said to show that humans evolved from apes. While the 98 per cent figure is debatable, it makes sense that apes and humans would have a high degree of similarity in their DNA. Both are mammals, so it is not surprising that their DNA is a closer match than when either is compared with, say, the DNA of a reptile.

But even given only a tiny percentage difference in the DNA between chimps and humans, the encyclopaedic information content of DNA means that a small difference would nonetheless require a lot of information to turn one creature into another. In the case of chimps and humans, even if we grant the ten million years of evolution asserted by evolutionists, bridging the gap would still be an impossibility.[24] It makes far more sense to see the similarities in DNA as evidence of a common designer, not a common ancestor.

Does 'embryonic recapitulation' prove evolution?

Embryonic recapitulation is the idea that the development of the human embryo in the womb is a rerun (or 'recapitulation') of the steps in humanity's alleged evolution from a primitive creature. It is also known as 'ontogeny recapitulates phylogeny' or the 'biogenetic law'. According to this notion, early in embryonic development you can see gill slits like those of a fish, and later a reptilian-like 'face', as the human embryo supposedly passes through the various stages from fish to amphibian to reptile and so on.

This idea was popularised by German evolutionist Ernst

Haeckel with a famous series of embryo drawings in 1868. Within months of the drawings being released to the public, however, leading anatomists had shown them to be fraudulent. They had been drawn to make the embryos of different species look alike. Haeckel was forced to issue a modest confession in which he blamed 'the draughtsman'—without acknowledging that he himself was that person.[25]

But did Haeckel have a point? Does 'ontogeny recapitulate phylogeny'? The answer is a resounding no—even from evolutionists. In 1980, the well-known evolutionist Stephen Jay Gould stated, 'Both the theory [of recapitulation] and [the] "ladder approach" to classification that it encouraged are, or should be, defunct today.'[26] In 1988, Professor Keith Thompson of Yale University said: 'Surely the biogenetic law is as dead as a doornail. It was finally exorcized from biology textbooks in the fifties. As a topic of serious theoretical inquiry, it was extinct in the twenties.'[27]

Sadly, the idea that embryonic development reflects evolution has remained widespread, with high school and university textbooks in the 1990s still using Haeckel's fraudulent drawings.[28] At least one recently-published encyclopaedia presents recapitulation as evidence for evolution.[29] So, despite the debunking by many high-profile scientists, it is not surprising that there are still many who believe, on the basis of the full range of Haeckel's diagrams, that embryos from widely differing creatures are similar in their early development, and therefore are evidence for a shared evolutionary ancestry.

However, the full extent of Haeckel's deception was made known to the scientific community in 1997, when a detailed study (by evolutionists), with actual photographs of a large number of different embryos, showed that embryos of different kinds are not only very distinct from one another, but also utterly different from Haeckel's published drawings.[30] This study was widely publicised in science journals,[31] so really there should be no excuse for anyone teaching science today to be unaware that the idea of embryonic similarities is outdated and based on fraud.

Even so, Haeckel's legacy is still evident in the popular imagination, with many clinging to the idea that early in development, the human embryo has gill slits like a fish. Abortion clinics have used the idea to ease their clients' consciences: 'We're only taking a fish from your body.' But it has been known for many decades that human embryos never have 'gill slits'. There are markings on the human embryo called pharyngeal clefts, which delineate throat pouches. No slit or opening develops. They have no relation to breathing, but develop into the thymus gland, parathyroid glands and middle ear canals.

What about 'vestigial organs'?

Many evolutionists argue that certain animal or human organs that seem to have no function must therefore be vestiges or 'leftovers' of an evolutionary heritage. Certain organs present in an ancestor creature are no longer needed, but are still present because natural selection has not eliminated them yet. Examples have included pigs' toes, flightless birds' small wings, so-called 'vestigial hind legs' in whales, and our own coccyx and appendix.

It is impossible, however, to prove that an organ is useless—the function may simply not yet have been discovered. This is why the list of supposed 'vestigial organs' has greatly decreased with increasing knowledge and scientific study. For example, 100 organs in humans were once considered useless leftovers from evolution but are now known to be essential.[32] The human appendix, once said to be a useless evolutionary relic, helps control bacteria entering the intestines. Similarly, our tonsils, also once thought useless, are now known to suppress throat infections.[33]

Even if it could be shown that a supposedly vestigial organ had no function, this would prove devolution, not evolution. For example, where the 'useless' wings of some flightless birds were derived from birds that once could fly, this is entirely consistent with the biblical account—a once-perfect creation which, since the Fall, has been 'in bondage to decay'. It is an example of the *loss* or *deterioration* of a feature, which occurs relatively easily through natural processes. In contrast, the acquisition of new features, requiring specific new DNA information, is impossible.

In any case, the muscles on the wings of many flightless birds are functional, allowing the birds to actively flap their wings during mating rituals and when scaring predators. Given such behaviour, perhaps these wings cannot be dismissed as 'useless' after all.

What about the alleged 'vestigial hind legs' within the flesh of whales, which evolutionists use as evidence that whales evolved from some cow-like land creature? Far from being useless, we now know that these so-called 'remnants' help strengthen the reproductive organs. (The bones are different in males and females.)

Just because we might not know the function of something does not mean it has no function. Take the case of pigs having two toes that do not reach the ground. In muddy conditions (much frequented by pigs) the extra toes probably make it easier to walk—a bit like the rider wheels on some long trucks which only touch the road when the truck is heavily laden.

As for the human coccyx (popularly known as the 'tailbone'), is it really a useless vestige of a tail inherited from our evolutionary ancestors? Absolutely not! The coccyx is a crucial point for attachment of certain muscles, tendons and ligaments. Without the coccyx, we would need a radically different support system for our internal organs.

This certainly fits with us having been created by an allknowing, loving God—we would expect all parts of the body to operate in harmony and have their own particular function.

Conclusion

The claimed biological evidence for evolution just does not stand up to scrutiny. Evolutionary arguments about vestigial organs and embryological development have been thoroughly discredited, while the evident similarities between living things speak of a common designer, not evolutionary relationships.

The biblical account of God having created the different basic kinds of organisms to reproduce 'after their kind' fits with what we observe. The basic kinds were capable of adapting to different environments by sorting the original created genetic information (reshuffled by sexual reproduction) via natural selection. It is evident that natural selection is not molecules-to-man evolution because it can only select from *existing* genes—it cannot generate new genetic information. Evolutionists look to mutations as the source of new genes, but mutations result in degenerate changes involving loss of genetic information—consistent with the biblical description of a fallen world 'in bondage to decay'.

The notion that natural processes have generated the vast amounts of complex coded information in the diversity of earth's life forms just does not fit the evidence. In contrast, the biblical account of creation logically explains the origin of that information, and the degeneration evident in living things today.

9

The Geological Evidence for Creation

Andrew A. Snelling

Fossils—creatures by design

What is immediately striking about fossils is that they are so similar to their modern counterparts. Most of them are therefore easily recognisable and thus identifiable for what they are. Some fossil creatures, such as the trilobites, are extinct of course; however, the classification system that works on all the living creatures today also works on all the fossils, including those that no longer exist. Furthermore, it is immediately apparent that fossilised creatures bear all the same attributes and qualities as modern creatures, which implies that they were designed as integrated working 'machines' that functioned perfectly while living in their respective biological communities.

The trilobites serve as a potent example of design features in fossil animals. Trilobites were jointed-limb animals (arthropods) with hard external skeletons, and they are found early in the fossil record. They are thought to have been marine creatures that lived on the ocean floor because their fossilised remains are commonly found with fossilised creatures whose descendants still live on the ocean floors today. The trilobite's shell is usually divided into three sections: the head shield (cephalon), a series of segments making up the thorax, and the tail shield (pygidium). Each shell's shield is made up of a raised central ridge or spine, and a wide flange on each side, constituting the three lobes that give the animal its name. Trilobites' jointed legs mean the animal must have had a complex muscle system, and they are thought to have had a circulatory system, including a heart, as well as a very complex nervous system.

However, it is the trilobites' eyes that are a marvel of design. Indeed, some scientists believe that the aggregate (schizochroal) eyes of some trilobites were the most sophisticated optical systems ever utilised by any organism.

A trilobite's eye was a compound eye made up of more than a single lens. Each lens 'saw' through only a very narrow window, which means its compound eye was ideally suited to detect motion. Furthermore, each one of the lenses in the trilobite's compound eye was specially designed to correct for spherical aberration so that each lens constructed a clear image. Even more amazing, some of the lenses were positioned so as to provide the trilobite with a zone of overlapping vision, thus allowing for depth perception as well. Thus, incredibly, trilobites' eyes allowed for vision in virtually all directions—up, down, forwards, backwards, sideways—at the same time! This elegant physical design employed Fermat's principle, Abbe's sine law and Snell's laws of refraction, and compensated for the optics of the birefringent calcium carbonate crystals of which the lenses were composed. Thus trilobites could see an undistorted image underwater in all directions, being able to determine distance in part of that range, while at the same time having the optimum sensor for motion detection.

Such an amazing, incredibly complex vision system has all the evidences of being constructed by an exceedingly brilliant and intelligent designer.

The trilobite's extraordinary complexity presents an impossible dilemma for those who claim, despite the obvious evidence to the contrary, that the trilobite with its eyes evolved by chance from some supposed primitive ancestor. The trilobite was not the type of creature that evolutionary theorists would have predicted would stem from the earliest multicellular animals with hard parts that are found fossilised in the lowest strata. Most evolutionists would have to believe that something resembling a sponge would have evolved before the trilobites. If evolution were true, then we would expect such a complex creature as the trilobite to have documentable ancestral fossils, preserved in the strata below where the trilobites are now found, showing how that complexity arose. However, no such fossils have ever been found. Instead, the trilobites with their amazing complexity appear abruptly in Cambrian strata deep in the Grand Canyon and many other places around the world, fully-formed and fully-functioning, with no hint whatsoever of how their exquisitely designed complexity evolved.

Thus fossil creatures such as the trilobites with their amazing eyes argue powerfully and overwhelmingly for design and fiat creation by an all-wise and all-knowing creator. And what is so patently true of the trilobites is similarly evident in the design and abrupt appearance of many other amazing creatures in the fossil record.

The fossil record and the Flood

Although it cannot be categorically said that no fossils are now being formed in sediments accumulating across the earth's surface today, it is nevertheless emphatically true that there are no modern parallels for the formation of the fossil deposits found in great numbers at various levels in the global geological record. Actually, the formation and preservation of the fossils required special conditions. After death the soft parts of organisms rot or are eaten, and any hard parts may be dissolved by water, or broken and scattered by scavengers or storms and so on. Obviously, to be fossilised, animal and plant remains must be buried quickly, entombed and sealed so that seeping groundwaters and even bacteria cannot subsequently destroy them. Just how quickly organisms were buried can only be fully appreciated when the details of the state of preservation of so many abundant fossils on an enormous scale globally are understood.

There are many instances where preservation has been remarkably complete. Even delicate structures such as flowers, insects and the soft tissues of animals have been fossilised meticulously, as well as footprints and burrows. Examples include soft tissues of fish such as muscle fibres, with even the cell nuclei still able to be studied; pieces of skin, developing eggs and even stomach walls and stomach contents; and dinosaur soft tissues such as skin impressions and organs that are able to be CAT-scanned to study their structure. These and many other examples require that burial was instantaneous, and preservation and fossilisation extremely rapid. Indeed, some fossilisation processes that have preserved soft tissues have been mimicked rapidly in the laboratory, demonstrating that such processes only require weeks at most.

Furthermore, pristine fossil shells supposed to be 165 million years old have still retained even the iridescence of their shells, leaving them indistinguishable from their modern counterparts. Other shells in the same rock layers still have their organic ligaments. All these and many other examples not only demonstrate that these creatures must have been catastrophically buried alive and rapidly fossilised, but are consistent with this having occurred only thousands, rather than millions, of years ago.

Such beautifully preserved fossils are not isolated specimens. They are found by the countless thousands over vast areas in what are known as 'fossil graveyards'. Several examples illustrate the scale and magnitude of the catastrophic deposition that had to be responsible for burying alive so many organisms over such vast areas of the earth's surface.

Within the Soom Shale in the Cedarberg Mountains of South Africa, for example, are found thousands of exceptionally preserved fossils of brachiopods, nautiloids, arthropods, conodonts and other organisms at several locations hundreds of kilometres apart. The sensory organs, walking appendages, fibrous muscular masses and even gill tracts of some of the arthropods are remarkably preserved, as are the complete feeding apparatuses of the conodonts. Countless thousands of these organisms had to be catastrophically buried alive over

Geological Evidence for Creation

thousands of square kilometres within this thinly-laminated shale unit, which also had to be catastrophically deposited and buried before burrowing organisms could destroy the fine laminations.

The shales of the Cow Branch Formation in the Virginia–North Carolina border area are another fossil graveyard. These shales contain an abundance of complete insects and preserve even the soft-part anatomy of some vertebrates, along with an unusual diversity of plants. Microscopic details of the insects are preserved with great fidelity, and many articulated specimens of aquatic reptiles have been fossilised complete with ghosts of the muscles on the tail and ligaments in the webbed hind feet. This mixture of terrestrial, freshwater and marine organisms buried together, fossilised and so well-preserved, is again consistent with catastrophic deposition and burial, as insects do not simply die, fall into a body of water and sink with dying fish to be gradually covered up by slowly accumulating sediments.

Many fossil graveyards of dinosaurs are found in the Morrison Formation and its equivalents that stretch from New Mexico in the south to Canada in the north, over an area of 1.5 million square kilometres. Most of the dinosaur remains are well-worn bone fragments and relatively pristine, semi-articulated skeletal segments; but they are often buried with the fossilised remains of crocodiles, turtles, lizards, frogs and even clams in a conglomeratic sandstone deposited by water moving at velocities of one to two metres per second. A spectacular fossil dinosaur graveyard is similarly found in the Cretaceous Djadokhta Formation of the Nemgt Basin of Mongolia, where in one area of four square kilometres hundreds of well-preserved articulated skeletons of dinosaurs have been found, together with associated skeletons of mammals, lizards and even birds. Many skulls are virtually complete with lower jaws still in articulation, and tympanic rings and ear ossicles well preserved. All of these are found in a distinctive sandstone layer that obviously resulted from catastrophic deposition and burial.

Even limestones should be regarded as fossil graveyards, because so many of them in the geological record are largely composed of the broken remains of countless billions of marine organisms that were destroyed then transported by fast-moving water before being dumped and buried. For example, the Thunder Bay Limestone of the Michigan Basin is at least four metres thick, stretches laterally for hundreds of kilometres, covers many hundreds of square kilometres and consists of the broken, jumbled and fossilised remains of all types of corals, bryozoans, crinoids, brachiopods and a profusion of other shallow marine creatures.

But perhaps the most extensive fossil graveyards are the Cretaceous chalk beds that are exposed in north-west Europe, but which stretch from Ireland across Europe to Turkey, Egypt and Israel, as well as being found in Texas, Kansas and other US states, and also on the coast of Western Australia. This is a global distribution of uniform beds up to 100 metres or more thick composed of the microscopic remains of countless trillions of coccolithophores and other tiny creatures, along with the larger remains of other marine creatures such as ammonites.

And finally, another impressive and distinctive fossil graveyard are the coal beds that represent trillions of tons of plant fossils all buried together with so few impurities. There is also a remarkable similarity in the plant fossils in the coal beds from continent to continent, such as in the British Coal Measures and the coal beds of the Illinois Basin. The features and contained fossils of these Carboniferous coal beds, for example, are essentially the same all the way from Texas across Europe to the Donetz Coal Basin north of the Caspian Sea. With so much evidence of the catastrophic burial of organisms in water-deposited sediments on such a vast scale throughout the geological record, it is totally reasonable to view this devastation as due to a catastrophic global Flood.

The order of fossils in the geological record is also highly significant. In the Cambrian strata, where so many fossils suddenly appear in the record, they are all of marine creatures. It is then only higher up in the sequence in the Devonian and Carboniferous that fish fossils are found in profusion, followed by amphibians and reptiles. In the Permian, and especially above in the Triassic, Jurassic and Cretaceous, are found many fossils of land-dwelling organisms. However, many marine fossils are also found in adjoining or associated strata, or even mixed in with these terrestrial fossils. This is consistent with the waters of a catastrophic global Flood first being active in the ocean basins and then rising up to eventually cover the continents. Indeed, many strata containing marine fossils are found up on the continents deposited on crystalline basement rocks, unmistakeable evidence that the ocean waters have flooded the continents in the past on a global scale.

Geological evidence for the Flood

As already described, the many fossil graveyards in the geological record, and the scale of them, are unmistakeable geological evidence for a catastrophic global Flood. The fact that the Morrison Formation and its equivalents, with their dinosaur fossil graveyards, stretch from New Mexico to Canada, and the chalk and coal beds can be traced from North America to Europe and elsewhere, is eloquent evidence of the catastrophic destruction of creatures and plants all across the earth's surface and their rapid burial in water-deposited sediments during a global flood that swept across the continents. Indeed, there are many examples of widespread, rapidly water-deposited strata in the geological record.

Traceable across North America are six recognisable strata sequence cycles known as megasequences, each of which represents a major cycle of ocean waters advancing right across the North American continent and then retreating. First and lowermost of these megasequences is exposed in the Grand Canyon as the Tonto Group. It comprises the Tapeats Sandstone, the Bright Angel Shale and the Muav Limestone, which represent sediments carried by ocean waters surging over the continent that were deposited laterally and stacked vertically at the same time, with the finest grain sizes at the top of the sequence in the limestone. At the base of this megasequence, the underlying crystalline rocks were eroded by these surging waters, which picked up boulders with diameters up to 4.5 metres and included them in the base of the Tapeats Sandstone.

The Tonto Group strata have an enormous horizontal

extent of many hundreds of kilometres; but it is also possible to map the occurrence of all the sandstone strata that correlate with the Tapeats Sandstone in this megasequence from southern California and Texas northwards across Montana and much of North Dakota through to Canada, and from southern California and Nevada right across to the Mid-West and New England, including Maine. The evidence indicates that this was a storm-driven, major, rapid inundation right across North America, exactly what would be expected to occur at the onset of a catastrophic global Flood.

The Triassic Hawkesbury Sandstone, higher up in the geological record, is a flat-lying sandstone layer up to 250 metres thick covering an area of about 20,000 square kilometres in the Sydney Basin of eastern Australia. Key depositional features, such as the frequent cross-bedding up to eight metres high, in places overturned, provide convincing evidence that underwater sand dunes or waves, 10–16 metres high, were moved along by water currents travelling at about 1.5 metres per second, with water waves as high as 20 metres and up to 250 kilometres wide carrying the billions of tons of sand required to deposit this sandstone over this vast area. Shale lenses within the Hawkesbury Sandstone preserve abundant fossils of many freshwater and marine fish, sharks, insects, freshwater-marine arthropods, crustaceans, amphibians, bivalves, gastropods and much plant debris, testimony to marine devastation of the land.

A similarly impressive example is the Coconino Sandstone and its equivalents that cover about 500,000 square kilometres across Arizona, New Mexico, Kansas, Oklahoma and Texas, representing an estimated volume of about 42,000 cubic kilometres of sand. Cross-beds in this sandstone indicate that water over 90 metres deep flowing at 1–1.5 metres per second moved sandwaves up to 18 metres high to deposit this sandstone layer over such an incredibly vast area. Yet at that rate of deposition, the 42,000 cubic kilometres of sand would have been deposited within a few days!

So it is not just the extent in area of these rock layers that is impressive evidence for a catastrophic global Flood, but the rate and scale at which they were deposited. The Triassic Shinarump Conglomerate has an average thickness of about 15 metres and covers more than 26,000 square kilometres in Utah and neighbouring states. Composed of sand and rounded pebbles, it is impossible to explain its deposition slowly and gradually by a network of streams, because no modern depositional environment produces such a vast conglomerate. It is far more realistic to explain this uniformly thick conglomerate as deposited by a massive sheet of rapidly flowing water that carried all the pebbles and sand en masse in what would therefore have been a catastrophic flooding event over this vast area in an extremely short time.

Though not as extensive in area, the Uluru Arkose and Mt Currie Conglomerate in central Australia are equally potent in demonstrating the devastating rate at which many strata were catastrophically deposited. The Uluru Arkose is a coarse sandstone consisting of poorly sorted, jagged grains of minerals and other rock types that is exposed as near vertical beds in the monolithic landmark known as Uluru (Ayers Rock). The total thickness of the arkose is estimated at almost six kilometres, while its extent is conservatively estimated at at least 30 square kilometres. The nearby huge, rounded, rocky domes known as Kata Tjuta (the Olgas) are composed of the six-kilometre thick Mt Currie Conglomerate that covers an estimated area of more than 600 square kilometres. This conglomerate is also poorly sorted and contains boulders up to 1.5 metres in diameter, as well as cobbles and pebbles of other rock types, held together by a matrix of finer fragments in cemented sand, silt and/or mud.

These two rock layers are regarded as being related, and the features within them are totally incompatible with any currently observable deposition mechanism. Instead, the evidence suggests deposition of this arkose and conglomerate concurrently as lateral equivalents by an amount and force of water sufficient to erode, transport and deposit at least 4000 cubic kilometres of boulders, pebbles, cobbles and sand distances of at least tens of kilometres in successive continuous pulses so as to stack the resultant beds to a thickness of 6000 metres over at least 600 square kilometres—all probably in a matter of hours or days at the very most. These are uppermost Precambrian strata that appear to represent the beginnings of this catastrophic global Flood event.

Another graphic example of catastrophic deposition at this same time is found in the Kingston Peak Formation in southeastern California. This rock unit is dominated by very coarse fragments of other rocks, ranging from pebble size to enormous blocks greater than 1.5 kilometres wide. In fact, much of this rock unit consists of rock fragments of all sizes and conglomerates up to 1000 metres thick. The observational data support a catastrophic deposition model in which gravitational collapse of earlier rock units was caused by sudden, massive earth movements that generated mass-flow deposits, enormous rock fragments and high-energy water currents that moved at rates of 15–30 metres per second all mixed together, resulting in flows of fluidised rock masses cascading downslope at speeds of 50–100 metres per second. Most of this Kingston Peak Formation would thus have accumulated within one hour. Other similar and related formations are found from Mexico northward through the western United States and Canada up to at least Alaska. Similar formations have also been recognised in at least 100 other locations in at least fifteen countries around the globe, including parts of southern and central Australia and southern Africa.

This demonstrated catastrophic deposition of so many rock layers in the geological record suggests that much of this record could have accumulated within a year-long global flood. Even at the present measured average sedimentation rate, the rock record would have accumulated in only a small fraction of the oft-claimed vast available timescale. Thus it is normally claimed that the geological record consists of either brief periods of catastrophic sedimentation separated by long periods of inactivity, or of long periods in which sediments are deposited and then eroded before the next sediments arrived.

But where is the evidence for long periods of time between rock layers? In the well-studied area just to the north of the Grand Canyon, more than 60 per cent of the Cambrian to Recent geological record is claimed time gaps, but there is virtually no erosion evident at the boundaries between strata where the claimed time gaps are supposed to be. This general lack of evidence of erosion, soil formation or burrowing animals suggests this claimed time never existed. Within the walls of the Grand Canyon itself there are nine boundaries between the ten major rock units, five of which are claimed to represent time gaps, but along which there is little or no evidence of any erosion, and certainly no evidence of soil formation or animal burrowing. In places the Carboniferous Redwall Limestone lies directly on the Cambrian Muav Limestone without any trace of significant erosion and thus no evidence for the claimed time gap of more than 140 million years! There is supposed to be a rock layer missing between the Coconino Sandstone and Hermit Shale, but the boundary between them is absolutely flat and knife-edged, while the Coconino Sandstone has preserved in it many footprints of animals that one would have expected to be also at that boundary and they are not. Where there is localised erosion at the boundaries between some rock units, it is totally consistent with the continuous catastrophic deposition of the whole rock sequence, just as would be expected in a catastrophic global Flood.

Finally, there are also many evidences where rock layers had to be bent and folded while still soft, but this is claimed to have happened many millions of years after the rock layers were supposed to have hardened—again situations that rule out the claimed millions of years. For example, when the plateau through which the Grand Canyon is cut was uplifted, the Cambrian Tapeats Sandstone was bent and folded in places almost at right angles without any signs within the sandstone of fracturing and breaking up. Instead, the sandstone appears to have been soft and pliable when the bending occurred, so at that time the sandstone could not yet have been hardened rock. However, the bending and folding occurred supposedly 470 million years after deposition of the Tapeats Sandstone, and after deposition of the 1200 metres thickness of rock layers on top of the Tapeats Sandstone in the Grand Canyon area. Instead, all these layers of sediment had to have been deposited rapidly and then bent while still soft as the plateau was uplifted, all in a brief time span before the layers fully hardened. Again, the scale of this evidence is consistent with a brief catastrophic global Flood.

Global tectonics in the Flood and the subsequent Ice Age

Tectonics refers to the development and relationship of the larger structures and broad features of the outer part of the earth. Today the earth's crust is still broken into a number of different pieces or plates that exhibit imperceptibly slow movement along their boundaries relative to one another, generating earthquakes and volcanic activity. Plate tectonics theory describes and explains the present and past motion of these plates, successfully building a consistent explanation of most of the earth's current surface and internal structural and geophysical features.

However, computer simulations have shown that, where the earth's crustal plates have been sinking into the mantle in what are known as subduction zones, slow plate movement cannot overcome the resistance of the mantle rock, so that the only way the process could occur would be at a catastrophic rate known as 'thermal runaway'. In this scenario, as the leading edge of the crustal slab began sinking into the mantle, the friction would generate heat, which would help overcome the friction to make the sinking go faster. This would then generate even more friction and more heat, so the sinking would go faster and faster until the crustal plate was sinking into the mantle at a rate of metres per second. The consequences would be a global catastrophic plate tectonic event in which the continents would be flooded, with copious volcanic and magmatic activity and torrential global rain as continents were split and moved rapidly across the earth's surface.

If this global catastrophic upheaval occurred in the past, all that would have been required to start it would have been upwelling heat and steam within the mantle that triggered the rifting and breaking up of the ocean floor and continents into crustal plates. This heat would have caused partial melting of the uppermost mantle, resulting in upwelling of the ocean floor along an extensive rift zone to form a mid-ocean ridge system all around the earth, much as there is today. Lavas would have been profusely extruded and the ocean floor pushed aside along this rifting ridge system. The molten rock in contact with the ocean water would have combined with the steam coming from inside the earth to produce supersonic steam jets that catapulted steam high into the atmosphere, where the steam would have eventually condensed as it spread around the earth to form intense global rain. The upwelling mid-ocean ridge system would also have caused the ocean floor and the sea level to rise, the ocean water thus rising up onto the continental land surfaces flooding them.

The accompanying earthquakes with all these earth

movements would have destabilised sediments on the ocean floors that would then have been carried landwards by the landward-flowing and flooding ocean water. These sediments would then have been progressively deposited up on the continents, along with the sediments generated by the erosion of the land surfaces due to the torrential rainfall.

Where the ocean floor abutted the continental crust, fracturing would have occurred so that the lighter continental crust would float and the denser ocean floor would sink into the mantle, the cold ocean floor rocks being denser than the warm mantle rocks beneath, thus triggering the thermal runaway subduction of ocean floor slabs previously described. The subducting ocean crust slabs would have sunk faster and faster into the mantle so that they could have reached the core/mantle boundary within a matter of weeks. As confirmation of this, there is seismic evidence of unmixed (undigested) cold dense slabs still at the core/mantle boundary today, suggesting that this global catastrophic tectonic upheaval occurred only relatively recently. The mantle rock displaced by these subducting slabs would have caused mantle-wide flow, with mantle rock upwelling to complete convection cells within the mantle. Thus mantle plumes would have risen to the earth's surface to produce prodigious outpourings of lavas that are found today at the earth's surface among all the sedimentary rock layers and are referred to as flood basalts. In other places the heat and instabilities would have caused local melting of the lower continental crust to produce granitic and other magmas that intruded into the catastrophically accumulating sediments and their entombed fossils.

With the earth's crustal plates moving at rates of metres per second, where continental plates eventually collided, one plate often overriding the other, the resultant forces would have buckled the strata and thrust them upwards to produce mountain chains. At today's plate motions of only millimetres per year, there would not be enough energy in these collisions to produce mountain belts. But during this catastrophic plate tectonic global Flood event, the metres per second plate motions would have had enough energy to buckle and push up strata to produce mountains. Sedimentary strata with marine fossils in these mountains, such as we see in the Himalayas today, indicate the sediments were originally deposited on the ocean floor.

Once all of the original ocean floor had subducted and sunk into the mantle, the mantle convection and plate movement would have begun to rapidly slow, so that today's residual plate motions could be considered the remnants of this catastrophic event that only lasted about a year. Indeed, today's earthquake and volcanic activity could likewise be regarded as residual effects, a position that is consistent with the evidence of the declining scale and rate of volcanic eruptions in the history of the earth since this global Flood event.

Further confirmation that this catastrophic plate tectonics upheaval was the driving force of a global Flood event is seen in the features preserved within the fossiliferous sedimentary strata that record the directions in which the water currents were flowing when they deposited the sediments and buried the fossils. Because of the earth's rotation and therefore the daily motion of the ocean tides, it would be predicted that as the ocean water flooded onto the continents during this global upheaval, the water currents would have persistently flowed in the same general direction during accumulation of most of the fossil-bearing geological record. If, on the other hand, these tectonic and sedimentation processes had been occurring at slow-and-gradual rates over millions of years, it would be predicted that the directions of water currents would have changed as the different groups of sedimentary strata were progressively deposited at different places and different times around the earth's surface. It is thus a startling revelation to find that the current indicators recorded in the sedimentary strata show the same consistent direction of water flow on every continent as the sediments were deposited, a unidirectional flow that fits with the global catastrophic plate tectonics Flood event. Such water flows could easily have been sustained for a year, but not for millions of years.

An immediate consequence when this catastrophic global Flood was over would have been ocean waters that were much warmer than they are today. This is confirmed by the temperature indicators in tiny shellfish buried in the sediments today on top of the new ocean floors generated from the upwelling lavas during the Flood. From pole to pole, the ocean water would have been 15 °C warmer than today's ocean water. Such warm ocean water would have more readily evaporated, so that in the early years after the Flood event there would have been much higher rainfall than today, and therefore conditions conducive to the rapid revegetation of the denuded earth.

However, at the same time, because of the tilt of the earth's axis and the seasonal variations, the polar and inland mountain regions would have gradually cooled to produce contrasting climatic zones. Thus eventually during the winters the rainfall in polar and mountainous regions would have begun to fall as snow. In mid to high latitudes the summers would also have cooled, so that eventually the winter snows would not all have melted through the cooler summers, the net result being thick accumulations of snow that compressed to form ice. Within a few centuries the ice would have accumulated so thickly that it began to flow outwards from the mountainous regions to eventually form huge continental ice sheets that covered North America, Siberia, northern Europe, Greenland and Antarctica–southern Australia–southern Africa. This is the period known as the Ice Age, with the maximum extent of these ice sheets being achieved within 500 years after the Flood.

Coastal and equatorial regions would have remained mild to warm and wet because of the warm nearby ocean waters. With so much water locked up in the continental ice sheets, the sea level would have been lowered by as much as 70 metres, producing land bridges, for example, across the Bering Strait and the English Channel, thus enabling the migration of animals and people. Even the coastal regions of Alaska and Siberia, including those along the Arctic Ocean shores, would have been ice free, with forests and grasslands capable of supporting large animals such as mammoths.

However, as the centuries passed the ocean waters would have gradually cooled, and so the evaporation rate also would have declined, meaning a diminished supply of snow to maintain the continental ice sheets. Once the ocean waters reached a temperature close to their temperatures today, the snowfalls would have rapidly diminished and a sudden climate switch would have occurred with warmer summers and colder winters. The coastlines of Alaska and Siberia would no longer have been protected by the warm ocean waters, and so the ground would have frozen to form the permafrost trapping the mammoths that didn't escape; and the freshwater from the melting continental ice sheets that flowed into the oceans and initially floated became frozen around the Arctic Ocean and Antarctica.

It is estimated that this deglaciation period, when most of the continental ice sheets melted and the sea level rose to essentially today's level, would have occurred rapidly in about 200 years. The contrast between a timescale of a total of 700 years for this Ice Age compared with the 1.8 million years in conventional thinking is stark, but the evidence and computer modelling heavily favours a short, rapid Ice Age generated as a consequence of the heat released during a global catastrophic plate tectonics Flood event.

The age of the earth

The geological evidence of rapid deposition of sedimentary strata over vast areas, the catastrophic burial and preservation of so many creatures and plants in fossil graveyards, and the evidence for catastrophic plate tectonics overwhelmingly point to a catastrophic global Flood that reshaped the earth's surface in about a year. This automatically eliminates more than 500 million years in the conventional timescale for the earth's history. However, it is usually claimed that radioactive methods have successfully dated the earth's rocks as millions and billions of years old. So just how reliable are these radioactive dating methods?

These methods are based on the radioactive decay of a parent element into a daughter element. The rocks to be dated are analysed for these parent and daughter elements and then, based on the quantities of each, a date is calculated. What is not always evident is that in this process three main assumptions are made:

1. The initial conditions under which the rock formed are known, so that either the elements were homogeneously distributed or there were no daughters to begin with, implying that all the daughters now measured have been produced from the parents by radioactive decay.
2. There have been no changes in the parent and daughter concentrations in the rocks since they formed apart from radioactive decay, so that neither groundwaters nor any other environmental influences have added or subtracted any of these important elements.
3. Rates of radioactive decay of the parent elements have remained constant during the earth's entire history.

The reality is that none of these assumptions are, strictly speaking, provable. This is because no human observers were present when the geological record was forming to verify the initial conditions, to verify that these elements in the rocks were undisturbed, or to verify that the radioactive decay rates have remained constant throughout the earth's history.

As a consequence, anomalous radioactive dating results are abundant, though this is not well known because such results are not always published. Groundwaters are known to dissolve or precipitate various elements, while surface weathering of outcrops, which are where samples are usually collected for radioactive dating, can subtly perturb element concentrations. The heat from magmatic intrusions has been demonstrated to cause migration of elements in the adjacent rocks, so that the rock dates are much lower than what they are expected to be for up to several kilometres away from the intrusions. Even the constituent minerals in some rocks yield highly variable dates, with results differing by hundreds of millions of years on different faces of the same mineral crystals.

But perhaps the most perplexing and revealing recent discovery is that recent and historic lavas on ocean islands and from active continental volcanoes yield radioactive dates of between one and two billion years. This has led to the recognition that, because these lavas are the result of melting in the mantle, these radioactive dates reflect inheritance of these radioactive elements from the mantle where mixing has occurred in the past. This has been recognised as due to the stirring of the mantle caused by plate motions, with past ocean crust subducted into the mantle and mantle material added to the continental and new ocean crust in lavas and intrusions. Of course, as already explained, such plate motions and therefore radioactive element mixing could only have occurred due to catastrophic plate tectonics during the year-long Flood, so these radioactive dates are not the true ages of the rocks.

However, notwithstanding all these anomalous results and problems with the radioactive dating methods, many of the radioactive dates obtained fit with conventional expectations of the ages of the rocks and correspond in sequence to the inferred order of the strata, namely, the oldest dates for the oldest rocks at the bottom of the geological record and dates getting younger up through the sequence to the youngest dates for the youngest rocks at the top. This is why there is usually so much confidence that the earth's rocks have been dated as millions and billions of years old, in spite of the increasing recognition that the radioactive elements and their daughter products have often been inherited from the mantle, and crustal, sources of the lavas and intrusions in the geological record. On the other hand, this systematic trend would also easily be explained by accelerated radioactive decay during the catastrophic global Flood. After all, if plate tectonics, erosion, sedimentation, fossilisation and other geological processes were occurring at catastrophic rates during this Flood, then it is consistent for radioactive decay to also be occurring at catastrophic rates, thus yielding radioactive dates much greater than the true ages of the rocks.

This has been confirmed by several lines of evidence. First, when uranium decays in tiny zircon crystals in granites, it produces both lead and helium; but while the lead gives conventional old dates, the equivalent helium has only partially leaked out of the crystals, yielding young dates that indicate more than a billion years of radioactive decay occurred rapidly only recently. Second, the radioactive decay has caused physical damage in crystals in granitic rocks. These radioactive halos are produced for uranium and daughter products such as polonium, the uranium halos supposedly taking hundreds of millions of years to be produced in contrast to polonium halos that form in minutes and days. Yet in these granites these halos could only have formed at the same time when the granites had cooled, which again means that hundreds of millions of years of radioactive decay of uranium must have occurred within days while the polonium halos formed too. Third, when specific rocks are dated using all four main radioactive dating methods, a procedure not usually used, the four methods give vastly different dates for the same rocks. However, the pattern of dates corresponds to the type and rate of radioactive decay as predicted by accelerated decay theory. So the evidence points to both catastrophic mixing of elements and accelerated radioactive decay during the Flood, resulting in both anomalous and systematic, erroneous, far-too-old dates for the earth's rocks.

However, there is also much evidence that the earth and its strata are young. Radiocarbon is used to date organic materials at only thousands of years old. Thus coal beds in the geological record, conventionally regarded as hundreds of millions of years old, should not have any radiocarbon in them. Yet recent analyses with sophisticated equipment, able to literally count radiocarbon atoms, has always found radiocarbon in coal samples, dating them at only thousands of years old. This is consistent with the vegetation being buried and fossilised to form the coal during the recent catastrophic global Flood. And even diamonds, supposedly over a billion years old, have yielded radiocarbon and a young age.

Measurements made over the past century-and-a-half have revealed that the earth's magnetic field is steadily decreasing in strength and losing its energy. At the measured rate of decrease the earth's magnetic field cannot be any older than 6000–7000 years. Furthermore, evidence has been found that the earth's magnetic field has rapidly reversed in the past, particularly during the catastrophic plate tectonics of the global Flood and its aftermath. Only freely decaying electric currents in the earth's core can explain the generation and rapid reversal of the earth's magnetic field and its young age. It follows that if the earth's magnetic field is only 6000–7000 years old, then the earth itself is that young.

There are numerous other evidences of the earth's youthfulness, such as the accumulation of salt in the sea, helium in the atmosphere and sediments on the ocean floor, as well as the erosion rate of the continents themselves. Even at today's rates these processes indicate an earth very much younger than conventional thinking, but in the context of a catastrophic global Flood these process rates would have been greatly accelerated, and this coincides with the geological, fossil and global tectonics evidence for a recent catastrophic Flood on a young earth.

Answers to other alleged problems with the Flood

If the waters of the Flood had to cover the mountains, where did they come from and then where did they go afterwards?

In asking this question many incorrectly conceive of the Flood waters having to cover Mt Everest, the summit of which is more than eight kilometres above the present sea level. However, as explained previously, the Himalayas are a product of the catastrophic collision of the Indian and Asian Plates during the catastrophic tectonics of the Flood, so Mt Everest didn't exist when the Flood began.

The description in the Bible is that the earth's crust broke open and fountains of water issued from inside the earth for

150 days, while there was global torrential rain for forty days. Thus most of the water for the Flood came from inside the earth. If the earth's present topography on both the continents and the ocean floors were levelled out, then the water in the current oceans would cover the entire earth's surface to a depth of more than 2.7 kilometres. As for where the waters went at the end of the Flood, they went into today's newly formed ocean basins, and the earth's surface is still 70 per cent covered by water.

There is a certain irony in scientists' recognition that a global flood must have occurred on Mars because of the landscape features observed there, even though there is no liquid water on Mars' surface today, yet many scientists still won't accept that there was a global Flood on the earth when the earth's surface is still 70 per cent covered with water!

Another alleged problem with the Flood is the biblical account of animals embarking with Noah and his family on the ark, which it is claimed could not have housed so many animals (and Noah and his family could not have fed them all anyway). However, only the air-breathing, land-dwelling animals (mammals, birds and reptiles) are described as being taken aboard the ark. Sea creatures, including whales, would have been able to survive in the floodwaters, so they did not need to be taken aboard. Similarly, many plants would have survived as seeds and others as mats of tangled vegetation, floating on the Flood waters, while many insects and other invertebrates were small enough to have survived on these floating vegetation mats.

Furthermore, it was only two of every 'kind' (not species) of land vertebrate animals that went aboard the ark. The members of the biblical 'kinds' are described as being able to interbreed with one another, so the fact that horses, zebras and donkeys can interbreed, as can dogs, wolves, coyotes and jackals, as well as all kinds of cattle including bison and water buffaloes, and all kinds of cats including tigers and lions, means that the original biblical 'kinds' probably equate to the genera of the classification system used by modern taxonomists. Thus only one pair each of horses, dogs and cats, for example, would need to be included in the representatives of about 8000 genera, including those that are now extinct, that needed to go aboard the ark, a total of about 16,000 individual animals. This estimate is generous, because among the extinct animals such as the dinosaurs, fossil evidence is so fragmentary that new finds are often designated as new genera, when really there are only about twenty-four well-established dinosaur genera.

As for size, young animals would be required to go aboard the ark so that they would reach maturity and therefore begin breeding after the Flood. Thus small juvenile dinosaurs, for example, would have gone aboard the ark, not their huge adult counterparts. In any case, there would have been plenty of room on the ark for all these animals, because the ark was approximately 137 metres long, 23 metres wide, and 13.7 metres deep, a total volume of 43,200 cubic metres—the equivalent volume to 522 standard railway stock carriages each able to hold 240 sheep. Thus if the average size of the animals on the ark was that of a sheep, then 16,000 animals of that size would have only occupied 1200 cubic metres, less than three per cent of the volume of the ark. This would have left plenty of space for food and water storage, for exercise and for Noah and his family on the floor space of the three decks of the ark. In any case, fresh water was available from the rain falling during the Flood. Furthermore, the space, feeding and other requirements for all these animals would have been greatly reduced if most of them hibernated during much of their time aboard the ark, a highly likely scenario given the latent ability in most animals to hibernate if required.

But how could Noah have possibly collected all those animals from all around the world and get them to the ark? In fact, the scriptural account nowhere says that Noah had to collect the animals but that they came to him.[1] Besides, if, as is supposed in all global tectonics models for earth history, there was originally one supercontinent, then all the required animals would likely have been living close by where Noah built the ark. Then after the Flood the animals and their descendants migrated across the post-Flood world using the land bridges that were exposed during the lower sea levels of the post-Flood Ice Age, and floating vegetation mats as rafts. So kangaroos came to Australia after the Flood (not before), which is why their fossils are only found in post-Flood sediments in Australia where kangaroos live today. If rabbits can spread right across Australia in less than 200 years after being introduced, then there is no problem with kangaroos migrating from the Middle East to Australia in 700 years or less during the Ice Age.

Where unique organisms are found today, that is either where their ancestors arrived and established themselves, or this is the remnant of a much wider distribution in the past, or both. And animals that today are specialised due to mutations, such as the koala with its diet, would originally have been less specialised and thus more able to migrate to where they are found today.

Alleged ape-men, cave-men and the true history of man

Because Noah and his family were the only human survivors of the Flood approximately 4500 years ago, then all people today are descended from them. This is consistent with population statistics. Indeed, starting with Noah's family and calculating the population growth using a generation span of forty years and an average of 2.5 children per family, the present world population of more than six billion people would be obtained in those 4500 years. Since the parameters used are conservative, this would allow for even more people to have lived and then died early in wars or of disease.

On the other hand, if human beings supposedly evolved even as late as 100,000 years ago, where are all the graves and the remains of the countless billions of people that would have lived and died throughout that period, and where are all their artefacts?

But what about the alleged ape-men? The hominid fossils that are claimed by evolutionists to be our ancestors are found in localised sediments at the earth's surface that clearly post-date the Flood. The different 'kinds' (genera) of animals that came off the ark must have rapidly diversified into today's species as they migrated across the earth's surface in the early centuries after the Flood and become established in separate breeding populations. Among the primates minor physical differences would have developed during such speciation. The hominid fossils are often fragmentary, with critical skeletal pieces missing, but from the available evidence most of the remains would appear just to be variations of the pygmy chimpanzees and their relatives. This includes the australopithecines, whose name anyway simply means 'southern apes'. The remains of a few of these creatures became locally buried and fossilised primarily in parts of Africa.

Meanwhile, instead of spreading out across the earth's surface, Noah's descendants, who still had the one language and culture, stayed together in the Tigris-Euphrates area and built a city called Babel. Genetic variations in the population meant that through intermarriage there was a complete gradation of external human features, such as colour shades of skin and hair, eye shapes and so on. The biblical account then describes how God directly intervened to confuse the people's language, thus creating turmoil and causing all the family clans to break up along language lines and migrate in all directions away from the Babel area. By now the Ice Age was well advanced, so some of these new tribal groups migrated across the land bridges into the Americas, while other groups headed into Africa. Those that went to parts of northern Europe and elsewhere could only find shelter in caves because of the severe weather associated with the Ice Age. The tribal groups that remained in the Babel area continued on with the civilisation they had already developed, but those tribal groups that moved into distant lands had a harsh time first of all using whatever they could, such as stone tools, to hunt and eventually clear land to develop agriculture again, before finding metals to again rebuild civilisations. Some tribes didn't progress through all these stages, so (even as today) there were tribal groups with stone tools living contemporaneously in adjacent areas with those in metal tool civilisations.

Not only does the evidence fit this model for the true history of man, but it has numerous implications. First, because the tribal groups that migrated to Africa, for example, arrived there after hominids had been locally buried and fossilised, this would give the false impression of an evolutionary progression, particularly as the tribal groups may have looked different from other human tribal groups and have initially used stone tools.

Likewise, the supposed 'Stone Age' to 'metal ages' to civilisation evolutionary progression in humanity and its culture is merely a figment of interpretation, rather than being supported by the evidence. And people such as the Neanderthals—who like the Cro-Magnons were our human relatives, because even though they may have looked slightly different they had even larger brain capacities than we have on average today—were artistic and buried their dead, but often lived in caves because that was convenient in the climate where they lived. Meanwhile, the tribal group that became the Australian Aborigines migrated south-east across Asia, crossed the land bridge down through the Indonesian islands, and eventually used rafts to paddle across a few remaining stretches of water to reach Australia, bringing the dingo with them, as some of them recall in their 'dreamtime' stories. They therefore would have arrived here only about 4000 years ago, or perhaps even later, because they have always practised circumcision, which was first instituted with Abraham.

Finally, unlike the evolutionary view of human history that

regards some people groups as less advanced and more primitive, and others as more advanced and civilised, the true history of man knows nothing of racial prejudice because all people, whether dead, fossil or living, are equal and of one blood, having all descended from Noah and his family after the Flood.

Conclusions from the evidence

The evidence in the earth's rocks and surface features bears unmistakeable testimony that this world and all the creatures in it were created by an intelligent designer, and this happened only thousands of years ago. A catastrophic global Flood then swept away most of the life on the earth's surface, burying many creatures in rapidly deposited sediments to produce the fossil record as the earth's surface was reshaped by catastrophic plate tectonics. Crustal rocks were recycled and mixed in the mantle, while radioactive decay was accelerated so that the radioactive elements in the older rocks and in new lavas and intrusions now yield anomalous and old ages, contrary to so much other evidence in the rocks and in the earth, the oceans and the atmosphere for a young world.

The animals and people that survived this catastrophic upheaval in the ark repopulated the world after the Flood. In the first few centuries after the Flood, a rapid Ice Age gripped the earth, lowering sea levels to provide land bridges that further aided human and animal migration. Many organisms migrated by drifting across the oceans on floating vegetation rafts. People originally settled in the Tigris-Euphrates region with the one language and culture, but divine intervention to confuse their language caused them to migrate as tribal groups with different languages and cultures into the four corners of the earth.

So when examined thoroughly, the geological and related evidences totally agree with the biblical account of the history of the earth and of man, causing us to acknowledge our creator God, who not only shaped the earth and created all life on it, including us, but has been intimately involved in our history and therefore our lives. How can we not seek to know him?[2]

10

Where Do Thoughts Come From?

Charles Taliaferro

The existence of consciousness has been employed in arguments against naturalism (the view that only the physical world exists) and for theism. This chapter begins by (I) clarifying the nature of consciousness and (II) bringing to light the problem it creates for naturalism, before (III) exhibiting some of the ways in which the emergence and existence of consciousness can support theism.

(I) Consciousness

'Consciousness' is currently one of the most intensive subjects of philosophical and scientific inquiry. There is no uniform, widely agreed definition of consciousness. For present purposes we may delimit consciousness in terms of our beliefs, desires, sensations, intentions and emotions. It may be that we have beliefs and so on that are unconscious or subconscious, but the ideal or evident cases of beliefs, desires and the like occur under conditions of conscious awareness. Other terms for 'consciousness' are the psychological or mental, the experiential and subjective. In ancient philosophy, Plato and Aristotle talked of consciousness in terms of the soul and its states and activities.

Although denying the existence of consciousness would seem (to many of us) sheer madness, some contemporary philosophers have done so. To some extent, it is not surprising that advances in philosophy and science should compel us to revise our ordinary, 'common sense' view of consciousness. Marvin Minsky warns about the limited use of mental terms when it comes to constructing 'powerful theories' about the world.

> . . . though prescientific idea terms like 'believe', 'know' and 'mean' are useful in daily life, they seem too coarse to support powerful theories; we need to supplant rather than to support and explicate them.[1]

He goes on to argue that as real as 'self' or 'understand' may seem to us today, they are different from objective things (like milk and sugar) that our theories must accept and explain.

But can we ever come to a point when we would have to give up the belief that we are conscious? The following philosophers have each come to propose (in varying degrees) that there are no such things as beliefs, desires or conscious, subjective experiences: Willard Quine, Richard Rorty, Paul and Patricia Churchland, Stephan Stich and Daniel Dennett. They raise (in Paul Churchland's words) this question:

> Is our basic conception of human cognition and agency yet another myth, moderately useful in the past perhaps, yet false at

edge or core? Will a proper theory of brain function present a significantly different or incompatible portrait of human nature?[2]

Indeed, Churchland, who is inclined to a positive answer to these questions, has argued that talk of our having beliefs and conscious states is a myth.

The problem with this full-scale rejection of the mental is that it involves an inevitable contradiction. The original book jacket of Stephen Stich's book, *From Folk Psychology to Cognitive Science*,[3] brings this out: Stich is described as believing that there are no beliefs. (In a private communication, he told me that he was outraged at this.) To accept a theory of the world inevitably involves believing it, and to argue for a theory presupposes that you can persuade others to accept your beliefs based on reasons. The charge of self-contradiction has been powerfully argued by Lynn Baker, William Hasker, Arthur Danto and others.[4] And, more generally, philosophers have come to abandon the project of eliminating consciousness.[5] The denial of our mental life simply flies in the face of every waking moment.

The subsequent philosophical task has been to explain, rather than to explain away, consciousness. It is at this point that naturalism is challenged. Jaegwon Kim, a senior philosopher of the highest reputation, offers this overview of the project facing the naturalistic philosopher:

> The shared project of the majority of those who have worked on the mind-body problem ... has been to find a way of accommodating the mental within a principled physicalist scheme, while at the same time preserving it as something distinctive.[6]

Kim is himself a naturalist, so he sees the task whereby we do this, without losing what we value or find special in our nature as creatures with minds, as vital.

(II) Naturalism in trouble

Given that the cosmos is fundamentally a material one—in origin and basic structure—why suppose that consciousness would emerge? Naturalism seems either to have to reduce consciousness to physical states or to recognise the emergence of something radically distinct from the spatio-temporal physical world. Let's consider each option.

On the side of reduction, Paul Churchland expresses well the motive behind seeing consciousness as a thoroughly physical matter:

> Most scientists and philosophers would cite the presumed fact that humans have their origins in 4.5 billion years of purely chemical and biological evolution as a weighty consideration in favor of expecting mental phenomena to be nothing but a particularly exquisite articulation of the basic properties of matter and energy.[7]

The problem is that conscious mental states appear (by their very nature) to be distinct from the basic and non-basic properties of matter and energy. Is my thinking 'I am puzzled by your gaze' the same thing as electrochemical processes in my brain? Is observing my behaviour, brain and nervous system the very same thing as observing my puzzlement? My puzzlement (and all my thoughts and emotions) may be brought on by brain and other bodily states, but it is not the very same thing as these states. The interaction between the mental and physical, in fact, presupposes a distinction between the two. There is indeed interaction, but the point needs to be made that interaction is not identity. It has been forcefully argued by a range of philosophers that knowledge of the physical does not (taken alone) constitute knowledge of consciousness.[8]

Those who seek to tame consciousness by seeing it as a merely physical affair face a problem similar to those who try to eliminate consciousness. For example, it has been pointed out that Dennett's attempt at getting rid of conscious appearances (it only appears that there are appearances) retains appearances.[9] If someone claims that subjectivity does not exist because you only subjectively think it exists, they are locked in a contradiction in much the same way that someone contradicts themselves when they claim to know that nobody knows anything.

A growing number of philosophers have recognised the difficulty of seeing consciousness as an exclusively biological process.[10] One philosopher, John Searle, has taken the route of expanding the concept of biology simply to cover consciousness. But one wonders whether this isn't a way simply of re-naming or labelling a mystery.[11] Is conscious thinking on the same level (or the same kind of thing) as the functioning of one's nervous system? The thinking carried out by the brain (or the thinking carried out by a person with his or her brain) seems like a different kind of thing from the heart pumping blood. After all, you can measure the size, location, colour and weight of the heart. But to claim to measure the size, location and colour of a thought like '1 + 1 = 2' seems bizarre at best.

And the issue is not merely a matter of the physical and mental merely appearing to be different; the physical description of the brain is not *ipso facto* a description of the person's thoughts. In a physical description of my brain and body, you will not get a physical description of my thoughts and emotions unless these are somehow brought in by way of laws involving physical–mental interaction—in which case the resulting descriptions would not be purely physical.

Some naturalists who do recognise the radical distinction between the conscious and the physical see consciousness as a 'supervenient property', a property that emerges or necessarily grows out of the physical.[12] The problem with this, however, is brought out by the naturalist David Armstrong. He suggests that it is quite reasonable that, when the nervous system reaches a certain level of complexity, it should evolve new properties or should affect something that was already in existence in a new way. He goes on to point out:

> But it is a quite different matter to hold that the nervous system should have the power to create something else [mental entities], of a quite different nature from itself, and create it out of no materials.[13]

Colin McGinn captures the problem of consciousness for naturalism as follows:

> We have a good idea how the Big Bang led to the creation of stars and galaxies, principally by the force of gravity. But we know of no comparable force that might explain how ever-expanding lumps of matter might have developed an inner conscious life.[14]

(III) Consciousness and theism

Theism has been employed at many points philosophically when it comes to accounting for our contingent cosmos. Consciousness has a role especially in design or teleological arguments. Why is there a cosmos which is governed by laws of nature and that has included the emergence of conscious, living, moral and religious beings? And consciousness has had a role in other arguments as well, such as arguments from religious experience (the apparent consciousness of God), knowledge or cognition (epistemological arguments), and beauty (aesthetic arguments).

A range of philosophers have argued that the existence of conscious beings is best accounted for by an overriding divine consciousness, thus explaining the existence of minds in terms of a divine mind.[15] The argument from consciousness has been developed in different ways, but the essential heart of the argument is that the emergence of consciousness (in humans and nonhuman animals) and the cosmos as a whole is best accounted for by an over-arching powerful consciousness (or mind) rather than mindless, non-conscious forces.

I seek to bring out the plausibility of this line of reasoning by replying to a series of objections.

Objection 1: Mysteries. Isn't 'explaining' the emergence of consciousness in terms of a supreme consciousness replacing a minor mystery with a colossal mystery?

Reply: Every philosophy will have to recognise a basic level of explanation. For example, in naturalism, physical causal relations involving the most fundamental constituents of the cosmos will have to be acknowledged as foundational, and not explained by some deeper, more basic relationship. The question 'Why?' will have to come to an end. The theistic philosopher is raising the following question: What is more basic—explanations involving consciousness or explanations lacking consciousness? If naturalism is accepted, the emergence of consciousness is a mystery, whereas the emergence of mind from mind is intelligible and has greater coherence.

Objection 2: Still mysterious. But isn't the creation of matter and energy by a divine consciousness still a mystery?

Reply: If by 'mystery' one means 'ultimate' or 'not further explainable', the appeal to God will retain a mystery. If one were to remove mystery by 'explaining' God in terms of physical laws, one would presumably have treated God as simply one among other items in nature. But, speaking directly to the question, we may see in our own powers of creative imagination the ways in which (we) conscious beings can create whole worlds with different laws and creatures (think of Tolkien's creation of Middle Earth in his trilogy *The Lord of the Rings*). In our own case we may catch a glimpse of what would be infinitely magnified in a divine consciousness.

A key element in conscious agency is that meaning and purpose can come into play in a way that they do not in non-conscious causation. In non-conscious processes the end or goal of the process does not enter into the cause; in other words, the processes do not work with any prevision of the end being brought about. In theism, there would be an ideal case of a causal power (God) creating the cosmos intentionally for the good. Otherwise, the cosmos and the emergence of consciousness are more mysterious.

Objection 3: Anthropomorphism. Why locate the cause in consciousness? Isn't that a base projection of humanity? David Hume in the seventeenth century put the objection in the following terms: why suppose that the creator of the cosmos bears a resemblance to mind (which Hume described as an agitation of the brain)?

Three replies: First, I have argued above that there are abundant reasons not to identify thought with brain process and states ('agitations of the brain').

Second, this objection diminishes the radical difference between conscious and non-conscious explanations. The first involves a full range of powers, especially conscious purposive activity. The choice between appealing to purposive or non-purposive causation goes far beyond a concern with anthropomorphism and human projection.

Third, thinking of God as supreme consciousness is not a matter of attributing an exclusively human characteristic to God. There is ample evidence of actual nonhuman consciousness (in some nonhuman animals) and we can readily imagine levels of nonhuman consciousness which far surpass our feeble endeavours.

It should also be noted that the concept of God is the concept of a supremely excellent, good reality. If we thought that God *surpasses consciousness*, would we be attributing an excellence to God? If 'surpassing consciousness' merely means being greater than human or animal or angelic consciousness then there is no problem. But if 'surpassing consciousness' means losing consciousness, presumably this would be an imperfection. A divine being lacking in consciousness would lack knowledge, power of agency, love and the like, all of which would remove from God key divine attributes of wonder and worship.

Objection 4: Future science. What if future science were able to produce consciousness from non-conscious causes? Imagine, for example, a supercomputer that was conscious, or a thinking humanoid.

Reply: Initially one should note that claims concerning computers and humanoids out of science fiction often involve the *simulation* of consciousness, not actual consciousness. But even if non-conscious processes produce or wind up causing conscious processes, there remains the fundamental question of whether or not this is because we live in a cosmos created by God who wills that consciousness emerge in nature.

In our own development, non-conscious elements (egg and sperm) come together and biological functions are carried out until there is a conscious human subject. Christians have differed in their account of when there is a soul or ensoulment (for consciousness). Some hold that God creates each person or soul directly; others (such as myself) hold that the emergence of consciousness is the outcome of God's comprehensive creative will. That is, I hold that when you came into being this was indeed a divine creation, but rather than being a special divine act (like a miracle) it is God's comprehensive will that when there is a level of physical complexity and coherence then there emerges consciousness. In the general original use of the word 'miracle'—an object of wonder—each bringing-into-being of consciousness is a miracle, but not in the sense of being a violation of the laws of nature.

Back to future science: just as human beings can use and misuse God's laws, there may be scientific use or abuse of God's will that our cosmos is one in which consciousness emerges. If God has made the cosmos such that consciousness emerges when certain physical conditions obtain, then presumably this emergence would take place if or when scientists re-create the very same physical conditions.

The argument from consciousness, like most theistic arguments, is best seen as part of a cumulative case for theism. See my book *Consciousness and the Mind of God* for some of the salient moves and counter-moves in the debate.[16]

11

The Question of Moral Values

Steven B. Cowan

Jeffrey Dahmer, the infamous serial killer, grew up in a fundamentalist Christian home. When he became an adult, he forsook the religious beliefs of his family and embraced atheism and the Darwinian theory of evolution. By his own admission, it was this new belief system that ultimately led him to commit his horrible crimes. After his arrest and trial he said:

> If you don't . . . think that there is a God to be accountable to, then . . . what's the point of trying to modify your behaviour to keep it within acceptable ranges? That's how I thought anyway. I always believed the theory of evolution as truth, that we all just came from the slime. When we . . . died, you know, that was it, there was nothing . . .[1]

After he was arrested, Dahmer had an overwhelming sense of shame over what he had done. He knew that his actions had been dreadfully wrong. Yet he could not explain this sense of right and wrong against the background of an atheistic worldview. He realised that only the existence of God can make sense of the existence of an objective moral order. So Dahmer came to embrace the Christian God. In his final public interview, he said: 'I've since come to believe that the Lord Jesus Christ is truly God, and I believe that I, as well as everyone else, will be accountable to him.'[2] He came to grips with the fact that there truly are objective moral values, and that such moral values depend for their existence on the existence of God.

What Dahmer discovered was the *moral argument* for God's existence.

Objective versus subjective moral values

Before we go on, it might be helpful to explain what we mean by the term 'objective moral values'.

Many people today believe that morality is a purely individual matter, or perhaps a cultural matter. That is, they believe that morality is a matter of individual or cultural preference. Morality is like matters of taste. I say that chocolate ice cream tastes best. You say that vanilla ice cream tastes best. Is one of us wrong and the other right? Of course not. We are both right. Which flavour of ice cream tastes best is a matter of one's own subjective preference. In the minds of many people today, principles of morality are just like ice cream: matters of subjective preference. This idea that morality is subjective is called *moral relativism*.

Moral objectivism is the opposite of moral relativism. By 'objective moral values' we mean moral values that do not depend for their validity upon whether or not they are accepted by any person or group of persons. To say, for example, that an action is morally right or wrong, in the objective sense, is to say that that action is right or wrong for anyone and everyone in relevantly similar circumstances. Its rightness or wrongness is not a matter of an individual's or a culture's subjective preference.

The moral argument

Now that we have a basic idea of what we mean by objective moral values, we may summarise the moral argument as follows:

Premise 1: If God does not exist, then objective moral values do not exist.

Premise 2: Objective moral values exist.

Conclusion: God exists.

This, or something like it, is the sequence of reasoning that Jeffrey Dahmer used to reach the conclusion that God exists. Is it a good argument?

I believe that Dahmer's reasoning was sound. The moral argument gives us powerful reason to believe in the existence of God. In order to show this, however, we must show that the premises of the argument are true or at least highly probable. That is, we must show that there are indeed objective moral values and that these objective moral values depend upon the existence of God.

Are there objective moral values?

That there are objective moral values is intuitively obvious. All of us, or at least most of us, have an innate and ineradicable knowledge of basic moral truths. We intuitively know that some things really are right and some things really are wrong. That people sometimes doubt this is due, in part, to the fact that they are distracted by extreme and borderline cases. They see the difficulty in adjudicating disagreements over matters such as capital punishment, physician-assisted suicide and affirmative action, so they wonder if there is any objective truth in morality at all.

But the places where people disagree strongly on moral issues is not the place to begin a discussion of whether or not there are objective moral values. We should start, rather, with clear-cut cases. How about this one: *it is wrong to torture children for the fun of it*. This statement expresses an objective moral value. We all simply *know* that torturing children for the fun of it is morally wrong. The person who disagrees is simply mistaken, and (like the colour-blind person who cannot see a certain shade of colour) is sadly defective. As Francis Beckwith and Gregory Koukl explain:

> Those who deny obvious moral rules—who say that murder and rape are morally benign, that cruelty is not a vice, and that cowardice is a virtue—do not merely have a different moral point of view; they have something wrong with them.[3]

Later in this chapter I will address a potential objection to the existence of objective moral values. For now let us grant that premise (2) of the moral argument is true—there are objective moral values.

The question before us now is: How do we *explain* the existence of these values? What we are asking is *what else* would have to be true given the fact that objective moral values exist? Just as a scientist, in trying to explain some particular phenomenon, will construct a background theory that allows him to make sense of his data, we are asking what background theory or worldview would allow us to make the most sense of objective morality. There are three possibilities.

First, there is a worldview known as *naturalism*. This worldview asserts that all that exists is the natural, physical world. Everything that exists, according to naturalism, is the result of chance, time and evolution.

A second worldview is *pantheism*. The pantheist holds a view that is the polar opposite of the naturalist. Whereas the naturalist believes that everything that exists is material and physical, the pantheist believes that everything is spiritual and divine. Somewhat more technically, the pantheist believes there really exists only one thing, and that one thing is an eternally existing and limitless Spirit or Mind. This means, of course, that the physical world that we experience through our five senses is an illusion (the Hindu term is *maya*).

The third possible worldview is *theism*. According to theism, the physical universe is the creation of God, an eternal, omnipotent, omniscient and omnibenevolent *personal* being. God is distinct from the created world, not part of it or identical to it. Yet the physical world is not all there is, and the world depends for its continued existence on God's sustaining power. Moreover, the world and all it contains were created for an intelligent purpose.

It is the contention of this chapter that the existence of objective morality makes perfect sense if theism is true, but cannot be adequately explained if naturalism or pantheism is true.

Does morality depend on God?

Premise (1) of our moral argument is tantamount to the claim that only theism (not naturalism or pantheism) can provide an adequate explanation for the existence of objective morality. That premise claims that if God does not exist, there can be no objective moral values. In defence of this claim, let us first see the inadequacy of atheistic naturalism for explaining objective morality.

It is worth noting at the outset that premise (1) of the moral argument has been widely accepted in the course of history by both theist and atheist alike. The great Russian novelist and philosopher Fyodor Dostoyevsky, a Christian, once wrote, 'If there is no God, everything is permitted.' And J.L. Mackie, one of the most prominent atheist philosophers of the twentieth century, agreed. He said, 'Moral properties constitute so odd a cluster of qualities and relations that they are most unlikely to have arisen in the ordinary course of events without an all-powerful god to create them.'[4] Richard Dawkins, an evolutionary biologist and outspoken atheist, explains it this way:

> In a universe of electrons and selfish genes, blind forces and genetic replication, some people are going to get hurt, other people are going to get lucky, and you won't find any rhyme or reason in it, nor any justice.[5]

Dawkins goes on to claim that the universe we observe has

precisely the properties we should expect if there was no design, no purpose, no evil and no good.

Why does it seem to these thinkers that apart from God there is no right and wrong? As the above quote from Dawkins suggests, it has to do with the atheist's naturalistic picture of the universe. According to naturalism, all that exists is the physical universe. Moreover, everything that exists sprang from a cosmic accident—the big bang—several billion years ago. The universe was not created by an intelligent being for some grandiose purpose. The big bang occurred simply as the result of unguided natural processes.

It follows from this that human beings are nothing special. We, too, are cosmic accidents who happened to evolve from the slime in a remote corner of the universe. And the history of evolution—which exhibits an incredible indifference to the birth and extinction of millions of entire species—shows us that the human race is destined to perish like the dinosaurs. In fact, the atheist-naturalist story has a very predictable outcome. The story will end with the heat-death of the entire universe as all the stars burn out, all the planets turn cold and every living thing dies.

Given this picture of the world, what basis is there for affirming the existence of objective moral values?

In order to attribute objective moral value to anything we have to have some reason to believe that something has *intrinsic* value—that something is valuable for its *own sake*. That is, if we are going to be able to say that it is wrong to take human life without a just cause, then we have to have some reason to believe that human beings are *special*—that there is something intrinsically valuable about human beings that sets them apart from dirt or dandelions or cockroaches. Yet naturalism seems to provide us no basis for saying that human life is special.

Of course, an atheist might claim that morality itself is the result of evolution. That is, he can say that human beings developed a sort of social instinct for morality in order to help them survive. In fact, this is exactly how Darwin explained the origin of morality.[6] The human species developed a belief in moral principles like justice and mercy because having those beliefs increases our capacity to survive.

This evolutionary account of morality may explain why we *have* moral beliefs. However, it does not provide a basis for affirming the existence of *objective* moral values. Philosopher Mark Linville notes that 'an evolutionary account of morality is essentially committed to *moral subjectivism*, the view that *all moral judgments are expressions of the sentiments or tastes of the speaker*'.[7] This is because, on the evolutionary account, moral values come into existence through natural selection as simply a means to help organisms better survive. They do not arise because there is anything intrinsically and objectively good about sympathy, mercy, justice and so on.

Many evolutionary thinkers admit this. James Rachels, for example, writes:

> Man is a moral (altruistic) being, not because he intuits the rightness of loving his neighbour, or because he responds to some noble ideal, but because his behaviour is comprised of tendencies which natural selection has favoured.[8]

Or as Michael Ruse puts it, 'Morality is just an aid to survival and reproduction . . . and any deeper meaning is illusory.'[9] It should be clear, therefore, that the naturalistic-atheistic worldview cannot provide a ground for objective moral values.

What about pantheism? Can it provide a basis for objective morality? No. Indeed, most pantheists are open in affirming moral relativism. Remember that pantheism is the view that there exists one eternal, limitless and divine Spirit. For the pantheist, this means that *I*, ultimately, *am* that one divine Spirit—though somehow (according to most pantheists) I have become ignorant of that fact. And since I am God, I may make my own rules.

Moreover, this Spirit exists without change, without any distinctions in its being, completely incomprehensible to human logic or language. None of our concepts or terms can adequately describe this being. We can say neither that it is personal nor impersonal, good nor evil, or anything else. All one can say about God, in Hinduism's ancient language of Sanskrit, is '*neti, neti*' (not this, not that). This means that words like 'good', 'evil', 'right' and 'wrong' have no meaning when applied to God/Ultimate Reality. For the pantheist, concepts like good and evil are aspects of *maya* (illusion) and are thus unreal. This is why the Hindu philosopher Shankara said that 'to him who has obtained the highest aim, no obligation can apply'.[10]

So like naturalism, pantheism cannot explain the existence of objective moral values.

Theism, however, does make sense of objective morality. For if God exists, then that means he has created us for a good purpose and has endowed us with intrinsic value as creatures made in his own image.[11] And since human beings are created in

God's image and have intrinsic value, we can honestly say they are truly special and morally significant. More than that, it makes sense that God's image-bearers would be created with a conscience—a moral sense that recognises and intuits the objective moral values with which the universe is furnished.

So unlike atheism and pantheism—in which moral values are strange and out of place—objective morality is perfectly at home in a theistic universe. Christian philosopher J.P. Moreland expresses this truth succinctly:

> Morality is more at home and less ad hoc in a theistic universe than in an atheistic universe. This is because God is a postulated entity who is himself good. He has the property of goodness.[12]

Moreland goes on to point out that human beings are made in God's image, and that to be a human being is to be defined not merely biologically but also in terms of moral properties. We have intrinsic value or worth as we reflect the intrinsic value of God and his worth.

Therefore, in conclusion, if we know that objective moral values exist, and we know that the existence of moral values depends upon God, then it follows by irresistible logic that God exists. Before we rest content with this conclusion, though, we will take a look at a few objections that may be raised to our defence of the moral argument.

Three objections to the moral argument

Objection 1: Different moral codes

We have defended premise (2) of the moral argument—that

The Question of Moral Values

there are objective moral values—by appealing to our moral intuitions. However, a critic of the argument may attempt to provide an argument of his own to show that moral relativism is true after all. The standard argument for moral relativism goes like this:[13]

1. Different cultures have different moral codes.
2. Therefore, there is no objective truth in morality.

In defence of the premise, the objector may point to various societies around the world which seem to have contradictory beliefs about morality. For example, most people in Western cultures believe that cannibalism is wrong. Yet there are peoples in the world who believe that cannibalism is morally permissible. So despite our defence of objective moral values, there may be some reason to call their existence into question.

There are three things we may say in response to this argument for moral relativism.

1. *The argument for moral relativism is invalid.* The conclusion of the argument does not follow from its premise. Just because people disagree about morality does not by itself prove that no one is right and no one is wrong. It may simply be the case that some people's views on morality are false!

2. *Moral relativism has unacceptable implications.* Let us suppose that moral relativism is true. What follows?

First of all, we could never judge the behaviour of any other culture to be morally inferior to our own. Moral relativism implies that all moral views are equally valid and that none is inferior or superior to any other. Therefore, moral relativists have to admit that even the culture of Nazi Germany, which killed six million innocent Jews, did nothing wrong! And they would have to admit that Jeffrey Dahmer did nothing wrong, either. Clearly this result is unacceptable.

Second, the concept of moral progress would become meaningless if moral relativism were true. Moral relativism entails that all behaviour is morally equal, so any change is just change, not change for the worse or for the better. This is yet another unacceptable consequence of moral relativism. We believe that we *have* made moral progress at times. The United States, for example, made moral progress when it abolished slavery in the 1860s. If moral relativism is true, however, the US did not get better, it just became different.

Third, if moral relativism were true, there could be no moral reformers. A moral reformer is someone who works to make society better. William Wilberforce and Martin Luther King are examples of highly respected moral reformers. However, if moral relativism is true, then morality is, at best, something decided by the majority of a given society. And that means that whatever the society says is right, *is* right; and whatever it says is wrong, *is* wrong. This implies that moral reformers who call into question the consensus of society would, by definition, be wrong! Again, moral relativism upsets what we deeply believe about morality.

3. *The premise of the argument for moral relativism is false.* The defender of moral relativism bases his argument on the claim that there is widespread disagreement around the world on what is right and wrong. This claim is not true. There are some superficial differences in the moral codes of various cultures. Nevertheless, all cultures have the same basic moral

principles underlying their moral codes. All cultures have laws, for example, against unjustifiable homicide, stealing and incest. All cultures value human life (at least of those humans who are members of the society), the nurturing of children, truth-telling, justice and so on.

Most apparent differences on morality usually boil down to *factual* differences, not moral differences. The disagreement lies not at the point of some basic moral principle, but at the point of whether or not some non-moral fact is true or false. To illustrate this, imagine an African tribe that believes it is morally permissible to feed physically deformed children to the hippopotamus. Our culture would see this as immoral. Is there a moral disagreement here? Only until you ask the question of *why* this tribe believes it is permissible to feed a deformed child to the hippopotamus. They believe that the deformed child is the *property of the hippopotamus*. They would reason like this:

1. It is right to give another being what belongs to it.
2. Deformed children belong to the hippopotamus.
3. Therefore, it is right to give deformed children to the hippopotamus.

We disagree with the derivative moral principle expressed in (3). But this principle is a conclusion drawn from the *general moral principle* expressed in (1) combined with the *factual claim* expressed in (2). We do not disagree with (1). It is on the factual claim (2) that we believe they have gone wrong. So there is no deep moral disagreement between our culture and this tribal culture. We share the same basic moral principles. It is only when we come to believe different non-moral facts that disagreements appear in moral behaviour. Therefore, the argument for moral relativism fails and our intuition that there are objective moral values remains unscathed.

Objection 2: Moral properties as brute facts

Another objection to the moral argument admits that there are objective moral values, but attacks the premise which asserts that morality depends on God. It might be asked why objective moral values cannot simply exist as brute facts. There are many strange and mysterious things that exist in the universe—why cannot moral properties be just another mysterious part of the furniture of the universe which came into existence accidentally and randomly like everything else?

The problem with this suggestion is that it cannot explain why moral values—as mere brute facts—would have anything whatsoever to do with us. Why should I or anyone else care about these brute moral values? Why would they have any more to do with me than a particular grain of sand on the far side of the moon?

To understand this point better, we need to note that one of the important characteristics of morality is *prescriptivity*. Moral values prescribe behaviour. They have the weight of *law*. Moral values, understood objectively, bind our consciences. But why should moral values, if they are simply brute facts that evolved without purpose or cause, bind my conscience? Moral prescriptions/laws make sense only if there is a moral prescriber/law-giver.

Moreover, the idea that moral values are just brute facts about the universe implies that moral values and properties could exist even if no persons (human or otherwise) ever existed. But this is absurd. Since moral values are prescriptive and thus bind our consciences, this means that there are such things as *moral obligations and rights*. But moral obligations and rights are had by people. Therefore, as philosopher Paul Copan declares, 'If no persons existed, then no moral properties would exist.'[14] Since this objection implies that moral values could exist without persons existing, we should not accept the view that moral facts are simply brute facts.

Objection 3: The Euthyphro dilemma

The last objection to the moral argument is also directed at the claim that morality depends on God. It is called the 'Euthyphro dilemma'.

This objection goes back to the Greek philosopher Socrates. On one occasion, Socrates had a debate on the nature of morality with a man named Euthyphro who maintained that what is holy (that is, moral) is whatever the gods like. Socrates then asked Euthyphro a difficult question: 'Is something holy because the gods like it, or do the gods like it because it is holy?' In a similar vein, someone today might ask, 'Is something good because God commands it, or does God command it because it is good?'

This question poses a dilemma for the theist who wants to ground morality in the existence of God. On the one hand, if we say that something is good because God commands it, then it would seem that good and bad are entirely arbitrary. Why? Because then if God had said that adultery and murder are good, they would be good—by definition! Yet we believe that morality is deeper and more significant than that. On the other hand, if God commands something because it is good, then it would seem that morality is independent of God. When God makes a command, he is merely recognising a moral standard which exists quite apart from him.

The Euthyphro dilemma poses a challenge to the moral argument because it forces the theist to choose between two unpleasant options. Either morality is dependent on the arbitrary will of God, or morality is independent of God.

The problem with the Euthyphro dilemma, however, is that it fails to consider a third alternative. The theist needs to say neither that God commands something because it is good, nor that something is good because God commands it. Rather, he can say that something is good insofar as it reflects *God's necessarily good nature*. So what ought Euthyphro to have said to Socrates' difficult question? Mark Linville suggests that perhaps the answer is found in the possibility that we are faced with a false dilemma. Much of Christian theological teaching maintains that it is not God's arbitrary will that is the standard but rather God's fixed nature. Therefore an appeal to God's nature rather than his arbitrary will allows us to maintain that God wills right actions because they are right. Linville writes:

> *[God's] will expresses his nature and his nature is the source of morality.* And because morality finds its source in God's nature, such a reply will not thereby commit us to saying that morality rests upon something independent of God.[15]

So the Euthyphro dilemma does not pose an obstacle to affirming that morality depends on the existence of God.

From these arguments we can deduce that objective moral values exist. No rational person will deny this evident truth. Yet, as we have seen, the existence of objective moral values makes sense only if God exists. Realising these facts should compel us then, like Jeffrey Dahmer, to embrace the God who is and who has written his moral law upon our hearts.

12

The Problem of Evil

Jon Paulien

There is clearly something wrong with the world. Between acts of genocide, suicide bombers, widespread pollution, random street muggings, sexual abuse and smart bombs that stupidly kill children, we can all tell that some sort of pervasive evil has twisted the minds and hearts of human beings. We long to believe that the world and those who live in it are basically good, but most of the everyday evidence seems to run in the opposite direction.

Can God be good and yet allow so much pain and suffering into the world? Is there any reason to hope that something better lies beneath the surface of what we see and experience?

The Bible tells us that things were not always this way. Before there was an earth, before there was even a universe, there was an Eternal Lover, a Being whose very nature was and is love. 'I have loved you with an everlasting love,' this Being declares.[1] Before there was an earth or any human being, this loving God envisioned what it would be like to have a universe full of creatures that could love and be loved. Like a woman who falls in love with a baby before it is born, God loved the creation before it was created. 'God is love.'[2]

The Bible goes on to tell us that God prepared the way for the creation by filling it with innumerable tokens of his love. There are flowers, almost infinite in variety, with hundreds of shades of every imaginable colour and incredible perfumes running from light and delicate to rich and dusky. There are fruits, grains, nuts and vegetables with their infinite variety of smells and tastes.[3] There are animals ranging from the awesome and magnificent, like the lion, tiger and bull elk, to the unbearably cute, like the koala, kiwi, chipmunk and meerkat.[4]

The incredible delight one finds in the plants and animals is not a necessary feature of existence. We could live without a variety of colours and tastes. We could live without animals. But life would not be nearly as enjoyable. We could also live without the songs of birds, but who would want to? (Except perhaps the annoying screech of the sulphur-crested cockatoo!) And that is only the beginning of God's gifts.

I could speak about mountains and lakes, beautiful sunsets over the ocean, the smell of fresh-cut grass and many other delights. The Bible tells us that these unnecessary but enchanting features of our world are the gifts of an extravagant Lover, who wants to fill the lives of those he loves with exquisite joy.[5] In spite of the evil we experience in the world today, these tokens of God's love are still there to be noticed and enjoyed.

But if God's intentions were so good, why is there so much pain and suffering in the midst of this beauty?

It all goes back to a choice that God made. When it came

time to create beings, God had to decide whether these beings would be controlled by him or truly free. One wonders at times whether it would be better if human beings did not have free will. As 'robots' we could be programmed to be good and kind and to function in a way that enhances the good of the whole creation. In a world of such beings, things would never go wrong.

But there is a problem. Full robotic control leaves no room for love. Imagine your spouse was a robot with a computer for a brain. Imagine you could program him or her to have the perfect body and to respond with loving words and actions in all circumstances. While at first blush this may sound like the perfect partner, the delight in such an arrangement would quickly wear off.

'I love you so much,' you say to your favourite robot.

'I love you with all my silicon,' the robot responds.

When you realise the response isn't free, the words rapidly become empty. Genuine love requires free will. It is only meaningful when it is chosen and given as a gift to the other. Love occurs only when someone is also free not to love, or to love someone else. But when someone else is free to love you, they are also free to hurt you and reject you. The possibility of love requires the possibility of evil. Freedom is the greatest of all risks.

The bottom line is that love and freedom go together. In order to have one you have to have the other. So when the God who is love, who is the Eternal Lover, decided to create, he also decided to make himself vulnerable to the choices of his creatures. He made all things good,[6] but he also allowed his creatures the freedom not to love, the freedom to reject him. Ultimately, evil exists not because God is a tyrant, but because he is committed to openness and freedom. Evil exists in this world not because God is powerless, but because he wanted human beings to be powerful in ways that mirrored his own freedom of action.

So God created the world and filled it with loving gifts for the human race. He gave the original humans the gift of his love, but he also gave them the gift of freedom.[7] He placed his loving heart in their hands to cherish it or reject it. God opened himself to pain and suffering in order to experience the genuine love of his creation.

And, according to the Bible, things went terribly "wrong." First, in heaven there was a being who became enraptured with his God-given abilities and position and led an insurrection against the government of God.[8] Echoes of that insurrection can be found in the book of Revelation: 'And there was war in heaven. Michael and his angels fought against the dragon, and the dragon and his angels fought back. But he was not strong enough, and they lost their place in heaven. The great dragon was hurled down—that ancient serpent called the devil, or Satan, who leads the whole world astray. He was hurled to the earth, and his angels with him.'[9]

Second, Satan did not give up the conflict when he was cast out of heaven; instead he transferred the insurrection to earth by "enlisting" the support of the first members of the human race, Adam and Eve. In the primeval garden he succeeded in turning their allegiance away from God and to themselves.[10] In the process, their loving relationship with God was broken, and pain and suffering were introduced into the world, resulting in decay and death.[11] To make it even worse, what looked like greater freedom to them (rejecting God's authority over them) instead left them subject to the domination of Satan.

From that point on it could be said of every human being, 'Every inclination of the thoughts of his heart was only evil all the time'.[12] The world became a place of greed, exploitation, murder and chaos. From that time on the earthly evidence regarding God's nature was a mixed bag: tokens of love mixed with portents of suffering and death. And worse yet, the Bible tells us that the world has become the chief battleground of a universal civil war, and its citizens are held hostage by rebel forces. Evil does not exist in this world because God is evil; it exists here because the world is enemy-occupied territory.

Why didn't God stop evil?

A question arises at this point. Why didn't God simply put a stop to evil when it occurred? Why didn't he stop it in heaven before it ever got to earth? Why not just eliminate evildoers on the spot and give their squandered freedom to others more worthy?

Imagine a couple of angels in heaven having a whispered conversation just outside the pearly gates. One angel whispers to the other, 'You know, I'm not so sure anymore that God is as loving and kind as he makes himself out to be. You know what I just heard . . .?'

As the other angel leans forward to hear the juicy titbit, a lightning bolt flashes out of the sky and vaporises the complaining angel.

Stunned, the other angel seeks out an old friend. 'You won't believe what I just saw! Charleburt was just saying some negative stuff about God and got vaporised by lightning, just like that! You know, maybe he was right. Maybe God isn't so loving and kind as he makes himself out to be.'

And at that instant another bolt of lightning flashes out of the sky and vaporises the second angel.

If this kind of thing went on for long, what would all the angels be doing? Looking anxiously about for lightning bolts, worried that they would be next! It would be the end of love and the beginning of fear in their relationship with God. From that time on they would do the right thing and say the right thing, not out of love for God, but out of fear. So eliminating evil the instant it occurs was not an option for a God of love.

A second option for dealing with rebellion would be to sanction it. God could change his law and character to reflect the new realities in the universe. Everybody would be allowed to do whatever they wanted.

But this too would be the end of genuine love. It would result in anarchy, 'every man for himself'. Evil would become the reigning doctrine in the universe and a destructive chaos would be the result. Injustice would reach even greater proportions than we now experience as everyone sought to take what they could from others. Sanctioning rebellion, therefore, was not an option for a God of justice.

As powerful as God was and is, therefore, the options for dealing with the consequences of freedom were not many. What was God to do? The Bible offers the answer.

God decided neither to rule the universe by force nor to

sanction the evil that infected it. Instead, according to the great British scholar and novelist C.S. Lewis,[13] he did a number of things to gradually turn the tide away from evil and in favour of love and justice. These are also outlined in the Bible. First, he provided the conscience, an inner sense of right and wrong that few humans are without. Then he provided some, from Abraham to Moses to Paul, with visions and dreams that helped clarify the central issues of good and evil. Next he provided the story of a people (Israel, the Jewish nation) and the struggles through which he sought to teach them more clearly about himself.

And then God did the most amazing thing of all.

The greatest story ever told

In Bethlehem, just south of Jerusalem in the Middle East, a baby appeared, whose birth we celebrate every year at Christmas time. As the story goes, he was born in a manger, and visited by both shepherds and wise men. He was then forced to flee with his parents to Egypt because he was a threat to the reigning king.[14] The reason the Christmas holiday is the high point of the year in Western countries is the conviction that this man, this single, solitary man, was the most important person who ever lived. His name was Jesus.

When Jesus reached adulthood, he went about doing good. He had an amazing ability to heal the sick and, on occasion, even raise the dead. He brought delight to a wedding couple by turning water into wine. He fed thousands with a handful of bread and a few fish.[15]

He also taught some memorable things. There were great

one-liners like 'Do to others what you would have them do to you', 'If someone strikes you on the right cheek, turn to him the other also' and 'Love one another as I have loved you'. He told unforgettable stories like the Good Samaritan, the Prodigal Son and the Parable of the Sower. He had memorable encounters with people like Nicodemus, a Samaritan woman and a dead man named Lazarus.[16]

But none of that is the reason Jesus' life was the most important in the history of the world. It was the strange habit Jesus had of going around talking as if he were God. Others have healed people; some have even claimed to raise the dead. But Jesus went beyond that, claiming an eternal relationship with God and doing things that only God can do.

Jesus is often referred to as a good man, or even the best man who ever walked the face of the earth. But neither description is accurate. Jesus could not be simply a good man. If a mere man claimed to be God he could not be a good man. To quote C.S. Lewis:

> A man who was merely a man and said the sort of things Jesus said would not be a great moral teacher. He would either be a lunatic—on a level with the man who says he is a poached egg—or else he would be the Devil of Hell. You must make your choice. Either this man was, and is, the Son of God: or else a madman or something worse.[17]

If Jesus was merely another prophet, a man among many, he would be a fraud for claiming to be God. But if he was what he claimed to be, God himself taking on human flesh, then his life, death and resurrection are the greatest events that ever happened in the course of human history. And they are the key to explaining how a loving God, who is powerful enough to stop it, could allow so much pain and suffering in this world (more on this later).

The climax of the story took place one Friday in Jerusalem, a sequence of events dramatised in Mel Gibson's movie *The Passion of the Christ*. The story of that Friday actually began on Thursday night. Jesus celebrated the Jewish Passover with his disciples. He then walked with them down the steep staircase of a street that led from the upper part of Jerusalem to the Kidron Valley south and east of the city.

From there they headed north up the valley to a favourite spot for prayer, in an olive grove called the Garden of Gethsemane, just east of Jerusalem. When they arrived at the garden Jesus agonised in prayer over the events he was about to experience. According to the Bible, Jesus' agony had little to do with the physical suffering he would go through the next day. Rather, as the 'God-man', he was designated to experience all the consequences of human evil in his own person.[18] His death on the cross would sum up all the pain, all the suffering, all the regret and all the rejection that evil has caused the human race. He would suffer loss of meaning, loss of relationship and all the misery of human sickness and death.[19] His anguish was much more mental and emotional than physical (in contrast to Gibson's movie).

After his arrest, Jesus was taken for immediate trial before the high priests of the national religion, Annas and Caiaphas. Due process seems not to have been a concern at the trial of Jesus. False witnesses gave their 'testimony', although disagreements among them mitigated its value to the accusers. Torture was used to try to extract a 'confession' from Jesus. Over the course of his various trials, he was slapped in the face, beaten with rods, whipped with long cords, mocked and derided. They spat in his face. A 'crown' of thorns was pressed into his head. Today such a trial would attract major attention from Amnesty International.

The religious authorities in charge of the trial decided that Jesus should be put to death. But since they had no authority to impose the death penalty, they needed to convince the Roman authorities that Jesus was a serious threat to Rome. So Jesus' case was bumped over to the Roman governor. After some consideration, Pilate sent the innocent man to death by crucifixion.

Crucifixion was a peculiarly Roman form of execution. An individual was required to carry the heavy wooden crosspiece to the place of execution as a public warning to others. Some people were nailed to the cross, others were tied with ropes. The key element, however, was that in order to breathe, victims had to exert strength to raise their bodies somewhat. Death came by suffocation when they were no longer strong enough to raise themselves. The process was slow and painful. An additional element of torture was shame and exposure, being hung naked in front of family and friends and in all kinds of weather.

Arriving at Golgotha, the place of execution, Jesus was nailed to the cross through the wrists and ankles and put on display between two common thieves. Three hours later he was dead, more from emotional and spiritual anguish than from physical causes. Rich friends of Jesus then secured his body and placed it in a cave-tomb nearby, closed off behind a huge rolling-stone door.

The story reaches its climax about thirty-six hours later, early Sunday morning. Several women decided to visit the tomb and anoint Jesus' body with spices, to preserve it and show him honour, even in death. But when they arrived at the tomb, the stone had been moved away and the tomb was empty. One or two men were standing nearby in dazzling apparel (one witness called them angels).[20] The women were told not to seek the living among the dead. Jesus had risen from the dead and would appear to his disciples again.

The implications of the cross

What was the cross all about in God's purpose? What difference did it make? I'd like to highlight two things.

The first difference is that the cross changes the way we look at our personal lives, particularly our mistakes and failures. According to the Bible, human beings are not simply imperfect creatures that need improvement; we are rebels who must lay down our arms. Those who crucified Jesus acted no differently than we would have, given the same circumstances. In other words, the struggle to overcome evil is not, first of all, a social or political task. It is a struggle against the evil within.

This 'repentance' is not fun. Acknowledging failure is humiliating and repugnant. But it is the necessary path toward redeeming our lives from the downward spiral of the evil that besets us all. It is the only way to bring our lives into the sunshine of reality. Repentance is simply recognising the truth about ourselves. We will never change until we are willing to be changed—until we recognise that change is needed.

The wonderful thing about God's plan is that he understands what this struggle for authenticity is all about. In submitting himself to the humiliation of the cross, Jesus experienced the kind of surrender we need. In the Garden of Gethsemane he struggled to give himself up to God's plan. And the Bible teaches that if we follow him in his surrender and humiliation, we will also share in his conquest of death and find new life in our present experience.[21]

Tragedies like September 11 and the Holocaust are more than just the work of a few kooks and fanatics. They are symptoms of deeper issues that plague us all. The struggle to recognise the evil within is fundamental to the human condition, whether we acknowledge it or not.

A second difference the cross makes is, at first glance, the very opposite of the first. We all have a fundamental need to value ourselves and to be valued by others. But how can we value ourselves when we recognise that the seeds of evil are within? It seems that the better we know ourselves, the more we dislike ourselves and the worse we feel. How can we elevate our sense of self-worth without escaping from the dark realities inside? That's where the cross comes in.

How much is a human being worth? It depends on the context. If someone were to melt me down into the chemicals of which my body is made, I understand I would be worth about $US12 (make that $13—I've gained a little weight). But the average American is valued by his or her employer at a much higher level than that, something like $50,000 dollars a year.

Now suppose you were a great basketball player like Michael Jordan. Suddenly the value jumps to tens of millions of dollars a year. And if you were the designer of the software everyone in the world uses, you would be valued at tens of billions of dollars (Bill Gates)!

You see, we are valued in terms of what others see in us. But according to the Bible, human value is infinitely higher than the value we assign to each other. Jesus was worth the whole universe (he made it), yet he knows all about us and loves us as we are. When he died on the cross, he established the value of the human person. When the creator of the universe and everyone in it (including all the great athletes and movie stars that people often worship) decides to die for you and me, it places an infinite value on our lives. And since the resurrected Jesus will never die again, my value is secure in him as long as I live.

So the cross provides a true and stable sense of value. This is what makes the story of that Friday in Jerusalem so very special. The cross is not just an atrocity. It is about God's willingness to take on human flesh and reveal himself where we are. It is about the value that the human race has in the eyes of God. It provides hope for a better world. How?

The best hope for a troubled world is an authentic walk with God that not only takes the 'terrorist within' seriously but also sees in others the value that God sees in them. If every one of us is flawed yet valuable, all other seekers after God become potential allies in the battle to create a kinder and gentler world. Armed with a clear picture of reality and a sense of our value, we can become change agents in the world.

So God's gift of Jesus Christ provides the fundamental answer to the problem of evil in the world. But that still leaves us with the question, 'If God cared enough to send Jesus, why doesn't God intervene more often to prevent catastrophic loss of life? Why is God silent in the face of suffering? Where was God during the Holocaust or on September 11? Or was he there and we just didn't notice?'

When tragedy strikes

For many people on September 11, survival seemed to be an accident of location and timing. George Sleigh was a manager at the American Bureau of Shipping, on the 91st floor of the North Tower of the World Trade Center in New York. He was on the phone in his office when he heard the roar of jet engines. Looking out of his window, he had just enough time to think, 'The wheels are up, the underbelly is white and, man, that guy is low.' It was 8.46 a.m. and a Boeing 767 was headed toward him at 500 miles per hour, with ninety-two people and more than fifty tons of jet fuel aboard. The jet exploded into the building at floors 93 to 98 just above him. The walls, ceiling and bookshelves in his office crumbled.

Crawling out from under the rubble, Sleigh looked up at the exposed beams and concrete underside of the 92nd floor. What he didn't know at the time was that his concrete ceiling was the floor of a giant tomb for more than 1300 people. Not a single person survived on any of the floors above him. But on his floor and below nearly everyone lived to see another day. The line of survival was as thin as a steel beam and a concrete slab. All of those on the 92nd floor died and all those on the 91st floor lived.

Counting heads, Sleigh discovered that eleven of the twenty-two employees in his office were on duty at the time. None were injured. Other than Sleigh's area the office was largely intact. Sleigh went back to his area to get his briefcase. The closest stairway was blocked, but the second was open. Heading down for several floors, Sleigh and his colleagues found the going quite peaceful. Nobody was behind them. By the time they reached the middle of the tower, Sleigh's office was engulfed with flames. Fifty minutes later, having become separated from his colleagues in the increasing press on the staircases (more and more people were evacuating and room had to be left for the firemen who were charging up to fight the blaze), Sleigh left the building and was loaded into an ambulance—bruised, bloody and covered with dust.

'Get out! Get out!' a policemen yelled, 'the building is coming down!' It was 9.59 a.m. and the South Tower was collapsing. But the North Tower's highest survivor was on his way to Beth Israel Hospital.

'Sometimes I think it was God's providence that spared me,' Sleigh said later. 'Other times, I wonder why me and not others. I realise I am a very fortunate man.' [22]

Why George Sleigh and not 3000 others? Why did God seem to go out of his way to preserve one life when so many other people lost their lives that day?

The problem with miracles

Journalist and editor Stephen Chavez reflected on the arbitrary nature of the events of September 11 and what that had to say about God.[23] He stated that there are two problems with miracles. For one thing, it is hard to tell the difference between a miracle and a coincidence. If a commuter plane goes down and half the people are killed, how many of the survivors were saved miraculously and how many were saved simply because they were sitting in the 'right' section of the plane? No doubt those who survived would be inclined to consider their survival a miracle.

This raises a second problem with miracles. Why didn't God miraculously preserve everyone's life? Tragedy is difficult enough to take by itself. But the preservation of even one person in the midst of slaughter, as wondrous as that may be, serves as the frame for a giant question mark regarding the loss of so many.

In the tragedies of September 11, thousands were killed and even more thousands were spared. There is no detectible pattern among the saved or the lost that would offer any explanation. Sometimes it was as simple as who got up and who slept in, or who was located on the 91st floor and who was located on the 92nd. Chavez concluded that survivors need to be careful how they celebrate miracles: 'Not everyone survives a terminal illness or an automobile accident; not every lost child (tool, dog, wallet, watch) is found.'

The story of Job

The Bible does not leave the issue of personal tragedy and suffering unaddressed. In one of its earliest writings, the story of a man named Job is told in the genre of a Hebrew play. The story wrestles with the issue of why bad things happen to good people. (On the other hand, Psalm 73 addresses the issue of why good things happen to bad people.) This story has had such an impact on the world that even in today's secular environment, nearly everyone has heard of 'the patience of Job'.

Job was a very wealthy man, perhaps the richest in the world. But his greatest treasure was his children, seven sons and three daughters. Every morning before the sun rose he prayed that God would protect them through the day. But one day, while Job was praying, his case came up in the heavenly court, although he was not aware of it.

Satan, the prince of evil and darkness, sneaks into the heavenly court with a crowd of 'the sons of God'. After noting his presence, God offers Satan a challenge: 'Have you noticed my servant Job? He worships me faithfully and is careful to do nothing wrong.'

Satan counters, 'Big deal. He's into religion for what he can get. You've given him everything. No wonder he worships you. But mark my words. Take away all he has and he'll curse you to your face!'

God responds, 'OK, we'll see. Everything he has is in your hands. Just don't hurt Job himself.'

The scene moves back to earth, where one disaster after another falls on Job's estate. Bandits, fire, marauding armies and storms destroy Job's animals, servants and possessions. Even his children are gone. He is left destitute and childless in a moment. Job's response? He falls to the ground, worships God and says,

> Naked I came from my mother's womb,
>   and naked I shall depart.

> The LORD gave and the LORD has taken away;
> may the name of the LORD be praised.[24]

After this the scene in the heavenly court reconvenes. God challenges Satan, pointing out that Job's faithfulness has not diminished, in spite of the great losses he has experienced. But Satan isn't finished yet.

'Big deal,' he exclaims, 'Skin for skin! A man will give all he has for his own life. But stretch out your hand and strike his flesh and bones, and he will surely curse you to your face.'

God responds by placing Job in Satan's control, with only one limitation. Satan must spare Job's life. So Satan goes out and afflicts Job with loathsome and itchy sores from head to toe. Even his wife turns against him and urges him to 'curse God and die'. But Job is not left alone. Three 'friends' hear about his troubles and come to console him in his sorrows.

Most of the play is about the attempt of Job's friends to convince him that God is not arbitrary. If things have gone wrong for Job, he must somehow be to blame for it. God is trying to get his attention. So if Job would just turn to God and humble himself, things would get better. Great friends!

In response, Job denies the charges. He insists that he is an exception to the rules, that he is innocent of anything that would justify his great losses. Under harassment from his friends, he begins to accuse God of injustice and oppression. In the real world the wicked prosper and the righteous die. And God sits there and watches it all. Job wishes he had never been born.

After a lengthy and tedious debate the four men fall silent and a fifth appears, named Elihu. In pious pride he comes to defend God against both Job and his friends. Pain is one way God uses to get people's attention, he declares. God never does the wrong thing, he gives people only what they deserve.

God responds

God then surprises all five men, approaching in the midst of a mighty thunderstorm. He speaks out of the storm and addresses Job and Job alone.[25]

At first God seems to support all that Job's four companions have said to him. He accuses Job of questioning him with ignorant, empty words. Then he throws a series of unanswerable questions Job's way. 'Where were you when I made the world? You know so much, tell me about it. Have you ever commanded a day to dawn? Have you ever walked the floor of the ocean? Can you guide the stars from year to year, or change their orbits?' And so on.

After Job admits his ignorance for the first time, God pelts him with another series of unanswerable questions. 'Do you gather food for the lions? Did you teach the hawk how to fly? Can you tie up a whale like a pet bird? Are you trying to put me in the wrong so you can be right?' Job offers the only possible response to overwhelming rightness and power. He says plaintively, 'Surely I spoke of things I did not understand, things too wonderful for me to know. My ears had heard of you but now my eyes have seen you. Therefore I despise myself and repent in dust and ashes.'

Here's the catch in the story. God never answers any of Job's questions; instead he asks Job questions. Nevertheless Job's attitude is totally changed. Out of his new understanding and

The Problem of Evil

relationship with God he is satisfied that God is just. Knowing *about* God is not the answer to his questions. *Knowing* God is.

Anyone who comes to this biblical play expecting all the answers to the problem of suffering is likely to be disappointed. Job's friends are full of answers, many of which are still offered today. But all these answers get mocked at some point in the book. When God appears, he offers no answers but just a sense of his overpowering greatness.

Perhaps the main point of the book is that none of the general answers to the problem of suffering had anything to do with why Job was suffering. The real reason Job suffered had to do with a 'wager' between God and Satan in the heavenly court. No statement in the earthly part of the book (chapters 3–42) ever returns to that issue, not even the statements of God himself. So the point of the book seems to be that the limited context of human experience does not allow a satisfying intellectual answer to the problem of suffering. We just don't have the context to understand, even if the one doing the explaining is God.

Don't get me wrong here. I'm not saying that our lives on this earth are hostage to cosmic wagers God is making out there somewhere. After all, Job is only a play, not a detailed explanation of the ways of God.

But the point of the play is that our lives are affected by wider issues in the universe as a whole, things that don't make sense from the perspective of a single planet alone. God's will is not always done on this earth. As God asserts in the book of Job, to try to explain September 11, the Holocaust and similar events is like commanding a day to dawn, roping a whale or walking on the floor of the ocean. It is just not a realistic enterprise for humans confined to the limited context of this earth.

Why then was Job satisfied with God's answer, even though it was not a real answer? If nothing else, it was because God cared enough to answer. After all, if God answered all of Job's questions, how could the story be a comfort to those who don't get any answers to their questions? But as it stands, Job's story can be a comfort to those left in the dark. While Job doesn't get any answers from God, he does encounter him, and that is enough. To know God is to trust him.

Perhaps the best news in the book of Job is that undeserved suffering will not last forever. It ended for Job, and it will one day end for the human race as a whole. To paraphrase Shakespeare, 'The earth is like a stage and we are merely players.' One day a much bigger picture will be revealed.

The suffering God

But the book of Job is not the Bible's last word on the matter of suffering. The cross is the Bible's final answer. As Jesus was dying on the cross, his greatest suffering had little to do with physical pain from the spikes through his hands and feet, the thorns piercing his forehead, or the torturous effort to breathe enforced by crucifixion. His greatest suffering arose from the apparent absence of God.

Jesus knows from experience what it is like to suffer undeserved suffering and pain. He did not deserve to be whipped, beaten, slapped and spat on. He did nothing to deserve a sentence of death, a hateful mob or the torture of crucifixion. To the victims of September 11 and other tragedies the cross says:

'God knows. He understands. He has tasted what it is like to suffer without having caused it in some way.'

Like the book of Job, the cross offers up no definitive answer to the problem of unjust suffering. What it does do, however, is offer companionship in suffering. The times when we experience undeserved suffering and pain are like our own Friday in Jerusalem. We feel as if our experience is unique, as if no one has ever been more alone. But Jesus himself went there in depth on the original Good Friday. He understands what it is like to be totally alone, totally rejected and abused. He's been there and done that. And in a sense he tasted just a bit of everyone's experience.

But for Jesus the story didn't end on that Friday, although it seemed that it would. When he cried out to God, 'Why have you forsaken me?', he himself seemed to see no hope for the future. But his suffering and abandonment turned out to be a prelude to the incredible affirmation of Easter Sunday. When he was raised from the dead his acceptance with God was re-affirmed. In some sense the whole human race stands in a new place with God.[26] The cross has turned human suffering into a prelude.

What difference does it make to believe in the cross today? For me it changes everything about suffering. Some have used undeserved suffering as an excuse to disbelieve in the existence of God. But atheism has not lessened human suffering one iota. If anything it makes it worse, because one is all alone in the suffering and it has no meaning and no future.

But the cross demonstrates several things that make a difference. It tells us that we are not alone, even though it may feel that way. It tells us that suffering doesn't mean that God is heartless; he cares enormously, but he doesn't always intervene to avert pain. God's absence in suffering is not a hostile or helpless absence; it has a higher purpose. In the light of the cross we have a reason to endure, even though we may not know the particular reason why. When we suffer without deserving it, we share in the experience of Jesus. When we feel the absence of God in our pain, we share in the experience of Jesus. He went there before us and understands how we feel.

I remember a man—I'll call him Harvey—who always had a smile for everyone, an encouraging word, a pat on the back or a hug. There seemed to be no limits to his optimism and joyful spirit. You never heard him say anything bad about anybody. When people in his church got to fussing with each other, he just kept on smiling and praying through everything.

In his 70s, however, Harvey's heart began to fail. If anyone could do battle with heart disease through the medicine of laughter and optimism, he was the one. When I visited him in the hospital I somehow expected him to be weak but to still be his normal self in terms of optimism. But it wasn't that way at all. He poured out his anguish: 'Why me? Why now?' This stalwart Christian had lost track of his optimism. I was truly confused, wondering momentarily if I had visited the wrong person. It was unquestionably him, but his spirit had been broken and I could hardly recognise him as a result.

I wondered what I could possibly say that would make any difference in his great hour of need. I thought of the sentiments I mentioned above. I decided to try something.

I told him, 'Harvey, if I were in your situation, I'm sure I'd feel as upset as you do.'

'I'm sorry I've been such bad company today,' Harvey said. 'It really does feel better to share how I feel about it with you.'

'Harvey, did you know that Jesus felt the same way as you do when he was on the cross?'

'He did? How do you know?'

'Do you remember what he said on the cross? "My God, my God, why have you forsaken me?" That tells me he knows how you feel, and that he thought it was OK to challenge God a little. God would rather hear our honest anger than our silence or even sweet words that aren't heartfelt.'

'You really think so?' Harvey asked.

'That's what the Bible says,' I replied. 'It's OK to tell God how you feel, even when you don't feel so good. But you know what I think?'

'What?'

'Now you can also understand what Jesus experienced on the cross. He didn't deserve to suffer and die any more than you do. What he went through wasn't right, but he endured knowing it would turn out all right in the end. If nothing else you are having a share in the suffering he went through. You can understand him in a way that I can't, with all my study and training. If God is allowing you to go through this, then he must know that the ultimate good will outweigh the present evil.'

We continued the conversation for a while along the same lines, and then he said, 'You know what? Today you taught me something. I was thinking about me. But now I see a bigger picture. Now I understand what Jesus went through for me. This isn't fun, but if it helps me know God better it is worth it.' Harvey's perspective had changed. He relaxed against the pillow with a slight smile and a look of peace on his face.

You don't always have to know the 'why' of suffering as long as you know you are not alone.

The ultimate act of terrorism

For those of us who experienced it, September 11 was an unimaginable expression of evil at its worst. It fundamentally altered our perception of the world and our own role in it. But September 11 was not the most evil act of all time. The Holocaust, as chillingly brutal and unfair as it was, was not the most evil act of all time. The Inquisition, the Crusades, the genocides of Armenians, Russians, Rwandans and Cambodians in the twentieth century, the slave trade across the Atlantic—all of these qualify as acts of systematic premeditated evil. But none of them qualify as the most evil act of all time.

The cross was the most evil act of all time. When human beings, for temporary and limited political advantage, crucified the God who came down and lived among us, they acted in the most incomprehensible, unfair and evil manner possible. In rejecting him, they were doing more than just condemning an innocent man to death; they were destroying the source of their own life and rejecting their own place in the universe. The cross of Jesus Christ is an evil act of infinite proportions. If the human race is capable of such an act, no evil action is unimaginable.

But there is a silver lining to the dark cloud of human evil. God has turned the cross into a powerful act of reversal. The greatest evil ever done has been transformed by God into the most powerful act of goodness ever performed. By death God brings life. Through defeat comes victory. Through shame, humiliation and rejection come glory, grace and acceptance. Through the cross God has turned the tables on evil and death. The greatest evil has become the basis for the greatest good.

13

Who Is God? What Is He Like?

Steven Thompson

Is this question important?

We know the probability is very high that God exists; but is it important to know the truth about *who* God is? Yes, it is! The question of God's identity is just as important as the question of his existence because it hugely impacts our quest for meaning. Our answer to the question of who God is determines the answer to so many follow-on questions about who we are, what life is about and what our place is in the universe.

The term 'worldview' has come to express our set of answers to these big questions of existence. A worldview is comprehensive and satisfactory only when it answers questions arising from all four of life's dimensions: the personal, social, spiritual and cosmic. A worldview that leaves out one or more of these dimensions is unsatisfactory because it can cause a fragmented sense of the world, and a sense of alienation.

The missing spiritual dimension in the worldview of a large number of contemporary people is often related to abandoning belief in a God who is involved with his creatures. Early in my career as a pastor I met a gracious and hospitable retired scientist whom I visited several times. He enjoyed my visits and told me about his work on a top-secret project of World War II, the development of radar. However, whenever we discussed the spiritual dimension of life, or if I cited a Bible passage which referred to anything divine or supernatural, he would smile courteously but condescendingly and chant, 'Fairies at the foot of the garden!' By this he reminded me of his lifetime commitment to a worldview that ruthlessly eliminated God by subjecting any claims about spiritual reality to a strictly materialistic, scientific scrutiny. The result was predictable—such scrutiny convinced him there was no God, and that claims of the existence of spiritual truth were not believable, since to him they could not be tested using 'scientific' methods.

Scientific method on its own contributes little or nothing to our knowledge of God. So how else can we learn who God is? Are there other sources for such knowledge?

The Bible as source of knowledge about God

Arguably the best knowledge about God would be that which he himself provides, rather than speculation or guesswork about him. Has God ever introduced himself to humans, so to speak, in a way which provides such knowledge? Has a reliable record of any such introductions been preserved? If so, they would seem to be the best source of first-hand knowledge of who God is.

Before answering this question, it is important to further define which sort of 'introduction' qualifies for our purpose.

Many people are convinced they have encountered God, that God has met and communicated with them. However, not all such claims should be believed. Direct human encounters with God are powerful and often result in long-term life turnarounds. They are also rare. Therefore, a good test for the reliability of a person's claim to have met God is the resulting changed life.

The Bible, consisting of the Hebrew Scriptures (Old Testament) and the Christian Scriptures (New Testament), contains the largest and richest collection of testimonies by individuals and people groups to whom God has 'introduced' himself, and who experienced consequent life changes. These stories have collectively served as the sourcebook for beliefs about God, and have provided the spiritual dimension for two world religions, Judaism and Christianity. The Bible's approximately 775,000 words recount a range of such divine 'introductions' to people living at various times in different parts of the ancient world. Its stories focus especially on the Hebrew people, beginning with their founder, Abraham, who lived about 2000 BC, and ending with accounts of the actions of the followers of Jesus Christ, who during the first century of our era spread the Christian faith the length and breadth of the Roman Empire and beyond.

A couple of features set these accounts apart from other collections of stories about encounters with God.

First, they provide unvarnished, very human glimpses into the lives of the people who encountered God. Even Hebrew heroes such as Abraham, Isaac, Jacob and his twelve sons, and the shepherd king David have their weak and foolish sides exposed, along with their struggles to trust God. This adds to the authenticity and believability of these stories as they depict real people, some of them experiencing struggles very much like our own.

The second feature of the Bible's accounts is that they present a consistent picture of God. Even though the stories take place in diverse ancient cultures spread across the Near Eastern world, and are written by different authors over a time span of thousands of years, they present a consistent and believable picture of God. The Bible is thus an extremely valuable source concerning the experiences of individuals and groups who encountered God.

God according to the Bible

The Bible consistently attributes a set of divine characteristics to God in its episodes of divine interaction. The following four especially are foundational: creator, person, ruler, judge.

Creator. First, the Bible presents God as creator of the world, including humans. The first and most notable identifying mark of the God of the Bible is his separation from, and ultimate authority over, creation. God is not *in* nature, and even the most impressive natural phenomena are but reflections of God's greatness. The technical term for this divine separateness from creation is 'transcendence'.

God's transcendence, however, does not cut him off from his creation. Rather than remaining in splendid isolation, God balances transcendence with a special form of 'immanence' that allows him to relate to creation. From the absolute otherness of his transcendence, the Bible describes God's encounters with creation, which clearly have as their purpose establishing a relationship.

God has not abandoned creation, leaving it to look after itself. Ongoing personal involvement is part of his plan. He possesses ultimate insight into creation's design, function and purpose. God knows why we are here, and where we are headed. The universe has not escaped from its creator's control, despite appearances to the contrary. After a period of conflict, God will remove the source of conflict and elevate his universe to a state of complete harmony and peace.

Person. Second, the God of the Bible is a person. It is significant that the title of this chapter asks 'Who is God?' rather than 'What is God?' There is vast difference of meaning between the terms. If while sitting inside at night I hear scratching on the front door, I ask myself, 'What is that?' But if I hear a knock, my question is different: 'Who is that?' In other words, *who* implies a person, with the potential for a relationship.

God, according to the Bible, possesses personhood and is the source of all personhood, including my own. Key attributes of human personhood include the will, individuality, a sense of differentiation, and the capacity to form relationships. The Bible attributes these to God, and repeatedly records human interaction with him in relational terms. No part of the Bible is free from the language of relationship; it dominates the books of Moses, the Psalms, the prophets, the New Testament letters and especially the teaching of Jesus. The God of the Bible intends to relate.

Furthermore, God takes the initiative in establishing life-changing relationships. Bible accounts depict him approaching people to provoke a relationship. He does not overwhelm or neutralise the personhood of those with whom he relates, reducing them from subjects to objects. They retain their identity, their individuality and, most importantly, their freedom to enter, continue or end the relationship.

Ruler. Third, God is presented in the Bible as ruler of this world—an absolute ruler, yet one 'big enough' and tolerant enough to embrace and contain this world's plurality, rebellion, suffering and evil, as well as its beauty and goodness. The Bible makes clear that regardless of determined resistance and rebellion against him, God maintains sovereignty over creation. No force in the universe can separate his creatures from his rule against their will.

On the other hand, he is only content to rule subjects who freely choose to be ruled by him. He never employs coercion. Finally, after repeatedly seeking the allegiance of every creature, his rule will lead through a process of cleansing and removal of evil to a state of perfect unity and harmony through all creation.

Judge. Fourth, the God of the Bible is judge. The task of judges is to decide whether the conduct of free humans is lawful. Judges are only needed where there is law, and where subjects have both freedom and choice.

According to the Bible, God is both lawgiver and judge. This unusual combination signals that there is a moral centre to the universe, a universal standard of right and wrong, the knowledge of which is communicated to creation. As free creatures with some power of choice we are held responsible, within limits, for our conduct. According to the Bible, God will judge us for the use we make of our freedom to choose good conduct and reject evil—in other words, to relate to him.

Unlike human judges, whose knowledge of both the law and human conduct is incomplete, the God of the Bible has complete knowledge of both spheres. He has no need for eyewitness testimony or high-tech forensic evidence in helping him reach a judgment, nor is there the possibility for error, since he knows all. Finally, he is not subject to bias or corruption. He is therefore the only judge whose judgments are completely true and just, and for which there is no higher court of appeal.

Why is Jesus so important?

The Bible ends with a set of accounts which provide the ultimate example of how God relates to humans. These accounts concern the person of Jesus of Nazareth, whose life was a series of largely public events witnessed by considerable numbers of people.

The amazing events in the life of Jesus and those around him, especially his resurrection, convinced many of his contemporaries that through Jesus God began to relate to people in a new and unique way. In the light of Jesus' resurrection, even the pain and suffering of the crucifixion came to be seen as central to God's plan for this world.

So to answer the question 'Who is God?' the Bible directs us to look to Jesus, God's ultimate revelation of himself and his plan for his creation.

14

Will the Real God Please Stand Up?

Eric Svendsen

I am going to tell you something about God that may shock you. It is a truth that is at once so foundational that it is the absolute first principle for knowing God, yet so profound that the very concept has baffled some of the greatest minds in history. Understand and embrace this truth, and every decision God makes and every action he takes suddenly becomes palatable. Reject it, and we may as well resign ourselves to a life of incessant frustration—always questioning God and never being able to come to a full knowledge of the truth.

What is this shocking truth? Simply this: the God of the Bible is bigger than most of us think. He is subservient to no one else's standard of behaviour, he is accountable to no one but himself, and he does whatever he pleases:

Our God is in heaven; he does whatever pleases him.

I am God, and there is no other; I am God, and there is none like me. I make known the end from the beginning, from ancient

> times, what is still to come. I say: My purpose will stand, and I will do all that I please . . . What I have said, that will I bring about; what I have planned, that will I do.
>
> The Lord Almighty has sworn, 'Surely, as I have planned, so it will be, and as I have purposed, so it will stand' . . . For the LORD Almighty has purposed, and who can thwart him? His hand is stretched out, and who can turn it back?[1]

There is a word that sums up this attribute of God and allows us to put a label on it; it's called 'sovereignty'. Sovereignty means not only that God is in complete and utter control of all creation, events and circumstances, but also that he actively carries out his own will—what *pleases* him—regardless of what his creation might think or say about it. In his sovereignty, God has decreed what will occur and how it will play out; that is to say, he has rendered *certain* all things that he has determined beforehand shall come to pass. There is nothing that is exempt from God's decree. God, we are told, 'works out everything in conformity with the purpose of his will'.[2] 'All things', including people, places, actions, events, movements and decisions, come under this decree of God.

What standard does God use in determining this decree? We may be certain that it is not some arbitrary decision, for it is 'in conformity with the purpose of his will'. But we may be equally certain that it is not based on the whims of his creation, since this decree is also carried out 'according to *his* good pleasure'.

> 'I will have mercy on whom I have mercy, and I will have compassion on whom I have compassion' . . . So then he has mercy

on whom he desires, and he hardens whom he desires . . . Or does not the potter have a right over the clay, to make from the same lump one vessel for honorable use, and another for common use?[3]

Equally important is the *goal* of this decree. All God's planning, and everything he decrees and carries out, he does 'for the praise of his glory'.[4] Whatever God does, he does it so that he may receive all glory, honour and praise from every living being.

This may seem a form of 'selfishness' to some—and it certainly would be it we were discussing anyone else. But here the point we made earlier becomes clear, namely, that God is not subservient to the same rules that apply to us. It would be a sin for any of us to seek our own glory, honour and praise. Not so with the sovereign God. He is the only being who rightfully seeks his own glory and rightfully stands in judgment of all others who attempt to do the same.

God stands outside of the rules and laws he has determined humankind will follow. But he does so perfectly and to humanity's benefit. The same biblical passages that tell us God works out everything in conformity with his own good pleasure also tell us that he has done this out of his great love and grace which he has freely bestowed on those who will turn to Christ.[5] We who were dead in our sins but have been raised spiritually to life are the direct benefactors of his divine plan. He has 'blessed us with every spiritual blessing', chosen us 'in him before the foundation of the world', 'predestined us to adoption as sons' and 'seated us in the heavenly realms'.[6]

Although God does what he wants, his sovereignty is always guided by immeasurable love, grace and mercy.

What sovereignty means

Apart from the concept of sovereignty, then, it is impossible to understand the God of the Bible as he truly is. With it, everything else falls naturally into place.

I doubt very much that most people really understand the ramifications of this. When we speak of the sovereignty of God, what we usually mean is that God is in control. We pray that God will 'work everything out', but *only* if that somehow turns out to our advantage. We want God to be in control, but not to the extent that *we* no longer are—and we certainly don't want him to be involved in any kind of activity that has negative ramifications.

Ask nearly anyone convinced of God's existence whether he believes that God is sovereign and he is bound to respond with a hearty, 'Yes, of course. Absolutely!' But ask him whether he believes that God has decreed all things that come to pass, including the hurricane that wiped out the east coast, the ten-year drought that dried up the crops, the invasion and destruction of one country by another, and even the traffic accident that took the lives of several people, and you are not likely to be met with an approving answer.

Yet this is just what sovereignty entails; and without it, the word, as it pertains to God, is rendered meaningless. God, we are told in Scripture, is sovereign over nature:

> I will send you rain in its season, and the ground will yield its crops and the trees of the field their fruit . . . I will break down your stubborn pride and make the sky above you like iron and the ground beneath you like bronze. Your strength will be spent in

vain, because your soil will not yield its crops, nor will the trees of the land yield their fruit.[7]

We are also told that God is sovereign over disasters:

> When a trumpet sounds in a city, do not the people tremble? When disaster comes to a city, has not the Lord caused it?[8]

We are told that God is sovereign even over evil, using it at his own good pleasure as a means of attaining his own ends:

> The LORD works out everything for his own ends—even the wicked for a day of disaster.

> 'So now the LORD has put a lying spirit in the mouths of all these prophets of yours. The Lord has decreed disaster for you.'

> And if the prophet is enticed to utter a [false] prophecy, I the LORD have enticed that prophet, and I will stretch out my hand against him and destroy him from among my people Israel. They will bear their guilt—the prophet will be as guilty as the one who consults him.[9]

But that is not all we're told about God's sovereignty. God's sovereignty extends even as far as the very actions and decisions of man:

> The king's heart is in the hand of the LORD; he directs it like a watercourse wherever he pleases.

> In his heart a man plans his course, but the LORD determines his steps.

Indeed Herod and Pontius Pilate met together with the Gentiles and the people of Israel in this city to conspire against your holy servant Jesus, whom you [the Lord] anointed. *They did what your power and will had decided beforehand should happen.*

The beast and the ten horns you saw will hate the prostitute. They will bring her to ruin and leave her naked; they will eat her flesh and burn her with fire. For *God has put it into their hearts to accomplish his purpose* by agreeing to give the beast their power to rule, until God's words are fulfilled.[10]

There is much more that can be said on this topic; indeed, full-length books have been written on it. What I have included here may even have raised more questions in your mind than it has answered. Unfortunately, that is unavoidable in a chapter of this length. But as I've mentioned, unless we have, at the very outset, the right picture of who the biblical God is, we're likely to head down a wrong path and end up with a god of our own imagination.

It's time we started thinking of God as he truly is and not as some frail, grandfatherly figure who wrings his hands in heaven hoping everything somehow turns out the way it should. Such a 'god' is little more than an idol of our own making, and is certainly not worthy of our awe and adoration.

All-powerful, all-knowing, eternal

We have already seen that God's sovereignty means that he has the *right* and *desire* to do all his will. But we have not yet revealed *how* he carries this out. It stands to reason that a God who does 'all his good pleasure' must have some way of accomplishing this—and indeed this is just what the Bible confirms. God is often called 'the Almighty',[11] and Scripture often poses the question 'Is anything too hard for the Lord?', frequently supplying the answer in spite of the question's rhetorical nature.[12]

This attribute of God—what we call 'omnipotence'—means that God is all-powerful; that is, he is able to perform all things that are proper acts of his power. This is an important qualification, since Scripture itself confirms some impossibilities with God. For instance, we are told 'it is impossible for God to lie', that he 'cannot be tempted by evil' and that he 'cannot deny himself'.[13] Upon further reflection, however, we see that these 'impossibilities' do not constitute limitations of God's omnipotence, because each one would be an act of weakness, not an act of power. Similarly with questions such as 'Can God make a rock so big that even he can't lift it?' or 'Can God make a triangle with only two corners, or a round square?' Obviously, none of these would be a true act of power. They would all be acts of nonsense.

The God who is both sovereign (he has the *inclination* to do whatever he pleases) and omnipotent (he has the *ability* to do whatever he pleases) is also all-knowing—what we call 'omniscient'. God's omniscience means that God knows all actual and possible things. We are told specifically that God 'knows everything'. We are warned that 'nothing in all creation is hidden from God's sight', but that 'everything is uncovered and laid bare before the eyes of him to whom we must give account'. We are assured that in Jesus Christ are hidden 'all the treasures of wisdom and knowledge'.[14]

But notice that the idea of omniscience goes beyond knowing *actual* things (things that either *have* happened or *will* happen) to include all *possible* things based on contingencies (things that *would* certainly happen given hypothetical circumstances). One of the best examples of this is where Jesus tells us that if the miracles that were performed in Korazin, Bethsaida and Capernaum had been performed in Tyre, Sidon and Sodom (a hypothetical situation that did not in fact happen), they would have repented long ago in sackcloth and ashes.[15] In other words, Jesus affirms here that he knows not only *actual* events, but also events which certainly would have occurred had the conditions been right.

One might legitimately ask the question: Since God knew that these cities would have repented had he only sent a prophet to them to perform miracles, then why didn't he send a prophet? Now, perhaps, you see the value of addressing the issue of God's sovereignty (God does whatever pleases him) as a foundation to understanding all of his other attributes, including his omniscience. Without this foundation, it is impossible to make sense of such passages.

We must make one other qualification regarding God's omniscience—namely, that God can know only those things that are proper objects of knowledge. This will spare us additional nonsensical questions such as 'Does God know how to create a mathematical formula or a scientific hypothesis so complex that even he cannot understand it?' Obviously, something unknowable to an all-knowing being is not a proper object of knowledge, and is akin to asking whether God's omnipotence allows him to create a rock so heavy that even he can't lift it. Such questions are absurd on their face.

One final attribute that falls under this heading is that God is eternal. What we mean by God's 'eternality' is that God has neither beginning nor end, and is not subject to the limitations of time. Time does not apply to God since God stands outside time. Rather, all time is 'present' before him. The Bible affirms God's eternality: 'from everlasting to everlasting you are God'. God declares of himself, 'I am the first and I am the last; apart from me there is no God'; ' 'I am the Alpha and the Omega,' says the Lord God, 'who is, and who was, and who is to come, the Almighty'; 'I am the Alpha and the Omega, the First and the Last, the Beginning and the End'.[16]

Because God is eternal (without beginning or end, and standing outside time), it is reasonable that he also be 'immutable'—that is, he does not change in his nature or his promises. And indeed, this is just the affirmation we find in Scripture: 'I the LORD do not change'; 'you [God] remain the same, and your years will never end'; 'Jesus Christ is the same yesterday and today and forever'.[17]

Creator of all things

One of the things that God in his sovereignty decided to do, and in his omnipotence and omniscience carried out, was the creation of the universe and life itself. Before creation, there was no such thing as matter. When we say that God created the heavens and the earth, what we mean is that God created the very matter of the universe *ex nihilo*, which is Latin for 'out of nothing'. We are told of Jesus Christ, the Word of God, that 'by him all things were created: things in heaven and on earth, visible and invisible . . . all things were created by him and for him'.[18]

But anything God creates, he does so in perfection. After God, in six days, created the heavens and the earth, he pronounced all he had created was 'very good'.[19] Implied in this goodness is purpose and order. God, we are told, is not a God of disorder, but of peace.[20] Hence, a universe which God creates will naturally exhibit an inherent purpose and order; and this is just what we find when we examine both the biblical and scientific characteristics of the universe itself. Throughout the creation week, we find that there are 'days', each one defined by a repeated, orderly event, 'and there was evening, and there was morning'.[21] We are later told that the seasons themselves—along with cold and heat, day and night, and seed time and harvest—will continue as regular cycles for as long as the earth endures.[22] Science tells us that due to its orderliness there must have been a master designer of the universe, and the Bible identifies just who that master designer is.

Of course, it must be admitted that there is much *dis*order in the universe as well, not least of which is decay, corruption, sickness, disease and even death. All these are products of the principle of entropy (the second law of thermodynamics), according to which the creation moves from a state of order to disorder. But this was not part of God's *original* design; it is in fact a perversion of God's intent for the universe, introduced by man's sin. When the first man and woman wilfully disobeyed the creator, the very fabric of the universe was torn, resulting in some very undesirable consequences.

The apostle Paul explains this phenomenon in his letter to the Romans:

> For the creation was subjected to frustration, not by its own choice, but by the will of the one who subjected it, in hope that the creation itself will be liberated from its bondage to decay and brought into the glorious freedom of the children of God. We know that the whole creation has been groaning as in the pains of childbirth right up to the present time.[23]

Of note in Paul's explanation is not only his affirmation of the law of entropy over the universe ('bondage to decay'), but also his acknowledgement that God is the 'mediate' (or indirect) cause of this state of the universe: 'not by [the creation's] own choice, but by the will of the one who subjected it'. Yet, ironically enough, the purpose of God in subjecting his creation to this bondage is to redeem it from this state: 'in hope that the creation itself will be liberated from its bondage to decay'.

Another kind of disorder in the universe comes in the form of destruction, whether by 'natural' causes (hurricane, drought, tornado, flood, earthquake) or by humanity's own doing (wars, vandalism, murder). The former may be immediately (directly) due to perversions of weather patterns brought on by Adam's sin, but in many cases are mediately initiated by God as a form of judgment on sin.[24] The latter may be immediately caused by someone's decision (good or bad, initiated perhaps by pride, arrogance, lust or some other sin), or, again, mediately initiated by God as a form of judgment on sin.[25] In no case are these things outside the control of a sovereign God, who 'works out everything in conformity with the purpose of his will'.

We must avoid two potential errors in thinking here. The first regards God as the author of evil. Keep in mind that if we view God rightly, then we must affirm that he cannot do anything that is evil.[26] So coming to the conclusion that he does (or has) perform(ed) evil is the direct result either of misdefining that particular action as an evil action (we, as sinful human beings, are not always in the best position to define right and wrong since 'the heart is deceitful above all things and beyond cure'[27]), or of failing to distinguish between immediate causes (where culpability lies) and mediate causes (where sovereignty lies). The second error is the opposite of the first: we begin to detach God from everything in the universe that appears to us to be negative or evil, and conclude that God's 'hands are tied', or that he's 'done all he can' to prevent the disaster. Such a view is not only based on a mistaken understanding of God's sovereignty, but is the result of the same kinds of misdefinitions and failure to distinguish terms that results in the first error.

One final form of disorder is really nothing of the kind, but is in reality a *suspension* of the order. God has created an orderly universe—one that follows certain cycles, patterns, chemical compositions and physical laws (such as gravity and entropy). But since God created these laws, they are proper objects of his power (his omnipotence) and he is at liberty to suspend them as he sovereignly chooses. And, indeed, he *does* choose to suspend them at times to advance his overall purpose. We see examples of this when Jesus walked on water (a suspension of the law of gravity), when he turned water into wine (a change in the chemistry of a substance), when he raised others from the dead, when he himself was raised from the dead, and ultimately, when he accomplishes the resurrection of all those who have believed in him—the ultimate victory over entropy and death.[28]

Overt suspensions (or even reversals) of natural laws are what we normally call 'miracles'—phenomena which, by their very nature, are unusual and infrequent, and so not something we should expect to see as a matter of course. However, there are also patterns of miracles that occur daily, such as life, purpose, self-conscious and intelligent thought, a universally acknowledged standard of morality and so on—all of which are inexplicable from the standpoint of the known physical laws of the universe. To be sure, we do not usually call these things 'miracles'; but failure to do so makes them no less miraculous.

Perfect and holy, and the source of moral law

When we refer to the *perfection* of God, what we mean is that God is complete and lacks nothing desirable in any of his attributes. One way God's perfection is manifested is in his moral attributes of *holiness* and *goodness*. God's holiness means that God is separated from and free of all evil, and completely dedicated to his own glory. His goodness means that he himself sets the very standard of what is good and evil. God does not do something because it is somehow inherently good (that would imply that God is subservient to a higher standard than himself); rather, something is good because God does it. Because of these two attributes, God himself is the absolute standard of morality; of right and wrong. The Bible says of God: 'Your eyes are too pure to look on evil; you cannot tolerate wrong.'[29]

God also desires—indeed, requires—that we ourselves imitate his holiness and conform our behaviour to his standards; and he communicates this to us in both *special* revelation and *general* revelation. In special revelation (the Bible) he commands us regarding false religions, 'Therefore come out from them and be separate, says the Lord. Touch no unclean thing, and I will receive you', and 'Be holy, because I am holy'.[30] Using general revelation, God has placed his laws in our hearts:

> For when Gentiles who do not have the Law do instinctively the things of the Law, these, not having the Law, are a law to themselves, in that they show the work of the Law written in their hearts, their conscience bearing witness, and their thoughts alternately accusing or else defending them.[31]

Having this law in our hearts allows us to know right from wrong (generally though not perfectly), and renders us without excuse before God:

> For the wrath of God is revealed from heaven against all ungodliness and unrighteousness of men, who suppress the truth in unrighteousness, because that which is known about God is evident within them; for God made it evident to them.[32]

This innate sense of right and wrong is completely without explanation apart from the existence of God, transcending as it does the boundaries of culture, language and custom. Yes, there are a few individuals and cultures that seem not to be impacted by this standard in idiosyncratic ways (such as cannibals in the old South Pacific, or homosexual communities in San Francisco); but that is not surprising when we learn that once an individual or a people decides to 'suppress the truth', God in turn 'gives them over' to a 'depraved mind' and allows them blissfully to pursue their unbelief.[33]

Our responsibility to pursue truth and act on our innate knowledge of God is paramount. God requires both right actions and true belief in him; and he, in his great love, has ensured that we can know what he requires so that we will turn to him. Anything less than a full turning to God and rejection of sinful self-fulfilment places our conscience in peril of being seared to the point that we no longer have the capacity to respond to God.

Why can't God just leave me alone?

There are many who find it repugnant to commit their lives to God. Although it is nothing new, our modern focus on 'self' has climbed to the pinnacle of human aspiration. Phrases such as 'because I'm worth it' and 'you deserve it', combined with concepts such as self-love, self-esteem and self-actualisation, rule the day. We want to be the masters of our own destiny, and we don't easily yield the helm to another who demands our allegiance and obedience. Moreover, we view this trait as a *desirable* one, and we hold up as models those who have successfully carved out and accomplished their own purpose in life.

The Scriptures have a word for this; it's called 'rebellion'. When we decide that we want to live a life without God, we have missed the point of God's purpose for us. We have not been created to serve ourselves, but God. We have been placed here not to carry out our own purposes, but God's. And we are not obliged to live by our own laws and standards, but God's.

We have already noted that we have an innate sense of right and wrong, but we don't always follow it. When we ignore it there are consequences; certainly eternal consequences, but perhaps even temporal. For instance, if on our way to the top of the corporate ladder we decide we must cheat and steal in the process, not only have we abused others who have been made in the very image of God,[34] but we have very likely obtained a bad reputation in the process. The man who commits adultery with another man's wife must be concerned not only about the judgment of God in eternity but also the retribution of the offended husband in this life. The man who assaults or murders another man, rapes a woman or lives a promiscuous lifestyle will certainly have to face eternal consequences, but also temporal consequences in the form of capital punishment, a prison sentence or sexually transmitted diseases. All of these actions attain 'self-fulfilment' for the moment; but all of them result in undesirable consequences in the end.

This is not to say that God is out to get us or simply wants to put a damper on our fun. God has not given his laws to chain us down but to guide us into a right way of living for our own benefit. Those who pursue righteousness rather than their natural inclination toward self-fulfilment will be free from the temporal consequences that those actions bring.

But let me be very clear here. Following the laws of God does not mean that we are spared the eternal consequences of our sin. Why? Because the moral law of God has jurisdiction well beyond our actions; it also governs our deepest thoughts and desires. Jesus himself said that if we restrain ourselves from committing a physical act of adultery with someone we desire, it avails us nothing if we still find that we lust after that person in our heart. And even if we restrain ourselves from assaulting and murdering someone with whom we are angry, the very fact that we are angry with him to begin with is tantamount to murdering him in our heart.[35] Hence, although it will certainly aid in avoiding the temporal consequences of sin, no amount of following God's law externally will result in attaining eternal life.

But here is the really bad news. As much as we may be able to restrain ourselves from most external acts of sin, we cannot stop ourselves from committing their internal equivalents. To be sure, the law of God was placed here to guide us into right living; but it became for us a two-edged sword in that it actually increased our desire to sin. This is where the principle of rebellion mentioned earlier comes into play. Under ideal conditions, the law of God would aid us in doing the right thing. But we should not delude ourselves into thinking we are under ideal conditions. Far from it. Indeed, in our natural state we are said to be 'slaves to sin', and consequently 'objects of God's wrath'.[36] The primary thing a law or rule—any law or rule—accomplishes in our natural state is a disclosure of the true rebellion of our heart.

When I was a young child, no older than nine or ten, I remember walking through a department store parking lot with my mother and siblings. There was a huge sale occurring, and the store had set out a pair of roving spotlights to light up the sky and announce the sale. I barely noticed the spotlights, and would happily have gone about my business of ignoring them had it not been for the very next words that came out of my mother's mouth: 'Don't look at the lights.' She said this, of course, solely for my benefit, so that I would not damage my eyes, and not at all to place unnecessary and burdensome laws in my path. But what do you suppose was the very next thing I did? I was perfectly fine and minding my own business before the 'law' was issued from my mother's lips. But as soon as I heard it, sin stirred up rebellion in my heart as a fireplace poker stirs fire from hot coals, and my strongest desire at that point was to disobey that 'law' and look directly at the lights.

Sin is a disease. It lies ever dormant in us until a law activates it and results in rebellion. Hence, no matter how much we restrain our outward disobedience to God's law, we are powerless to cure the inward disease—the rebellion—that sometimes expresses itself in outward acts of sin.

> Those who live according to the sinful nature have their minds set on what that nature desires . . . The sinful mind is hostile to God. It does not submit to God's law, nor can it do so. Those controlled by the sinful nature cannot please God.[37]

We are, indeed, in a very real dilemma, and one which the Bible fully recognises:

> What shall we say, then? Is the law sin? Certainly not! Indeed I would not have known what sin was except through the law. For I would not have known what coveting really was if the law had not said, 'Do not covet.' But sin, seizing the opportunity afforded by the commandment, produced in me every kind of covetous desire. For apart from law, sin is dead. Once I was alive apart from law; but when the commandment came, sin sprang to life and

I died. I found that the very commandment that was intended to bring life actually brought death.[38]

God is just

But that's not the worst news of all. The smouldering coals of sin over which we are powerless require the righteous judgment of God. God is perfectly just, and cannot therefore turn a blind eye to sin and forgive us without a basis for that forgiveness. To ask God to enter into 'special arrangements' with us and allow us into heaven on our own terms is tantamount to asking him to relinquish his justice (an essential attribute of his deity) and cease being God. It's not something God *could* do, even in his great desire to save us, because God always acts in accordance with his justice. And when sin is in view, that justice is always accompanied by God's wrath.

Yet God's justice and wrath are never carried out rashly or in a cavalier way, but are always accompanied by great patience: 'Do I take any pleasure in the death of the wicked? declares the Sovereign LORD. Rather, am I not pleased when they turn from their ways and live?'[39] In the days of Noah, when God saw how corrupt man had become, he determined to 'wipe mankind . . . from the face of the earth'.[40] But it was only after waiting more than a century—and after Noah (whom Peter calls a 'preacher of righteousness'[41]) had warned the inhabitants of the world of impending judgment—that God brought the flood waters upon the earth.

The patience of God is the very reason sin is allowed to exist in the world today. Many people ask me: 'Why, if God exists, does he allow evil in the world? Why doesn't he just obliterate it all?' My response is always the same: 'Would you like him to start with you?' The problem is, we want the world to be free from all *other* evil, not the evil we ourselves enjoy. But therein lies the dilemma once again. We want justice in the world—just not against *us*. We want to be in right standing with God to avoid his wrath, but we find that we ourselves are incapable of accomplishing it. The only thing that is capable of turning away God's wrath is full payment for that sin; only then is God's justice fully satisfied.

God is love

The good news is, as much as the Bible warns of God's justice and wrath, it also assures us of his love. Indeed, the Bible tells us that 'God is love';[42] and because of his great love, God himself has provided for us a way out of this dilemma, without compromising the requirements of his justice:

> You see, at just the right time, when we were still powerless, Christ died for the ungodly. Very rarely will anyone die for a righteous man, though for a good man someone might possibly dare to die. But God demonstrates his own love for us in this: While we were still sinners, Christ died for us. Since we have now been justified by his blood, how much more shall we be saved from God's wrath through him! For if, when we were God's enemies, we were reconciled to him through the death of his Son, how much more, having been reconciled, shall we be saved through his life![43]

The death of Christ brought full satisfaction to God's justice for those who have been 'justified' by his death. Justification is a *legal* pronouncement whereby God *declares* us righteous and in good standing with him for all eternity. How may we avail ourselves of this state of justification? Remember, it cannot be earned through good deeds. We have already seen that we are in a dilemma in this regard. As much as we may *want* to do good things, those actions are always tainted by sin. And so, God does something that no one would expect—he justifies us while we are in a state of ungodliness, through belief in his Son and *without consideration of our deeds*:

> Now when a man works [that is, attempts to earn salvation through what he does], his wages are not credited to him as a gift, but as an obligation. However, to the man who does not work but trusts God who justifies the wicked, his faith is credited as righteousness.[44]

God 'imputes' (legally attributes) the righteousness of Christ to us as one would transfer funds from a full bank account to an empty one. Our 'bank account' of righteousness is completely empty; but for those who believe in Christ, God transfers Christ's righteousness to their empty accounts. The result is that God's justice is fully satisfied toward those who believe in Christ, and his wrath is turned away from them: 'Therefore, there is now no condemnation for those who are in Christ Jesus.'[45] God no longer holds his justified ones against the standard of the law, but has freely forgiven them all their sin: 'we have been released from the law so that we serve in the new way of the Spirit, and not in the old way of the written code.'[46]

Moreover, he has accomplished this justification all on his

own, with no help from us. This is the pinnacle act of God's love:

> This is how God showed his love among us: He sent his one and only Son into the world that we might live through him. This is love: not that we loved God, but that he loved us and sent his Son as an atoning sacrifice for our sins.[47]

We didn't love God; he loved us, and he sacrificed his only Son while we were still in a state of hostility toward him. 'We love because he first loved us.'[48] That is amazing love indeed!

Part 2

God, History and the Bible

15

Can the Bible Be Relied On?

Stephen Caesar

One of the first tests for the reliability of the Bible based on internal evidence is whether or not the biblical documents have been passed down to us with accuracy and integrity. In other words, have the books of the Bible been changed over the centuries? Is the Bible we have today the same as the Bible that was written so long ago?

These questions can be answered with a resounding 'Yes', thanks to what is regarded as the greatest archaeological discovery of the twentieth century: the Dead Sea Scrolls. Discovered at Qumran near the Dead Sea in 1947, the Qumran literature (as scholars call the Scrolls) included complete scrolls or fragments of all the books of the Old Testament except for the book of Esther. Among them was a complete scroll of the book of Isaiah, while another contained chapters 53–60, as well as fragments of others.[1]

Before the discovery of the Scrolls, the two oldest copies of the Old Testament, the Aleppo Codex and the Leningrad

Codex, were dated to around 900 AD. The Dead Sea Scrolls, some of which dated to as early as 250 BC,[2] thus moved back the date of the earliest Old Testament manuscript by over one thousand years. There was more to the Qumran literature than that, however. It also demonstrated the loyalty and integrity with which the books of the Old Testament had been handed down through the generations.

James A. Sanders, President of the Ancient Biblical Manuscript Center, explains the importance of this find: 'Careful study of the biblical texts among the Dead Sea Scrolls has established that there was a period of intense stabilization of the text during the first century AD.'[3] He goes on to point out that all scribes and translators had responsibilities both to the past to transmit accurately the text being copied or translated, and to the present to update the ancient, often archaic, Hebrew text so that it could be understood by contemporary Jews. Up to the first century AD, translators and scribes focused largely on their obligation to the present, on the need for their readers to understand the text copied or translated. However, as Sanders has pointed out, during the first century AD a marked change occurred towards more accurate transmission and greater stability of the text as it was known at that point.

By the close of the twentieth century, nearly all of the Dead Sea Scrolls had been transcribed, transliterated, translated and either published or nearly published. They did not, as some early critics had speculated, reveal any 'lost books' of the Bible, nor did they alter our view of the Hebrew Bible. Instead, the Scrolls demonstrated the stability of the ancient biblical text, as Dr Sanders pointed out. This text, in turn, has been preserved in the standard Hebrew edition of the Old Testament, the Masoretic Text.[4]

The Masoretic Text is the basis for all modern translations of the Old Testament. The name comes from the Hebrew word for 'traditionalist', and it is a fitting title, for the Masoretes were Jewish scribes who were utterly dedicated to the meticulous preservation of the Hebrew Scriptures. Their exactitude has paid off. Professor Lawrence Schiffman of the Department of Hebrew and Judaic Studies at New York University notes that most of the differences between the Qumran biblical texts and the Masoretic Hebrew Bible (the standard, received text) are miniscule. For example, he writes, 'where the Masoretic and Samaritan texts have the Hebrew equivalent of "the cattle of Egypt" in Exodus 9:6, the Qumran text (4Qpaleo-Exod^m) and the Septuagint have "the cattle of the Egyptians".'[5] Eric M. Meyers, Professor of Judaic Studies at Duke University, similarly remarks: 'The Qumran copies of most of the books of the Hebrew Bible are surprisingly faithful precursors of the Masoretic text, the version of the Hebrew Bible that became canonical in Judaism at the end of the first century AD.'[6]

The Masoretes' duty was to 'build a hedge' around the Scriptures, protecting it from corruption or amendment. To do this they counted every verse, every word and even every letter in every book of the Old Testament. They recorded how often the same word occurred at the beginning, middle or end of a verse. They recorded the middle verse, middle word and middle letter of every biblical book.[7] This was actually a tradition begun by the scribes who left us the Dead Sea Scrolls. At the end of some of the Qumran scrolls, archaeologists found that the scribes had given the total number of words in the book, and recorded which word was the exact middle, so that later copyists could count both ways to be sure they had not left out so much as a single letter. It was this tradition that the Masoretes inherited.[8]

The Masoretes, the heirs of the Qumran scribes, were thus extremely loyal to their forebears. Emanuel Tov of the Hebrew University in Jerusalem points out that a good example of an early proto-Masoretic source is the shorter Isaiah scroll from Cave 1 at Qumran (1QIsa^b). Comparison of this scroll with the medieval text shows, Tov observes, that 'the consonantal framework of the MT [Masoretic Text] changed little after the first century BC. This is confirmed by the fragments found from Wadi Murabba^{c}ât south of Qumran, dating from about 130 AD, all of which are virtually identical with the medieval text.'[9]

Jewish scribes were astoundingly meticulous in their transmission of the Scriptures. Strict rules were applied to the copying of biblical texts: the scribe's ink had to be black and indelible, the parchment on which he wrote had to be from an animal edible under the Mosaic Law, and, according to the Talmud, he traced his lines of writing with a ruler and style in order to ensure that they would be straight and uniform. Also mandatory was exactness in spelling, crowning certain letters and dotting others, and following prescribed regulations regarding spacing for sections. The Talmud records the following adjuration by Rabbi Ishmael to a scribe: 'My son, be careful in thy work as it is a heavenly work, lest thou err in omitting or adding one iota and so cause destruction of the whole universe.' Erasures were forbidden; if three or more errors were found on a single page of Scripture, the entire scroll was not to be used.[10]

What about the New Testament?

When it comes to the New Testament, there has been no single find comparable to the Dead Sea Scrolls. Rather, the evidence for the reliable transmission of the New Testament documents has been gathered in fits and spots. Nonetheless, as is the case with the Qumran literature, the evidence is overwhelmingly in favour of reliable transmission.

By 1989 the Münster Institute for New Testament Textual Research had catalogued the number of manuscripts of the Greek New Testament at a total of 5488. There were 96 papyri, 299 uncials (manuscripts using large, rounded letters), 2812 minuscules (manuscripts using a small, cursive script) and 2281 lectionaries (containing selected passages arranged according to the liturgical year, for use in church services). Bruce M. Metzger, Professor of New Testament Language and Literature at Princeton University, points out: 'When compared to the numbers of existing manuscripts of ancient classical writers, these numbers are extraordinarily large . . . fifty-nine manuscripts contain the entire New Testament.'[11]

The oldest known New Testament manuscript is a papyrus fragment dated to 100–150 AD, containing five verses from the Gospel of John, chapter 5. It is currently housed in the John Rylands Library in England. The oldest manuscripts that contain substantial portions of the New Testament are the Bodmer papyrus of the Gospel of John and the Chester Beatty papyrus, which contains ten of Paul's epistles. Both date to around 200 AD. The oldest parchment New Testament copies are the Codex Vaticanus and the Codex Sinaiticus, both of which date to the 300s AD. About three hundred New Testament manuscripts from the period 300–1000 AD are in our possession today.[12] By the close of the twentieth century, the following New Testament manuscripts or fragments were in scholarly hands:[13]

Technical designation	Date of manuscript	New Testament portion
Papyri		
P38	c. 300 AD	parts of Acts 18:27–19:16
P45	3rd century	parts of four gospels and Acts
P46	c. 200 AD	ten Pauline letters
P47	3rd century	Revelation 9:10–17:2
P52	100–150 AD	John 18:31–33, 37–38
P66	c. 200	parts of John
P75	3rd century	most of Luke; 2/3 of John
Uncials		
—	4th century	entire New Testament
A	5th century	almost entire New Testament
B	4th century	four gospels, Acts, several epistles
C	5th century	four gospels and Acts
W	5th century	four gospels
0169	4th century	Revelation 3:19–4:3
0212	3rd century	Diatessaron,[14] small parts four gospels

Because of such powerful manuscript evidence for both the Old and New Testaments, Millar Burrows of Yale, former Director of the American School of Oriental Research in Jerusalem, concludes:

> On the whole such evidence as archaeology has afforded thus far, especially by providing additional and older manuscripts of the books of the Bible, strengthens our confidence in the accuracy with which the text has been transmitted through the centuries.[15]

Burrows acknowledges that there are many cases of minor variations in wording, but these changes do not affect the main facts of the history or the doctrines of the Christian faith. In fact, he points out that archaeological discoveries have not materially altered the text of the Bible in any way. Rather, they have confirmed that not only the main substance of what has been written, but even the words, aside from minor variations, have been transmitted with remarkable faithfulness, so that 'there need be no doubt whatever regarding the teaching conveyed by them'. Burrows asserts that if we consider what Amos, Isaiah, Jesus or Paul thought and taught, our knowledge is neither improved nor changed by any of the manuscripts that have been discovered.

Truth, not propaganda

One of the most notable features of the biblical writers is the fact that they don't gloss over the sins and shortcomings of their greatest heroes. They are interested in truth, not propaganda. This is true of both Old and New Testaments.

The Old Testament

The first example is where Genesis states that 'Noah found grace in the eyes of the Lord.'[16] So righteous was he, in fact, that he was specifically chosen by God to survive the worldwide

Flood that was sent to punish a violent world. Why, then, would the biblical authors so openly admit to Noah's great embarrassing sin, in which he got so drunk on wine that he passed out naked in his tent for all to see?[17] The only answer is that the incident actually happened, and the author of Genesis wanted only to record the truth, not propaganda that made Noah anything but what he really was—a flawed human being like the rest of us.

The same can be said about all three of the great Hebrew patriarchs, Abraham, Isaac and Jacob. Abraham was to be the 'father of many nations' (including the Chosen People, the Jews) and the entire world would be blessed through him.[18] Yet this colossus of faith blatantly sinned when he entered Egypt: lying to save his own skin, he told the Egyptians that his wife Sarah was his sister, which caused all sorts of problems.[19] Even worse, he told the same lie later to the Philistines, with the same disastrous results.[20] His son Isaac, through whom God's promises were to be carried out, committed the exact same sin, and for the same selfish reason: 'Lest I die for her.'[21] His son Jacob, the father of the twelve tribes of Israel, was worse than both his father and grandfather combined. His schemes and swindles are too numerous to mention in a limited space. Why would the holy book of the Chosen People admit that its founders were such scoundrels? Because they were, and the author of the patriarchal narratives was merely recording the truth.

The same thing can be seen in the character of the Old Testament's central figure, Moses. Moses was handpicked by God to deliver the Israelites from slavery in Egypt and to give them the divine Law on which the Jewish religion is based.

Moses was so righteous that God himself said, 'With him I will speak mouth to mouth . . . and not in dark speeches'.[22] So why does the Bible admit so brazenly to Moses' sin in front of the entire Israelite nation? At one point the Israelites were dying of thirst and were beginning to grumble against God. Moses sought God's counsel and was told to perform a miracle by standing before a rock and ordering it to produce water. He did so, but his faith failed him and he struck the rock twice with his staff, and water came out.[23] God severely punished Moses for his public misdeed: 'Because you did not trust in me enough to honour me as holy in the sight of the Israelites, you will not bring this community into the land I give them.'[24] This was quite an embarrassment for the great leader: his entire life centred on getting his people out of Egypt and into the Promised Land, yet he himself wasn't even allowed to enter it.

We find the same thing with King David, who is referred to as a man after God's own heart.[25] Yet the Bible openly admits to a horrible sin committed by this great man.[26] Despite the fact that he had numerous wives and concubines, he lusted after Bathsheba, another man's wife. Determined to have her, he impregnated her and then intentionally sent off her husband, the righteous Uriah the Hittite, to be killed in battle. The deed having been done, David married Bathsheba. In one of the most dramatic scenes in the Bible, the prophet Nathan confronted the king, rebuking him to his face for his wicked deed. The biblical historian made no effort to hide such a dastardly act on David's part. He could simply have neglected to add it to his account if he had not been interested in faithfully recording the truth.

We see this again with David's son Solomon, the king who brought the Israelite nation to the height of its wealth and power. Solomon was highly commended by God because, at the beginning of his reign, he prayed for wisdom and knowledge rather than for 'wealth, riches or honour'.[27] Subsequently Solomon was known all over the ancient Near East for his great wisdom and knowledge.[28] Yet the Bible also admits that, when he was old, Solomon permitted his many wives and concubines to turn away from God and embrace idols. So bad was his backsliding that he even built altars for Chemosh, the god of Moab, and Molech, the god of the Ammonites, right in Jerusalem.[29] Once again, the biblical historian could simply have left this embarrassing feature of Solomon's life out of the record, but he did not—because he was interested in recording the unvarnished truth.

The New Testament

When we move to the New Testament, we find a similar attitude: an all-consuming desire for truth, not propaganda.

Aside from Jesus Christ, the chief figures of the New Testament are the twelve apostles. These men were not pillars of righteousness. For example, Jesus consistently chided them for their lack of faith. When James and John requested that Jesus destroy a Samaritan village, he fiercely rebuked them. He had to rebuke the same two apostles again when they arrogantly asked to be placed at his right and left hand 'in your glory', causing much resentment among the other apostles. Peter, the leader of the apostles, was the worst of all. At one point Jesus rebuked him so severely that he called him Satan!

Later Peter egotistically boasted that no matter what happened, he would never desert Jesus. Yet when the chips were down, and Jesus was about to be crucified, Peter denied him three times, even using foul language.[30]

If, as some suppose, the New Testament documents were fictions invented to serve the desires of early church leaders (who claimed apostolic succession), why would they make the apostles look so awful?

These counterproductive features in both the Old and New Testaments make a strong argument for the general reliability of the Bible as a book whose authors were interested in only one thing—the truth. Their desire was not to make mere human beings into flawless superheroes, but to show the absolute perfection of God and his Son, Jesus Christ.

Eyewitness reports

But were the writers of the Bible credible eyewitnesses? Were they able and willing to tell the truth? This subject is closely related to the previous section. Rather than focusing on individuals, we shall home in on more general aspects of the Bible which show that the authors were credible eyewitnesses who were not bending reality to suit their purposes.

Among the most notable of the Bible's claims is that the Israelite people are descended from slaves in Egypt. The story is told in the book of Exodus. Is this rooted in eyewitness testimony? Professor Nahum Sarna of Brandeis University stated that Exodus 'cannot possibly be fictional. No nation would be likely to invent for itself . . . an inglorious and inconvenient tradition of this nature' unless it was true.[31] 'If you're making up history,' added Professor Richard Friedman of the University of California, 'it's that you were descended from gods or kings, not from slaves.'[32] Indeed, the Exodus story stands unique in the ancient world as a story of national origins. Across the globe, origin stories almost invariably have their own people specially created by the gods, superior to all others. Only the Old Testament is reliable enough to tell the historical truth about how the Israelites came into being.

Similarly, the Old Testament is historically accurate enough to be brutally honest about how poorly God's Chosen People stuck by him. From the very outset, while Moses was receiving the Ten Commandments from God on Mt Sinai, the Israelites fashioned a gold bull-idol and set it up as a pagan god.[33] From then on, the Old Testament is painfully forthright in its portrayal of the Israelites as constantly moving back and forth from worshipping God to embracing paganism.[34] This is especially evident in—but certainly not restricted to—the major prophets (Isaiah, Jeremiah and Ezekiel). Dishonest propagandists would not have been so open about the chronic apostasy of the Chosen People. An Israelite mythologist would rather have painted an artificially rosy picture of his countrymen as righteously and unfailingly basking in the light of the national God. However, because the Old Testament authors were credible eyewitnesses, they recorded for posterity exactly what they saw—and not what they wanted their readers to hear.

The same is true in the case of the New Testament, but from a different angle. Being part of the Roman world, with its myriad of contending philosophies, cults and religions, the New Testament authors were acutely aware of their need to provide a credible, sustainable, verifiable and objective presentation of the story of Jesus Christ. This is why both Paul and Peter exhorted Christians to earnestly defend their faith using facts and logic.[35] Historian Herbert Schlossberg writes: 'The New Testament writers insisted on the factuality of their message, rejecting a spurious spirituality that has dogged the church down to our own day.'[36]

Schlossberg gives several examples to illustrate this insistence. When John the Baptist was wavering in faith that Jesus was the true Messiah, Jesus vindicated his ministry by reminding John's disciples of the miracles he was performing.[37] Doubting Thomas refused to believe that the resurrection had occurred, and it was only after seeing the risen Christ and feeling his wounds that he was willing to change his mind.[38] The early church emphatically denied that its teachings had anything to do with 'cleverly designed myths' and insisted that they were as based on verifiable facts: 'we were eyewitnesses'.[39] Schlossberg further points out that Judas' replacement as an apostle had to be an eyewitness,[40] and when Paul spoke of the literal truth of the resurrection, he referred to the fact that there were about five hundred people, most of whom were still alive at the time, who could attest to the truthfulness of the account based on what they themselves had seen. He would not allow this cornerstone of Christian believe to be spiritualised but asserted that if it had not taken place as he was describing it, then Christian faith was in vain and believers were to be pitied above all others.[41]

Ralph P. Martin, Professor of New Testament and Director of the Graduate Studies Program at Fuller Theological

Seminary, similarly comments that 'the Gospel-writers were certainly not purveyors of myths, professing to extol the mighty deeds of a legendary figure, a cult-hero of a kind of mystery religion, to whom a name could hardly be given'.[42] He observes that a lot of the detail in the gospels would make no sense if the writers were simply making up stories. The impression most readers gain is exactly the opposite, namely, that 'the evangelists thought they were reporting solid history, and that the chief actor in their drama was a flesh-and-blood character, living a human life under Palestinian skies'.[43]

Most importantly, the apostles and other early Christian leaders were obliged to tell stories of Jesus that were, in their time, still very recent events. Norman Anderson, former Professor of Oriental Laws and Director of the Institute of Advanced Legal Studies at the University of London, wrote:

> We must always remember, and insist, that the preaching and teaching of the apostolic church must have been firmly based on a real, historical person who had actually—and very recently—lived a life and given teaching of the unique quality to which the preaching of the apostles called men.[44]

Brazen lies about Jesus' miracles would have been easily and quickly shot down, and Christianity as a religious movement would have faded quickly. Instead it triumphed, eventually becoming the official religion of the very empire that had tried so hard to destroy it. Why? Because the bedrock of Christianity, the four gospels, consisted of historical truths that could not be successfully denied or refuted by Christianity's opponents.

Evidence from outside the Bible

The Jewish historian Flavius Josephus, who lived in the apostolic era, mentioned the ministry of Jesus of Nazareth during the years that Judaea was under the Roman governor Pontius Pilate. He recorded Jesus as 'a wise man . . . a doer of wonderful works', who 'drew over to him both many of the Jews, and many of the Gentiles'. He also mentioned that Pilate condemned him to crucifixion.[45]

Another Jewish source dating from around the time of Jesus is the Talmud, a collection of religious writings and Bible commentaries compiled by Jewish rabbis beginning in the years after the Jewish revolt of 70 AD. The Talmud mentions Jesus on numerous occasions, and even though its authors were extremely hostile to him and the religion he founded, nevertheless they provide some surprising confirmations of the gospels. For example, the Talmudic tractate *Sanhedrin* (43a) states:

> Jesus was hanged [that is, hanged on a cross[46]] on Passover Eve. Forty days previously the herald had cried, 'He is being led out for stoning, because he has practised sorcery and led Israel astray and enticed them into apostasy. Whosoever has anything to say in his defence, let him come and declare it.' As nothing was brought forward in his defence, he was hanged on Passover Eve.[47]

This passage confirms several historical facts as recorded in the gospels, such as the assertion that Jesus was crucified on Passover Eve, that he performed miracles which the Jewish leaders attributed to the devil, and that no one came to his defence at his trial.[48] Immediately following this Talmudic passage is a comment by a rabbi known as 'Ulla, who lived around the end of the third century. He wrote of Jesus:

> Would you believe that any defence would have been so zealously sought for him? He was a deceiver, and the All-merciful says: 'You shall not spare him, neither shall you conceal him [Deuteronomy 13:8].' It was different with Jesus, for he was near to the kingship.[49]

'Ulla's comment about Jesus being 'near to the kingship' most likely refers to his royal descent from the house of David.[50]

Another Talmudic confirmation of the historical Jesus is provided by Rabbi Abbahu in the *Ta'aniyot* II, 65b. Abbahu is warning his followers not to heed the claims that Jesus made publicly and privately:

> If a man says to you 'I am God', he is lying; [if he says to you] 'I am the son of man' he will retract in the end. [If he says] 'I shall ascend to heaven', he will not fulfil what he said.[51]

This corroborates the four gospels' assertion that Jesus claimed that he was God, that he was the son of man, and that he would ascend to heaven.[52]

Neither Josephus, who was merely recording secular history, nor the Jewish authors of the Talmud, who vehemently opposed early Christianity, had any reason to make up myths and fairy tales about Jesus doing miracles or claiming to be the Son of God. If these occurrences, as recorded in the gospels, were actually fictional stories made up at a later date, then the Talmudists would not have bothered to include them as historical facts in their writings. The authors of the Talmud, being fervent opponents of Jesus and his followers, would *never* have bothered to make up stories about Jesus doing miracles (practising 'sorcery').

The earliest of the Talmudic rabbis lived among the very people who witnessed Jesus' life and ministry; they could have openly denied the stories contained in the gospels, but they did not. It was only in later centuries, and in parts of the Roman world far from where Jesus had lived, that denials of his miracles and resurrection began circulating—but in the time and place where Jesus lived, historians openly admitted that the miracles, as well as the resurrection, were true. There is only one answer as to why this should be: the authors of the Talmud admitted that Jesus performed miracles because these events actually happened, and any attempt to deny them to the reading public would have been shouted down as historically fraudulent.

Are the documents free of contradictions?

In dealing with a multi-authored book such as the Bible, it is extremely important to ask this question. If some books of the Bible contradict others, how can they all be divinely inspired? Critics claim the Bible contains contradictions, but careful examination shows that each one can be explained. In some cases, apparent contradictions are even found to be complementary.

Here is one example. Genesis 1:6 in the King James Version refers to the earth's atmosphere as 'a firmament'. This English word is obviously derived from the word 'firm', which suggests that our atmosphere was thought of by the biblical writer as a hard, solid mass—which of course it is not.

The problem lies not with the Bible, however, but with the English translation. The word 'firmament' was borrowed from the Latin Bible (known as the Vulgate), which used the word *firmamentum*, 'something made solid'. This in turn was based on the Greek translation of the Old Testament (known as the Septuagint). During the time the Septuagint was translated, the Greeks believed that the sky was a solid, crystalline sphere arching above the earth.[53] However, when the original author of Genesis wrote this verse under divine inspiration, he used the Hebrew *râqîya'*. This word is derived from the root word *râqa'*, meaning 'to expand' or 'to spread around'. Job states that God 'spread out [*râqa'*] the sky'; the same word appears when David states of his enemies that he 'spread [*râqa'*] them abroad'.[54] Thus the original authorship of the Bible rightly described our planet's atmosphere as an open, spread-out expanse, not as a solid mass. This is confirmed when Genesis mentions 'fowl that may fly above the earth in the open firmament [*râqîya'*] of heaven'.[55]

Now that we've established that Genesis views the atmosphere as a stretched out expanse and not a hard dome, we come across a seeming difficulty: Job 37:18 states that the atmosphere is as hard as bronze. As it turns out, both passages are scientifically correct. Lewis Thomas, formerly of Yale University and a member of the National Academy of Sciences, writes: 'We know that it [the sky] is tough and thick enough so that when hard objects strike it from the outside they burst into flames.'[56] In other words, from an earth-side point of view, our atmosphere is indeed a wide open, spread out expanse through which birds can freely fly; but from the point of view of outer space, our atmosphere is as hard as a dome of bronze. This causes what astronauts call the 're-entry problem'—the fact that a spaceship re-entering the earth's atmosphere will bounce off it as if it were metal, unless the craft approaches at precisely the correct angle. So two biblical statements that appeared to be contradictory are actually complementary.

Other apparent contradictions are due to cultural and linguistic issues that no longer apply in today's world. One example is an apparent disagreement about the time of Jesus' crucifixion. Mark states that Jesus was crucified at 'the third hour', whereas John says he was still with Pontius Pilate at 'about the sixth hour'.[57] Jack Finegan, Professor of New Testament History and Archaeology and Director of the Palestine Institute of Archaeology at the Pacific School of Religion in Berkeley, California, observes that on the basis of the common reckoning of time from daybreak or 6.00 a.m., the third hour was what we would call nine o'clock in the morning. If John was reckoning on the same basis as Mark, there would indeed be a contradiction. But if John was counting time by the Roman method—that is, from midnight—the time he gave was only six o'clock in the morning. This gives more than enough time for the various stages of the trial and abuse of Jesus, and for him to be sent off for crucifixion by nine o'clock.[58]

An even more blatant apparent discrepancy in the gospels involves the day on which Jesus was crucified. According to Mark, Jesus was crucified on the day after the Passover meal. However, according to John, he was crucified on the day before the Passover meal.[59] This is a very complicated problem, but not insurmountable. In the Jewish tradition each day begins in the evening so that both events (the last supper and the crucifixion) occurred on the same day. According to all four gospels, the last supper and the crucifixion took place just before Passover on the so-called 'day of preparation' (Greek *paraskeuē*), which in that particular year took place on a Friday. As a result, there was preparation for both the Passover and the Sabbath at the same time.[60] Given these facts, Professor Bo Reicke of Basel University in Switzerland has concluded: 'Contrary to what is often stated, the Synoptic and Johannine reports do not contradict each other in this point. In the Jewish calendar, the day of preparation for Passover to which all four gospels refer had to be 14 Nisan.'[61] The beginning of the lunar month of Nisan was established every year according to the first appearance of the crescent moon in March, and though exact timing was not possible in those days, modern research has shown that 14 Nisan fell on a Friday in two of the years in question: c. April 7 in 30 AD and c. April 3 in 33 AD. The political realities mentioned in the gospels support the dating of the last supper and crucifixion to an evening and the following day around April 3 of the year 33.[62]

Although this issue is complicated and highly technical, a careful, studious investigation of the subject shows that no contradiction exists as to whether or not Christ was crucified on the Passover. And so it is with the rest of the seeming contradictions in the Bible—careful scrutiny of the texts, combined with a keen knowledge of the ancient biblical world provided by leading scholars, will in the end iron them all out.

In fact, the occurrence of some minor discrepancies actually supports the reliability of the Bible, since it shows that the authors did not get together and try to eliminate these differences but rather wrote what they witnessed. Another important point is that in some apparent contradictions we may not understand the details of the particular circumstances. Unless there really is an obvious contradiction, we should follow Aristotle's dictum that the benefit of the doubt should go to the author, not the critic.

16

Historical Evidence for the Biblical Flood

Jerry Bergman

There is evidence which suggests that world creation and flood myths have a common origin in history.[1] Most creation and flood myths worldwide have a basic core of themes. This fact provides strong evidence of a common origin of these myths based on actual events. The Genesis account, however, stands out from all others. This is true partly because we have more knowledge of this account and partly because of its lack of the corruption common in other creation and flood stories. The original historical events can be seen through the modifications and embellishments that were added to these stories from generation to generation.

Almost every culture has a creation myth, yet most are basically variations of the core theme of the Genesis creation account. Although found all over the world, these myths have 'haunting similarities'.[2] This would be expected if the source of all creation myths stems back to a common human experience or some actual historical event.[3] If they came from a single early source, then oral transmission, time and local cultural circumstances could have embellished or at least modified them, sometimes greatly.

For this reason, we would expect that the details of the extant creation myths would vary but that the *basic outline* would be similar—or at least that most of the stories would have common elements.[4] Conversely, as we will document, the Genesis account stands in stark contrast to almost every other creation story.[5] Even Darwinian evolution, a view that has been called 'the newest of all the stories of the beginnings of life',[6] is classified as a creation myth by those who study primal myths. In contrast to most of the other creation myths, evolutionism is the only one that does not involve creation by outside intelligence.[7]

Basic classes of creation myths

Creation stories are commonly classified into a few basic groups, and many myths contain elements from two or more groups.[8] This is additional evidence that, although altered in time, most creation myths had their origin in an actual set of events or records.

1. *Creation from nothing.* This group involves the creator 'calling forth into being' the creation, which came into existence totally as a result of his will. Steindl-Rast notes that many creationist myths described the Supreme Being as creating the world literally by thinking it into existence, 'by a word of command, by singing or by merely wishing it to be', and it became reality.[9] The best example is Christianity, which has traditionally taught that creation was from 'nothing' or *ex nihilo,* and

several Scriptures support this view. Genesis states six times 'and God said . . . and it was so', indicating creation *ex nihilo*. (Some of creation was by way of modification of parts of existing creation, such as the creation of Eve from Adam; but ultimately, in the very beginning, all of creation was *ex nihilo*.)

2. *Creation from chaos.* Creation from chaos (producing a structure from undifferentiated material) is another common theme. These myths generally stress that creation involved the process of forming the earth and living things from a mass of unstructured elements.[10] From the original chaos, order occurred as a result of some activity, force or process. This is the theme of Genesis 1:1–2, which states that the earth was undifferentiated in the beginning ('And the earth was without form, and void' [KJV]).

Christie notes that, for the Chinese, 'creation was the act of reducing chaos to order, a theme which persists throughout Chinese thought'.[11] The Greek creation myth states 'in the beginning . . . was Nothing', then 'out of Nothing came' heaven, the earth, the sea and everything else.[12] Other examples include Babylonian, Mixtec, Finnish, Indian, Japanese and Egyptian creation myths.

3. *Emergence myths.* In this category, God creates materials *ex nihilo,* then other parts of creation *emerge* from these materials. Examples include the formation of man from the dust of the earth and of woman from Adam's side. The Scriptures often refer to God as a potter, moulding an existing substance into something else.[13]

Important examples of emergence myths include Acoma, Hopi, Navajo and most New World creation myths.[14] The

Navajo, the Aztecs and the Pueblos teach that all of life, as well as the first man and woman, was created from the soil or earth by God or the gods.[15] One Hopi myth teaches that at one time only water existed and two deities caused dry land to appear. Later the gods made birds out of clay.[16] Mbiti concludes that the 'metaphor of the potter is commonly used to describe God's creative activity' in ancient African creation myths.[17]

4. *Separation myths.* In many myths, division or a separation of 'parents' or something else occurs. Genesis contains several separation examples in creation, such as the division of the waters and of night and day. Hasel found that the 'idea of the creation of heaven and earth by division is common to all ancient Near Eastern cosmogonies' and in myths the world over.[18] One Polynesian myth teaches that darkness once rested upon the heaven and earth until the light and darkness were separated by God.[19]

5. *Earth-divider myths.* A similar major creation myth involves a divine being dividing the water by bringing the land up from the sea, permanently separating the two. In Genesis, an earth shrouded in darkness and 'without form and void' is first bathed in light and then divided into dry land and sea from which plant and animal life springs forth. Genesis 1:10 says that God divided the land and water by commanding 'let the dry ground appear [out of the sea]. And it was so. God called the dry ground 'land', and the gathered waters he called 'seas'.'

6. *Creation from a cosmic egg or seed.* Some creation myths include the concept of a cosmic egg, a 'germ' (a raw material such as water or clay) or seed out of which God formed

humans, animals, plants or even the earth.[20] Examples include Indian, Phoenician, Egyptian, Orphic, Chinese and other texts. One third-century AD Chinese myth taught that chaos 'was like a hen's egg' from which the parts separated, the 'heavy elements forming the Earth, and the light, pure ones the Sky. These were *yin* and *yang*'.[21] The concept of a world egg is found in many cultures; the classical Indian cosmogonies, for example, teach that the creator, Brahma, formed the heavens from the upper part of an egg and the earth from its lower part.[22]

The cosmological view currently in vogue among secular scientists, the big bang hypothesis (called the 'standard model' because of its wide acceptance), also postulates a 'cosmic egg' from which the entire universe sprang.[23]

7. *Light*. In addition, an 'enormous number of creation myths . . . involve the sun, and the life-giving, regenerative properties of light . . . Everywhere the sun or light plays an important, if not a central role' in creation.[24] The Bible often use the words 'sun' and 'light' in this sense, even stating that 'God is light'.[25] According to Genesis, the first act of God after the creation of the heavens was light: 'darkness was over the surface of the deep . . . And God said, "Let there be light", and there was light.'[26]

The term 'light' often refers not only to physical light, but also to knowledge and insight. The importance of light (meaning knowledge and wisdom) is reflected in virtually all non-biblical creation stories. Excellent examples include the Eskimo, Muslim and Zoroastrian creation stories.

The basis for creation myths

The essential categories of creation myths outlined above are directly taught, or at least clearly reflected, in Genesis. From the biblical account one could assume that Adam and Eve gave their immediate descendants information that became part of later historical records, parts of which later became incorporated in Genesis. As the descendants of Adam scattered, they would have carried what they remembered (primarily, the essential elements) of the history now recorded in Genesis.[27]

This history was oral in most cultures for years, and therefore could have been embellished or changed as a society developed.[28] Also, many of the mythologies would have made use of actual historical events known to the hearers/readers, often adding heroic, tragic, moralistic or other elements for a particular audience or a reigning king. The historical basis would be presumed by the storyteller but is often lost to modern readers. However, in the case of creation myths, many of the essential Genesis elements have remained the same.[29] All creation myths appear to be derived from the events that Genesis is based on, and in many cases, large remnants of the original story remain. Genesis, though, stands in contrast to all other creation accounts because it contains none of the embellishments common to the others, but only the bare outline of historical events.[30]

Flood stories are found in nearly every culture

Syrian, Sumerian, Greek, Babylonian, Chinese, Persian, Irish, American Indian, Toltec, Cholulan and even Estonian creation stories all include a variant of the biblical Flood story.

Strickling concludes from his study of world flood legends that 'nearly all' flood accounts 'are variations of the theme in the biblical account . . . however, a statistical analysis indicates the purity of the biblical account and reveals evidence of subsequent upheavals having corrupted in varying degrees all other accounts'.[31] Among the similarities, Strickling found a favoured family was saved in thirty-two of the flood accounts, and in twenty-one survival was due to a 'boat' of some type. Other similarities include: (1) a forewarning, (2) one flood only and (3) preservation of non-human types of life such as animals.

Typical are the many American Indian flood traditions, such as the Navajo flood that 'covered the first world' because the world grew extremely sinful, especially with sexual sins.[32] Regarding the great flood, Warshofsky notes that an account similar to the 'biblical account of a great, universal flood is part of the mythology and legend *of almost every culture on earth*'.[33] He adds that even people living far from the sea who would not be expected to have a flood account—such as the Hopi Indians in the American Southwest and the Incas high in the Andes of Peru—have legends of a flood so great that it covered the tops of the mountains and wiped out virtually all life on earth.

In a study of over two hundred flood myths, Morris found these similarities:[34]

Comparison of world flood myths

Event	Per cent that contain
1. Was the catastrophe a flood, not another type?	95
2. Was the flood global?	95
3. Was a favoured family saved?	88

4. Was the geography local?	82
5. Was the rainbow mentioned?	75
6. Did animals play any part?	73
7. Was survival due to a boat?	70
8. Were animals also saved?	67
9. Was the flood due to the wickedness of mankind?	66
10. Were they forewarned?	66
11. Did survivors land on a mountain?	57
12. Were birds sent out?	35
13. Did survivors offer a sacrifice?	13
14. Were specifically eight persons saved?	9

Morris concludes from his study that 'one of the strongest evidences for the global flood which annihilated all people on Earth except for Noah and his family, has been the ubiquitous presence of flood legends in the folklore of people groups from around the world'.[35] Although it is often claimed that the biblical account of the deluge was derived from a Babylonian source, I believe it is more reasonable to conclude that both accounts came from a still older source, possibly one of those that Moses used to write Genesis.[36]

Why the similarity of creation and flood myths?

Van Over, a leading myth researcher, concludes, 'The surprising and perplexing fact is that the *basic* themes for [creation] myths in widely different geographical areas are *strikingly similar*.'[37] The many scholars who have puzzled over this phenomenon include the renowned anthropologist Claude Levi-Strauss. After years of studying creation myths, he concluded that there exists an 'astounding similarity between myths collected in widely different regions' of the world.[38]

That world creation myths 'resemble one another to an extraordinary degree'[39] is not debated; *why* they are so much alike is the concern. The scholarly argument over why this similarity exists continues to this day. 'No definite answer seems yet to have developed, but theories abound.'[40]

One theory is that most of the extant creation myths arise from the need to understand our origins. But the fact that seeking an answer to our origins is common to humans does not fully explain why the many creation accounts are so similar. Another position—that argued here—is that the origin of these myths lies in actual historical events and that time embellished, romanticised and tailored the original story to local people's customs, needs and traditions. The tendency for time and culture to modify was an influence on most other historical accounts, and also supports the conclusion that the original source lies in actual historical events and the common human need to understand.

The problem of meaning in ancient creation myths

Freund notes that, in contrast to Genesis, many of the creation myths were written by 'philosophers and teachers' and only incidentally refer to creation.[41] Their primary purpose is not to discuss origins, and often they only indirectly refer to them as past events. Many ancient creation accounts are obviously didactic stories written not primarily to inform the reader about the means of physical creation, but to teach some moral principle or to instruct about some social tradition.[42]

One example is the epic of Gilgamesh. This account is

about a legendary king who travelled a great distance on a dangerous journey in search of the 'tree of life'. He finally found it after many adventures, but it was snatched out of his boat by a serpent. The biblical parallels here are striking, as are the differences. For example, Genesis talks about the tree of the knowledge of good and evil, and also about the antidote, the fruit of the tree of life.[43] Clearly, a strong similarity exists between this creation myth and the account in Genesis.

A major difficulty in understanding ancient creation myths is determining the degree that the ancients understood them as literal. If archaeologists ten thousand years from now unearthed certain remains of contemporary American civilisation, they could easily assume (based only on this evidence) that Americans believed in literal creatures called Santa Claus, Rudolph the Red-Nosed Reindeer and the Tooth Fairy. Few people today believe that the earth literally has four physical corners (as stated in Isaiah 11:12) or that the sun rises or sets. Expressions such as 'I could die of embarrassment' or 'I could kill him for doing that' are not literal, and no one (except possibly young children) interprets these common vivid figures of speech literally.

These examples illustrate the difficulty in understanding a culture from a few isolated artefacts, especially words.[44] Likewise, evidence exists that the ancients may not have believed that Zeus caused rain, or that the sun was a god, or that many of their other myths were literally true.[45] Better understanding of the ancients has altered our modern picture of them.[46] This new view argues that our modern understanding of the earliest religion stands in 'sharp contrast to the preconceived notions anthropologists had in the eighteenth and nineteenth centuries'.[47] These researchers took it for granted that religious beliefs and the human mind itself developed, in 'close parallel to physiological evolution', to ever greater refinement. Steindl-Rast, though, concludes that during the last century 'a wealth of objective material had been accumulated which proves that the most ancient cultural stratum to which we can penetrate by anthropological methods is simple but by no means "savage"'.[48]

Modern humans have no monopoly on wisdom, and the greatest of the ancient scientists were, 'considering the handicaps under which they worked, fully the equals of any in our own time'.[49] Plato's writings, Aesop's fables and other literary works clearly demonstrate that the ancients had a tremendous amount of insight into life and living—and indeed, if the reader could understand ancient Greek, he or she probably would feel at home in the company of the likes of Aesop, Plato, Aristotle and Socrates, and would no doubt learn much from them.[50]

To assimilate into our world, the ancients would have to adapt to our technology, but not necessarily to our 'worldly wisdom'. Levi-Strauss argues that the conclusion that these myths were only naive attempts to explain reality is incorrect, and that they were not trying to provide a scientific explanation for phenomena they did not understand. He reasons: 'why should these societies do it in such elaborate and devious ways, when all of them are also acquainted with empirical explanations?'[51]

Evolutionary assumptions dictate that the further back in time one travels, the more 'primitive', less sophisticated and more foolish human beliefs about the natural world become. In certain areas this is true, primarily because accumulated knowledge gives successive generations an advantage over their predecessors. But no evidence exists that the *brain* or *human intelligence* has undergone an evolutionary progression since the dawn of recorded history. An ancient Greek or Roman would probably feel fully at home in our culture, if he or she were raised in it.[52]

The benefit of accumulated knowledge of past generations, an advantage that has been especially true during the past several centuries in the West, tends to distort our evaluation of the ancients. The ancients had a tremendous amount of insight and knowledge, and we are selling them short by viewing their creation myths as the product of ignorance. Chiera notes that the Babylonian and Assyrian creation stories were 'ancient cosmogonies [with] sophisticated philosophical substratum'.[53]

This increase in knowledge notwithstanding, we still remain ignorant about much that exists. There are many problems that we are no closer to solving today than were the ancients. Speculations about the origin of the universe abound, and a study of many of the time-tested truths of the ancients helps us to realise that we have been meandering around the truth, and in some ways they were closer to it.[54]

The problem of understanding symbolism also exists in interpreting the Hebrew creation account recorded in Genesis. But we have a significant advantage in understanding ancient Hebrew because we have a huge body of literature about it compared to the mythology of dead cultures.[55] Furthermore, thousands of extant ancient writings discuss the various nuances and meanings of words, and these can be used to aid in understanding biblical manuscripts. This is not true for most of the other myths. Many are far removed from Western civilisation and culture, and in many cases their meaning was lost long ago.

Also, the extant ancient manuscripts for Genesis are far more complete than for any other creation account, and the record can be used as an historical outline to direct research. It has also been more extensively studied than any other ancient manuscript, enabling us to draw conclusions about the meaning of the Genesis account with far more confidence than the creation accounts of other cultures.

From this knowledge we can conclude that Genesis was not intended to be primarily a didactic story, but a brief, matter-of-fact summary of the creation of the heavens and earth. The well-known popular science writer, leading Darwinist and outspoken atheist Isaac Asimov, after noting that the biblical writers 'laboured to produce something that was as reasonable and as useful as possible', concluded that they 'succeeded wonderfully' and that 'there is no version of primeval history, preceding the discoveries of modern science, that is as rational and as inspiring as that of the first eleven chapters of the Book of Genesis'.[56]

Furthermore, the Genesis creation account was both validated and explained by Christ, the apostles and the early church (and they lived, not in a 'primitive' civilisation, but in cities much like ours). Christ and the New Testament writers cited the creation account and accepted it as true, stating it as fact and using the events to illustrate or teach some point about its meaning. Their many statements include these:

> All things were made through him [Christ], and without him was not anything made that was made.
>
> Sovereign Lord, you made the heaven and the earth and the sea, and everything in them.
>
> We are bringing you good news, telling you to turn from these worthless things to the living God, who made heaven and earth and sea and everything in them.
>
> Now what you worship as something unknown I am going to proclaim to you. The God who made the world and everything in it is the Lord of heaven and earth . . . he himself gives all men life and breath and everything else.
>
> Our Lord and God . . . you created all things, and by your will they were created and have their being.[57]

Nonetheless, as is true of all creation accounts, the biblical account uses certain figures of speech and allegories. For example, the reference to the earth's four corners in Isaiah obviously does not refer to a physical, four-cornered structure, but is an expression that was common then and is still so today. Our problem is to determine which statements are literal and which are symbolic.[58] Unfortunately, in order to reduce the credibility of the biblical record, many critics try to literalise portions that are obviously not meant to be literal or assume that certain statements refer to ancient myths, such as claiming that the Genesis 'firmament' is the metal dome that some ancients believed encircled the earth.[59] On the other hand, others try to ignore the clear meaning of the biblical text to fit the creation account into the latest secular ideas, sometimes with tragic consequences.

Some conclusions

Although all creation myths have similar themes and common basic skeletons, the Hebrew account stands apart in many major ways.[60] Virtually all non-biblical creation myths contain obviously false ideas such as seeing the earth as a large, flat disk sitting on a giant turtle that swims in a great sea of blood. The Genesis myth, in Hasel's words, is 'antimythical', meaning it is a simple description of events void of pagan embellishments such as this.[61] Hasel summarises his comparison of the Genesis 1 creation account with ancient Near Eastern accounts by pointing out that Genesis employs certain terms and motifs that were partly chosen in deliberate contrast to comparable ancient Near Eastern accounts, and 'uses them with a meaning and emphasis not only consonant with but expressive of the purpose, worldview and understanding of reality as expressed in this Hebrew account of creation'. He concludes that 'the Genesis cosmology represents not only a "complete break" with the ancient Near Eastern mythological cosmologies' but goes beyond this, with a 'parting of the spiritual ways brought about by a conscious and deliberate antimythical polemic' that actually undermined the prevailing mythological cosmologies.[62]

My experience from researching the material reviewed here suggests that a comparative study of creation myths can be a very beneficial part of the study of apologetics. Research of ancient cultures finds that stories attempting to explain the existence of humans, animals, plants, the world and the universe 'are found in almost every culture in the world, both in the religions of archaic peoples and in the greatest civilisation religions'.[63] The universality of creation myths points to a basic human need for a *causal* explanation of our world.

17

Archaeological Evidence for the Exodus

David K. Down

Getting the dates right

By the present chronological system of ancient history there are serious problems in synchronising Egyptian history with the events described in the biblical account of the Exodus. Most scholars have therefore concluded that the Bible record is unreliable or distorted. The Old Testament book of 1 Kings states: 'in the four hundred and eightieth year after the children of Israel had come out of the land of Egypt, in the fourth year of Solomon's reign over Israel . . . he began to build the house of the LORD.'[1] Most scholars would accept a date of about 970 BC for the beginning of Solomon's reign. So his fourth year would be 966 BC, and counting back 480 years this would place the Exodus about 1445 BC.

The reasons scholars advance for rejecting this date concern both Egypt and Israel. In Egypt the Eighteenth Dynasty reigned from about 1550 to 1320 BC. This was the most powerful, most affluent and best recorded dynasty that ever ruled the land of Egypt. During this dynasty there is no trace of any national disaster such as must have occurred as the result of the ten devastating plagues that fell on the land of Egypt prior to the Exodus. There is no sign of any military disaster such as befell the Egyptian army when it perished in the waters of the Red Sea.

It has been argued that the Egyptians only recorded their victories, not their defeats. True. But from the Eighteenth Dynasty have come so many inscriptions and papyrus documents that it is possible to map the movements of the whole dynasty, and there is no trace of a large number of Semitic slaves or of any national disaster such as would have resulted from the ten plagues and the loss of the entire army in the Red Sea. Moreover, all the mummies of the Pharaohs of that era are in the Egyptian Museum in Cairo. If Thutmosis III or Amenhotep II was the Pharaoh of the Exodus, his body should have been in the Red Sea, not in the Cairo Museum.

Another problem is the fact that the main centre of activity of the Eighteenth Dynasty was in Luxor, one thousand kilometres away from the land of Goshen, where most of the Exodus action took place. It is irreconcilable with the Exodus account to have Moses commuting back and forth from the Delta to Luxor to implore Pharaoh to 'let my people go', or even to Memphis, the political capital.

In the land of Canaan there is also a lack of evidence for an invasion by a foreign army. The biblical date of the invasion, about 1405 BC (forty years after the Exodus), falls in the middle of the Late Bronze Period, but there is no archaeological evidence during this period for the destruction of Jericho and Ai, the arrival of a new people with a different culture, or the annihilation of the old population. There is continuity of pottery styles and no indication that a dramatic change of population took place.

So, because of the lack of archaeological evidence both in Egypt and Israel to support a date of 1445 BC, most scholars have rejected the information supplied in 1 Kings 6:1 and accepted a date for the Exodus closer to 1200 BC, though there is only flimsy circumstantial evidence to support this. In the mid-1990s *Time* magazine carried the front cover heading IS THE BIBLE FACT OR FICTION?, and answered its own question by saying that there are 'parts of the Old Testament where the evidence is contradictory or still absent, including slavery in Egypt, the existence of Moses, the Exodus and Joshua's conquest of the Holy Land'.[2]

Concerning Palestine, the article said: 'Kathleen Kenyon, who excavated Jericho for eight years, found no evidence for destruction at that time' (meaning the Late Bronze period). The expression 'at that time' is most significant. Actually there is plenty of evidence for slavery in Egypt, the dramatic incidents of the Exodus and Joshua's conquest of the Holy Land, but not *at that time*.

Kenyon adhered to the traditional chronology, which would place a 1405 BC invasion of Palestine in the middle of the Late Bronze period, and there is certainly no evidence for it there. But as Kenyon herself noted, there is plenty of evidence for the destruction of Jericho and the invasion of Palestine by a new people at the end of the *Early* Bronze period. What if the traditional time periods as understood by archaeology were incorrect so that the end of the Early Bronze period actually fell around 1405 BC rather than earlier? Then the evidence of archaeology and the evidence of the Bible would line up.

Chronology is probably the hottest issue in archaeological debates, which sometimes bicker over decades, sometimes centuries, of error. And in fact there is growing concern in archaeological circles over certain dates in ancient history, especially in Egyptian history.

In 1992 archaeologist Peter James published a book, *Centuries of Darkness*, in which he claimed that the TIP (the Third Intermediate Period of Egypt, Dynasties Twenty-one to Twenty-three) was actually contemporary with other dynasties and that Egyptian dates for this period should be reduced by 250 years.[3] His book was a scholarly work, and he and the four other lecturers in archaeology who contributed were respected scholars. The book also carried a Foreword by Professor Colin Renfrew of Cambridge University, who himself wrote a scholarly book called *Archaeology: Theory, Methods and Practice*. In his Foreword Professor Renfrew wrote: 'This disquieting book draws attention, in a penetrating and original way, to a crucial period in world history, and to the very shaky nature of the dating, of the whole chronological framework, upon which our current interpretations rest.'[4] He went on to comment on the revolutionary suggestion that the existing chronologies for that crucial phase in human history are in error by several centuries, and that in consequence history may have to be rewritten. He felt that the critical analysis of the authors was correct and that a chronological revolution may be forthcoming.

Another scholar, David Rohl, who received his degree in

Egyptology and Ancient History from University College in London and was involved in excavations at Kadesh in Syria, has written numerous papers on the dating problem. In 1995 he published the book *A Test of Time*, which was subsequently put to air on BBC television. In this work Rohl also claimed that there has been a mistake in calculating the TIP, and that Egyptian chronology needs to be reduced by 350 years.[5]

As with James, Rohl is not trying to defend the Bible. He wrote, 'I have no religious axe to grind—I am simply an historian in search of truth.' He went on to acknowledge that he was prepared to accept that the Old Testament narratives are as valid a source for ancient history as any other ancient document.[6]

This reduction of dates would bring the Twelfth Dynasty down to the time of Joseph and Moses, and the Exodus in 1445 BC would have been in the early Thirteenth Dynasty rather than the Eighteenth or Nineteenth Dynasty as is generally believed. When this revised dating is adopted, remarkable agreement is found between the histories of Egypt and Israel.[7]

The story of Joseph

According to the Bible, Joseph, the son of Jacob, was sold into Egyptian slavery by his jealous brothers. In Egypt he was able to interpret one of Pharaoh's dreams to mean that there would be seven years of plenty and then seven years of famine. Pharaoh appointed Joseph as vizier of Egypt, and entrusted him with the task of collecting grain in preparation for the famine.[8]

This was an event of national importance, and there should

be some record of it in Egypt. In the time of Sesostris I, second king of the Twelfth Dynasty, there is such a record. Ameni was an official under Sesostris I. In his tomb at Beni Hassan he left an inscription that is very relevant:

> No one was unhappy in my days, not even in the years of famine, for I had tilled all the fields in the nome of Mah, up to its southern and northern frontiers. Thus I prolonged the life of its inhabitants and preserved the food which it produced. No hungry man was in it. I distributed equally to the widow as to the married woman. I did not prefer the great to the humble in all that I gave away.[9]

The Bible says that Joseph 'gathered up all the food of the seven years ... and laid up the food in the cities'.[10] Ameni could well have been one of Joseph's provincial governors who did his part in storing the grain for the coming famine.

Brugsch-Bey recognised the significance of the inscription, and if it had not been for his adherence to the traditional chronology would have readily identified it. He noted that the concluding words of this inscription, in which Ameni sings his own praises, gave rise to the idea that they contain an allusion to the sojourn of the patriarch Joseph in Egypt, and to the seven years of famine under his administration. However, Brugsch-Bey felt that there was the difference in time, which could not be made to agree with the days of Joseph. Freed of this limitation, the inscription assumes great significance.

The heart of the Twelfth Dynasty was in the Delta. There were palaces at Memphis, Avaris and the Faiyyum, which provide a suitable setting for the establishment of the Israelites in the land of Goshen. The religious centre was Heliopolis, the city of the sun then known as On. An obelisk still stands at Heliopolis, now a suburb of Cairo. This obelisk was erected by Sesostris I and is today known as the Pillar of On. If we understand that Sesostris I was the Pharaoh under whom Joseph was vizier, Genesis 41:45 is very meaningful: 'And Pharaoh . . . gave him as a wife Asenath, the daughter of Poti-Pherah priest of On.'

The Faiyyum is a vast oasis in the desert west of Meidum that today supports a population of two million people. It was developed during the Twelfth Dynasty. The whole area is watered by a canal dug during that period. This canal is today known as Bar Yosef or Joseph's Canal though no one seems to know why. It may have carried this name ever since it was dug in the Twelfth Dynasty. It could have been the work of Joseph in preparation for the seven years of famine.

As for Joseph himself, there was a vizier under Sesostris I who had extraordinary powers. His name was Mentuhotep. Of him Brugsch-Bey wrote: 'In a word, our Mentuhotep, who was invested with several priestly dignities and who was Pharaoh's treasurer, appears as the alter ego of the king.'[11] Brugsch-Bey noted that there was significance in the report that when Mentuhotep arrived, the great personages bowed down before him at the outer door of the royal palace. The Bible record states, 'Pharaoh took his signet ring off his hand and put it on Joseph's hand . . . and he had him ride in the second chariot which he had; and they cried out before him "Bow the knee!"'[12] This was an unusual honour to be bestowed on an Egyptian vizier, but that was the way it was with Mentuhotep. The fact that he is not named as Joseph is of little consequence. The Egyptians, as well as Bible characters, frequently had more than one name.

Following the death of Joseph, 'the children of Israel were fruitful and increased abundantly, multiplied and grew exceedingly mighty; and the land was filled with them'.[13] No doubt the heaviest concentration of the Israelite immigrants was in the land of Goshen in the Delta, but knowing the Hebrew capacity for industry, trade and enterprise, there is no reason to conclude that they would all confine themselves to the same location.

At Beni Hassan, 240 kilometres south of Cairo, is the tomb of a nobleman named Knumhotep of the Twelfth Dynasty. On the walls of his tomb are paintings depicting the visit of some Asiatic people to Egypt. At present this is placed closer to the time of the visit of Abraham to Egypt, but if the migration of Jacob and his large retinue took place during the Twelfth Dynasty, as is claimed here, then this wall painting should be related to this migration. It is indicative that such a migration did occur at this time and that the Hebrews had spread as far as Central Egypt. The Hebrews have always had a remarkable ability to maintain their identity, and this would also explain the foreign settlement at Kahun that archaeologist Sir Flinders Petrie investigated in 1889 (see below).

The oppression of Israel

The Bible continues, 'Now there arose up a new king over Egypt, who did not know Joseph.'[14] This does not mean he was ignorant of Joseph's services to the nation but that he wished to make no recognition of them, and as far as possible to bury them in oblivion. Josephus, the Jewish historian of the first century AD, wrote: 'having in length of time forgotten the benefits received from Joseph, particularly the crown being now come into another family, they [the Egyptians] became very abusive to the Israelites, and contrived many ways of afflicting them.'[15]

By the revised chronology, this Pharaoh would have been Sesostris III. His predecessor, Sesostris II, had no living sons at the time of his death. Sesostris III may have been related to this predecessor but he was not in direct line to the throne, so he could be classified as belonging to 'another family', as Josephus says. From his statues and inscriptions we may conclude that he was a harsh despot quite capable of inflicting cruelty on the Israelite slaves.

The last ruler of the Twelfth Dynasty was Queen Sebekneferu, who had no children. She would have been the daughter of Pharaoh who 'came down to wash herself at the river'.[16] She was not down there to get clean. She would have had a sumptuous bathroom at her palace for that. She would have been taking a ceremonial bath, praying to the fertility river god Hapi for a baby; and when she saw this beautiful Hebrew boy floating in a basket she would have regarded it as an answer to her prayers. How else can we explain an Egyptian princess adopting a slave child and planning to make him the next Pharaoh?

But when Moses was forty years of age he killed an Egyptian while defending one of his own people.[17] When Pharaoh heard of the incident Moses was forced to flee to the land of Midian, and when Amenemhet III died, Sebekneferu assumed the throne and ruled for four years. But when she died the dynasty ended and was succeeded by the Thirteenth Dynasty. 'Her reign, occasioned presumably by the absence of a male heir to the throne, marks the virtual end of a great epoch in Egyptian history.'[18]

From the historical records we also learn that Asiatic slaves were used during the Twelfth Dynasty. *The Cambridge Ancient History* notes that the Asiatic inhabitants of the country at this period must have been many times more numerous than has generally been supposed. It is not known whether or not this largely slave population could have played a part in hastening, or in paving the way for, the impending Hyksos domination of Egypt.[19] However, *Encyclopaedia Britannica* reports that Asian slaves, whether as merchandise or prisoners of war, were plentiful in wealthy Egyptian households.[20]

Another archaeologist, Sir Alan Gardiner, wrote that 'on stelae and in papyri Asiatic slaves are increasingly often mentioned, though there is no means of telling whether they were prisoners of war or had infiltrated into Egypt of their own accord.'[21] From the biblical records we can say that they did infiltrate into Egypt of their own accord, but were subsequently enslaved.

During the Twelfth Dynasty an extensive building program was carried on in the Delta where most of the Israelites were located. The temples of the Eighteenth Dynasty at Luxor were too far away from the Delta to have been built with Israelite slave labour, and they were built of stone. The buildings constructed in the Delta under the Twelfth Dynasty were made of mud brick. Mountains of such bricks went into the city of Avaris and nearby cities. Moreover, the pyramids of Sesostris III and

Amenemhet III were also made of mud bricks. The early dynasties' burial places were made of mud brick. The magnificent third and fourth dynasty pyramids were built of stone. For some strange reason, these Twelfth Dynasty rulers reverted to mud brick.

Josephus wrote that the Egyptians 'became very abusive to the Israelites, and contrived many ways of afflicting them . . . They set them to build pyramids.'[22] On the assumption that the oppression took place during the Eighteenth or Nineteenth Dynasty, this is regarded by scholars as a glaring blunder by Josephus, for by that time the pyramid age had ended. The Pharaohs of these dynasties were buried in the Valley of the Kings near Luxor. But maybe it is the scholars who have blundered, for the kings of the Twelfth Dynasty did build pyramids, and what is more, they built them of mud bricks laced with straw. 'Pharaoh commanded the taskmasters of the people and their officers, saying, "You shall no longer give the people straw to make brick as before. Let them go and gather straw for themselves".'[23]

Especially relevant is the research done by Rosalie David, Keeper of Egyptology at the Manchester Museum, whose book, *The Pyramid Builders of Ancient Egypt*, was published in 1986. David researched the work done by Sir Flinders Petrie in the Faiyyum in 1889 when he explored the pyramids of the Twelfth Dynasty and identified their owners. He also excavated the remains of a town that had been occupied by the workmen who actually built these pyramids. Petrie wrote: 'On the desert adjoining the north side of the temple, I saw traces of a town, brick walls, houses and pottery; moreover, the pottery was of a style as yet unknown to me.' He went on to describe how it dawned on him that this could hardly be other than the town of the pyramid builders, originally called Ha-Usertesen-hotep and now known as Kahun. A little further digging soon put it beyond doubt, and he found cylinders of that age and no other. He was convinced that he had actually discovered an unaltered town of the Twelfth Dynasty, regularly laid out by the royal architect for the workmen and stores required in building the pyramid and its temple.[24]

From the unidentified pottery and other evidence, Petrie concluded that the occupants had been foreigners. Expanding on this thought, David has an entire chapter entitled 'The foreign population at Kahun'. She wrote, 'From his excavations at Kahun, Petrie formed the opinion that a certain element of the population there had come from outside Egypt.' She went on to point out that it was apparent the foreigners were present in the town in some numbers, and this may reflect the situation elsewhere in Egypt at that time. She also observed that these people were loosely classed by Egyptians as 'Asiatics', although their exact homeland in Syria or Palestine could not be determined and the reason for their presence in Egypt was unclear.[25]

Neither Petrie nor David guessed that these Asiatics were the Israelites because they clung to the traditional chronology. But obviously the evidence fits the biblical records in a remarkable way. The book of Genesis tells how and why they got there, and what they were doing in Egypt:

> Then Jacob arose from Beersheba; and the sons of Israel carried their father Jacob, their little ones, and their wives, in the carts

which Pharaoh had sent to carry him. So they took their livestock and their goods, which they had acquired in the land of Canaan, and went to Egypt, Jacob and all his descendants with him.[26]

Evidence is not lacking to indicate that these Israelites, called by the Egyptians 'Asiatics', became slaves. David points out that a famous papyrus (the Brooklyn Papyrus) was left to the Brooklyn Museum. On the verso of this papyrus, a woman named Senebtisi attempts to establish her legal rights to the possession of ninety-five servants. A list of them is included which states their titles, names, surnames and occupations. Of the seventy-seven entries that are presented well enough to enable the individual's nationality to be read, twenty-nine appear to be Egyptians while forty-eight are Asiatics. David observes that although the foreign names were not precise enough to enable the exact homeland of these Asiatics to be identified, it can be said that they were from a 'Semitic group of the north west'. She continues:

> The Brooklyn Papyrus is important here because it shows that one household employed a large proportion of Asiatics and this household was situated in Upper Egypt (the south) and not in the Delta; therefore it is apparent that Asiatic servants were by now disseminated throughout the community.[27]

An intriguing aspect of Petrie's discoveries was the unusual number of infant burials beneath the floors of the houses at Kahun, a tragic reminder of the harsh edicts issued by the cruel tyrants of the oppression. Petrie found that beneath the brick floors of the rooms was the best place to search—not only for hidden things such as a statuette of a dancer or a pair of ivory castanets, but also for numerous burials of babies in wooden boxes. These boxes appeared to have been made for clothes and household use but were found to contain the remains of babies, sometimes buried two to three to a box, and aged only a few months at death. The remains were often accompanied by necklaces and other items. On the necklaces were sometimes cylinders with the kings' names, and Petrie concluded that these burials, and the habitation of the town, were of the Twelfth Dynasty, from Usertesen (Sesostris) II onward.[28]

David comments that internment of bodies at domestic sites was not an Egyptian custom. However, such practices did occur in other areas of the ancient Near East.

We have in the Bible what is probably a partial record of the efforts of the Pharaohs of the oppression to curb the growth of the Israelites:

> Then the king of Egypt spoke to the Hebrew midwives, of whom the name of the one was Shiphrah and the name of the other Puah; and he said, 'When you do the duties of a midwife for the Hebrew women, and see them on the birthstools, if it is a son, then you shall kill him.'[29]

These were probably only the midwives in the vicinity of the royal palace. Obviously a large population such as the Israelites then were, scattered all over Egypt, would require more than two midwives. These two midwives evaded their grim responsibility to Pharaoh by claiming that the Hebrew women gave birth before they arrived. But we do not know how many other midwives were obliged to carry out the edict.

Later, when Pharaoh found that these measures had not been effective, he ordered the Hebrews' Egyptian neighbours to see that the babies were killed: 'Pharaoh commanded all his people, saying, "Every [Hebrew] son who is born you shall cast into the river".'[30] Some parents managed to conceal their newborn babies for some time: 'when she [Moses' mother] saw that he was a beautiful child, she hid him three months.'[31] But many babies must have been drowned. Whether the parents retrieved the bodies, or whether some babies were put to death by other means, we do not know. There must have been many traumatic scenes as babies were torn from their mothers' arms by hostile neighbours. But this could account for the many infant burials at Kahun.

The son and successor of Sesostris III, Amenemhet III, was no less grim than his father. He built a mud brick pyramid at Hawarra in the Faiyyum and ruled for forty-eight years. He had two daughters, one of whom apparently died prematurely, and possibly one son, the shadowy figure Amenemhet IV. Some scholars place this son after Amenemhet III though he may have been a co-regent of Amenemhet III in his closing years. His fate seems uncertain, and Dr Donovan Courville of Loma Linda University, who wrote the two-volume book *The Exodus Problem and its Ramifications*, considers that he could have been Moses.[32]

Finally there is striking evidence pointing to the slaves' sudden departure. Up to the time of Khasekemre-Neferhotep I of the middle Thirteenth Dynasty there was evidence of continuous occupation of Kahun. Then it suddenly all stopped. David points out that there is every indication that Kahun continued to flourish throughout the Twelfth Dynasty and into the Thirteenth Dynasty. However, it is evident that the completion of the king's pyramid was not the reason why Kahun's inhabitants eventually deserted the town, abandoning their tools and other possessions in the shops and houses. While there are different opinions of how this first period of occupation at Kahun drew to a close, the quantity, range and type of articles of everyday use that were left behind in the houses may indeed suggest that the departure was very sudden and unplanned.[33] 'And it came to pass at the end of the four hundred and thirty years—on that very same day—it came to pass that all the armies of the LORD went out from the land of Egypt.'[34]

A period of instability followed the demise of the Twelfth Dynasty. Fourteen kings followed each other in rapid succession, the earlier ones probably ruling in the Delta before the Twelfth Dynasty ended. Khasekemre-Neferhotep I restored some stability, ruling for eleven years, and his was the last scarab to be found at Kahun before the city was deserted. He was the last king to rule before the Hyksos occupied Egypt 'without a battle', according to Manetho, the Egyptian priest who lived in the fourth century BC and who drew up a list of kings and dynasties and their length of reigns.

Without a battle? Where was the Egyptian army? It would have been at the bottom of the Red Sea.[35] Khasekemre-Neferhotep I was probably the Pharaoh of the Exodus. His mummy has never been found.

The plagues of Egypt

According to Exodus, Pharaoh at last yielded to Moses'

demands to allow his slaves to leave because of the ten devastating plagues that fell on Egypt.[36] The waters of the sacred River Nile were turned to blood, flocks and herds were destroyed, hail flattened the crops and wrecked the trees, lightning set combustible material on fire and the firstborn of every Egyptian family died.

The economy of Egypt must have been shattered and there should be some record of such a catastrophe. There is.

In the Leiden Museum in Holland is a papyrus written in a later period but which most scholars recognise as being a copy of a papyrus from an earlier dynasty. It could have been from the end of the Thirteenth Dynasty describing the conditions that prevailed after the plagues had struck. It reads:

> Nay, but the heart is violent. Plague stalks through the land and blood is everywhere . . . Nay, but the river is blood. Does a man drink from it? As a human he rejects it. He thirsts for water . . . Nay, but gates, columns and walls are consumed with fire . . . Nay, but men are few. He that lays his brother in the ground is everywhere . . . Nay, but the son of the high-born man is no longer to be recognised . . . The stranger people from outside are come into Egypt . . . Nay, but corn has perished everywhere. People are stripped of clothing, perfume and oil. Everyone says, 'There is no more.' The storehouse is bare . . . It has come to this. The king has been taken away by poor men.[37]

The invasion of Palestine

Under the revised chronology, the invasion of Palestine in 1405 BC occurred not during the Late Bronze Period, as currently believed, but at the end of the Early Bronze Period. With this placement we find the archaeological evidence that is so lacking during the Late Bronze era. In Jericho Professor John Garstang found walls that had toppled outwards, layers of ash and charred wheat and dates, all of which he dated to the middle of the Late Bronze Period, and therefore (in his thinking) to the Israelite invasion. Later, when Kathleen Kenyon re-excavated Jericho, she was obliged to redate Garstang's walls and ash to the end of the Early Bronze Age. By the traditional chronology this made it much too early to be attributable to the Israelites. But with the revised chronology it once more becomes the obvious evidence it is, and visitors to Jericho today can still see a one-metre layer of pink ash from this period.

And it is not only at Jericho that the evidence is found. In her book *Archaeology in the Holy Land*, Kenyon reports that the final end of the Early Bronze Age civilisation came with 'catastrophic completeness'. She describes how the last of the Early Bronze Age walls of Jericho appears to have been built very quickly, using old broken bricks, and was probably not completed when it was destroyed by fire. Little or none of the town inside the walls has survived. Excavations show that there was an absolute break in the archaeological layers, and that a new people took the place of the earlier inhabitants. Kenyon found that every town in Palestine that has so far been investigated shows the same break. She wrote: 'The newcomers were nomads, not interested in town life, and they so completely drove out or absorbed the old population . . . that all traces of the Early Bronze Age civilization disappeared.'[38]

An absolute break; a new people; every town in Palestine; nomads who completely drove out or absorbed the old population. Could we find a more apt description of the Israelite invasion: nomads from the desert who initially were not interested in living in the cities?

James Pritchard, who excavated in Gibeon in 1956, found the same types of evidence. He describes how the relics of Middle Bronze Age I people (immediately after the end of the Early Bronze era) seem to indicate a fresh migration into the town of a nomadic people who brought with them an entirely new tradition in pottery forms and new customs in burial practices. Pritchard suggests that these nomads may have come into Palestine from the desert at the crossing of the Jordan near Jericho and may then have pushed on to settle eventually at places such as Gibeon, Tell el-Ajjul and Lachish, where tombs of this distinctive type have been found.[39] Nothing could more aptly fit the biblical record of the Israelites coming in from their desert wanderings, crossing the Jordan at Jericho and occupying the Promised Land.

Strange to say, Pritchard uses these very facts to try and prove the unreliability of the book of Joshua. When the Gibeonites defected to Joshua, the other five prominent cities of Palestine were apprehensive: 'they feared greatly, because Gibeon was a great city, like one of the royal cities . . . and all its men were mighty.'[40] Pritchard did indeed find that Gibeon was a great city, but *not* during the Late Bronze Period when he thought that the Israelite invasion took place.[41] Extraordinary! He found the biblical Gibeonites but did not recognise them.

Until recently Dr Rudolph Cohen was the Deputy Director of the Israel Antiquities Authority. He spent twenty-five years excavating in the Negev (southern Israel) including Kadesh Barnea, where the Israelites spent a lot of time.[42] That was where Moses sent out twelve spies to reconnoitre Canaan. We should expect the Israelites to have left a lot of broken pottery in the vicinity and Dr Cohen identifies this as the Middle Bronze Age I pottery found there.[43] He claims that he can trace the route the Israelites took from the pottery they left behind—but this is at the end of the Early Bronze period, not during the Late Bronze period. Cohen wrote: 'The similarity between the course of the Middle Bronze I migration and the route of the Exodus seems too close to be coincidental.'

So there is plenty of evidence for Israelite slavery in Egypt, the sudden disappearance of these slaves, the devastation of Egypt by the ten plagues, the destruction of the Egyptian army, and the invasion of Palestine by a nomadic people replacing the preceding population—if we look for it at the right time.

18

The Historical Reliability of the Old Testament

Paul Ferguson

Long ago the French archaeologist Andre Parrot discovered thousands of tablets at Mari, an important capital city located in the middle Euphrates area. Parrot was very optimistic about witnesses to culture from the time of the Hebrew patriarchs found in these tablets. He suggested that it would not be too much of a surprise if a letter were found from Abraham's father, Terah, requesting permission to cross a boundary and settle in Haran.[1] Such detailed information, however, is yet to be found.

Nevertheless, if someone discovered Terah's post office box in Haran with a clay tablet from his son Abraham mentioning he had just gotten a promise from God, that would not prove that God talks to people. Finding a personal relationship with Abraham's God in, through and according to the Scripture would validate the existence of a personal Deity. Finding out that promises in the Bible are realised in our lives by faith would demonstrate its full reliability on spiritual matters.

Yet it has been found that in areas that can be checked by

archaeology, the Bible proves to be accurate. Whenever all the normal historical facts are known and properly understood about a biblical account, it has been found to be correct and trustworthy. Therefore we can be sure of what the Bible says about unseen, spiritual realities only discerned with the 'eyes of our heart'.[2]

Space does not permit a detailed analysis of all the historical phenomenon in the Old Testament. We have selected some test cases from the historical books and the prophet Jonah to illustrate this claim. We start with Jabin, king of the Canaanite city of Hazor.

King Jabin of Hazor

The city of Hazor is about ten miles north of the Sea of Galilee. It was one of the largest cities in Israel. It covered 200 acres and may have had a population of about 30,000.[3] It was located at a major junction on the international trade route and rivalled the greatest cities of the Near East in size and commercial importance. In the Mari tablets, Hazor is referred to as a centre for caravans in the metal trade that travelled from there to Babylon.

The king of Assyria gives instructions to the king of Mari to host diplomats from Hazor. Another letter mentions the hosting of Babylonian officials at this city. These letters speak of considerable consignments of tin and other valuable items connected with Hazor. The city is mentioned in a Babylonian dream book as the termination point of the author's journey to the west. It is the only city in Palestine so clearly referred to in Mesopotamian archives.[4] The Mari material comes from around the mid-eighteenth century BC. There are fourteen documents from Mari attesting ties with Hazor—not, as Abraham Malamat says, an inconsiderable number.[5]

According to the Amarna tablets from fourteenth-century BC Egypt, the king of Hazor was accused of rebelling against Pharaoh by the king of Tyre. He had supposedly joined the enemy and was taking over Egyptian territory. The king of Hazor countered that he was only holding these captured cities for the Pharaoh. In these letters he is the only monarch of the many city-states in Canaan who uses the Babylonian word for 'king' (*sharru*) of himself and is referred to as king by his peers.[6] Hypercritic Goesta Ahlstrom notes, 'The impression one gets [from this] is that Hazor is a powerful and aggressive north-Palestinian state.'[7]

The Bible perfectly describes this situation by stating that 'before that time Hazor was the head of all those kingdoms'.[8] It mentions King Jabin of Hazor as heading up the northern coalition against Joshua.[9] The book of Judges mentions a Jabin as being the 'king of Canaan'.[10]

Jabin, king of Hazor, is repeatedly found in archaeological records

The name of the king of Hazor is mentioned at Mari as Ibni-Addu. Malamat states that in the west this would have been pronounced Yabni-Addu.[11] It means either 'son of Hadad', the storm god, or 'Hadad has built [the dynasty]'. 'Yabin', from an older Hebrew word 'Yabnu', is the form found in the Hebrew Bible. In the early 1990s this name was found in a clay tablet from the city of Hazor itself. It was published by Wayne

Horowitz, my former teacher at Hebrew University.[12] About eight different clay tablets written in the Babylonian language have been found at Hazor. At no other city in Palestine have this many tablets been found.

Some critics have wondered, if Jabin was killed by Joshua, how he ends up threatening Deborah and Barak in the book of Judges.[13] The current chief excavator at Hazor, Amnon Ben-Tor, explains this by the fact that 'Jabin' was a dynastic name used over and over in this city.[14] This is confirmed by the fact that the name Jabin was found in eighteenth-century Mesopotamian records and in a fifteenth-century BC Egyptian listing of Canaanite officials.[15]

Archaeological evidence coincides with the biblical description of Hazor's destruction

Joshua states that Hazor alone among the northern cities was burned with fire. It states further that this destruction was done exactly and completely as Moses had commanded.[16] This command was stated as follows:

> But this is how you must deal with them: break down their altars, smash their pillars, hew down their sacred poles, and burn their idols with fire.[17]

Exodus 23:24 adds that they must 'utterly demolish them [the gods of Canaan] and break their pillars in pieces'.

Obedience to this command was especially noticeable at Hazor since the largest temple and idol in the land of Canaan were found here. According to Ben-Tor, this statue was three foot tall and must have come from a block of basalt weighing more than a ton. It had been smashed into nearly a hundred pieces, which were scattered in a six-foot-wide circle. The head and hands of the statue, and of several others, were missing, apparently cut off by the invaders.[18] Ben-Tor describes a thick layer of ash and charred wood—in places three feet deep—found on the mound from the time of Joshua's invasion. Jars of oil created an inferno estimated to have been 2350°F. This intense heat turned mud bricks to glass, cracked basalt slabs and melted clay vessels.

The excavator notes that the largest numbers of Egyptian statues ever found at one site in Israel were discovered here. None were intact. Three had chisel marks showing that someone had deliberately chopped off their arms and heads. Some statues were of Egyptian Pharaohs (who, by the way, claimed to be gods).[19] Ben-Tor's article shows pictures of these broken statues fully blackened by carbon from the burning.[20]

Pottery found among the debris dates the destruction to the fourteenth or thirteenth century BC, the time of Joshua. The only people at this time who could have destroyed Hazor were other Canaanites, Egyptians or Israelites. Canaanites would not have obliterated idols of their own gods. Egyptians would not have defaced statues of their own Pharaohs. This leaves only Joshua and the Israelites as the invaders. These statues are, so to speak, Joshua's calling card.

No rebuilding—and no pigs

Another startling fact is that after the temple at Hazor was destroyed, nothing was built on this site for over three hundred years. Ben-Tor observes that this is especially surprising since city residents did not normally leave prime real estate unoccupied. Ahlstrom states:

> From what we have seen at Hazor ... and at other places ... when temples were destroyed they were rebuilt on the same spot ... recognition of a site's holiness ... the realm of the god(s) had to be rebuilt on the same place.[21]

Yet in Hazor, and indeed all over Palestine, the populations after the destruction of various cities were replaced by a semi-nomadic culture. They had very crude pottery and did not have temples or idols. They lived in unfortified and crude settlements with poorly constructed houses. These people are seen as culturally less advanced than the previous population.

At this time we also encounter a mysterious disappearance of pigs. At Ashkelon, Ekron and Timna, three Philistine sites of this period, between eight and 18 per cent of the excavated bones belonged to pigs. At Ebal and Raddana, two Israelite sites, the percentage of pig bones was zero, and at Shiloh, 0.1 per cent (someone sneaking a ham sandwich?). Pigs disappeared from the faunal assemblages of Canaan's hill country. Obvious conclusion: Israelites first settled in the central highlands in Iron Age I (1200 to 1000 BC). Pigs were widely cultivated by the Canaanites yet virtually disappeared when a new population, the Israelites, appeared in the period of the Judges—another calling card![22]

The language of Joshua 11
Yet another piece of archaeological evidence further confirms the accuracy of the biblical account of Jabin. Scholars have found that the language and style of Joshua 11 matches the way

governmental documents in the second millennium BC recorded their victories.

Egyptologist James Hoffmeier summarises research done on military records of the Egyptians, Assyrians and Hittites from the time of Joshua. This has shown that 'Near Eastern scribes used similar, if not identical, theological perspectives and literary conventions in their military writing'.[23] While there are parallels in military writing in the first millennium BC, Hoffmeier concludes those in Joshua's time are more compelling. Elsewhere he concludes that Hebrew scribes borrowed the style of the Egyptian daybook tradition. He notes that this style especially fits the way military campaigns were recorded in the annals of Thutmose III, who lived in a time period close to the Joshua campaigns (c. 1450 BC).[24]

All of this adds up to an extremely compelling case for the Bible's account of Jabin. Norman Gottwald is a critic who has completely denied the conquest model of the occupation of Canaan by the Israelites; he believes they were there all the time as lower class Canaanites who eventually revolted against elite aristocrats. Yet in commenting on the people who occupied Hazor and other sites after their destruction, he states that 'it is tempting to see them as the technically impoverished seminomadic Israelites'.[25] He continues, 'The evidence from Hazor suits Joshua 11 well.'

The evidence found at Hazar is so strong that it actually tempts critics to believe the Bible![26]

The gates of Solomon

Israeli archaeologist Yigael Yadin excavated Hazor in the 1950s and '60s. He discovered there a six-chambered gate that was built exactly the same way as other gates at Megiddo and Gezer. Each had three guardrooms on each side with two towers flanking the passageway.

Yadin claimed that both the stratigraphy and pottery assigned these gates to the time of Solomon.[27] 1 Kings 9:15 states that Solomon fortified these three cities. Yadin held that the gates' identical architecture is a 'bold illustration of a centralised, royal building operation attributable to Solomon on archaeological grounds'.[28]

An early excavator at Gezer had dismissed the gate there as a Maccabean castle. Therefore he did not completely uncover the gate. Yadin studied the site maps and predicted exactly what would be unearthed—a gate like the ones at Megiddo and Hazor. On the basis of following Solomon's unifying building scheme set down in the Bible, Yadin knew exactly what would be uncovered before the digging took place. His ability to predict what would be found mystified many of the Arab workers.[29]

According to Israeli archaeologist Gabriel Barkay, the Solomonic gates at these three cities 'have become ornaments of biblical archaeology and prime examples of the correlation between the biblical narrative and archaeological record'.[30] He notes that despite some differences in building styles, the gates can still be attributed to the architects of Solomon. Negev and Gibson have concluded they are nearly the same.[31] Graham Davies of Cambridge University notes the identical plan and similar measurements of the three gates makes them all Solomonic.[32] Carol Meyers believes the gates at the three cities are close enough in size and proportion to indicate a common architectural plan, with differences being accounted for by variations for local conditions.[33] It is important to note that none of these critics are conservatives.

Davies believes the building program mentioned in 1 Kings 9:15 suggests a definite plan on Solomon's part to ensure the security of his kingdom.[34] Hazor is in the north. Gezer is in the south near the coast. Megiddo is at the top of the north central Jezreel valley. The similarities of the gates indicate a central administration of Israel in the time of Solomon.

Desperate down dating

Tel Aviv University archaeologist Israel Finkelstein has tried to 'down date' the entire chronological system in order to put these gates into the time of the later Israelite king Ahab.[35] Ben-Tor vigorously denies this, as do most other Israeli archaeologists.[36] Even Finkelstein's co-director at Megiddo has written a book refuting his colleague's down dating of the Solomonic gate there.[37]

In an article on Megiddo, Yigael Shiloh noted that each of the cities of Gezer, Megiddo and Hazor 'exhibits identical architectural and archaeological elements, characteristic of the major centres in Solomon's kingdom'. He connected the building of these three centres with the building of Solomon mentioned in 1 Kings 9.[38]

The mound of Megiddo is located on a strategically important trade route called the Via Maris ('the sea road'). It protects the northern top of the valley of Jezreel, Israel's 'bread basket'. The mound rises a hundred feet above the valley and contains the layers of over twenty ancient cities. Many decisive battles have been fought there. An Egyptian Pharaoh who captured it right before the Exodus said, 'The capturing of Megiddo is as the capturing of a thousand towns.' The book of Revelation declares it to be the sight of the final end-time battle.[39]

From the twentieth century BC through to the twelfth century BC, Megiddo was a Canaanite city. In both Joshua and Judges it is listed as one of the cities the tribe of Manasseh could not take.[40] It is not mentioned as an Israelite possession until 1 Kings identifies it as one of the main administrative centres in Solomon's empire.[41]

Archaeology bears out this picture. In the time of the Israelite judges (eleventh century BC), there was a flourishing, wealthy, densely built city. Pottery and cultural items show abundant Canaanite and Philistine influence. This is called city VI. At the time of David it was destroyed in a great conflagration.[42] There followed a poor, small village until the time of Solomon. Then massive fortifications sprang up. Two elaborate, monumental palaces were built at this time along with storage facilities and other buildings of a public nature. This fits the biblical descriptions perfectly.[43]

Finkelstein's theory, however, disputes all this. His down dating results in removing the monumental city from Solomon and giving it to Ahab or someone later. Since Finkelstein is one of the three dig directors at Megiddo, the signs on the site now reflect this view. They state that the six-chambered gate may be from Solomon but note other views dating it much later. Tour groups are doubtlessly puzzled by these signs.

This has been partly due to the fact that the records about

city VI up till now have been very sparse. In the summer of 1988, University of Chicago archaeologist Doug Esse was getting ready for a seminar on Megiddo. He discovered that the records of this city had not been completely published by the University of Chicago team that excavated it during the 1930s. He set about publishing them. When he died in 1992, his doctoral student Tim Harrison took up the project. At about that time I took a course on the archaeology of Megiddo from Harrison at the University of Chicago. Although this was a very liberal setting, no credence was given to the down dating views of city VI during this class.

Harrison's work is about to be released. At the end of 2003 he wrote a summary of it in the *Biblical Archaeology Review*.[44] He notes that during 1200 to 1000 BC, the period of the judges, the hill country of Canaan experienced an explosion of small sedentary settlements (250 to 300 of them). According to the Bible, this was the territory first occupied by Israelites.[45]

In contrast, city VI contains a diverse mix of imported pottery traditions—Cypriot, Phoenician and Philistine. Large buildings were discovered, one with a Philistine seal in it. Some of this building style resembles Philistine building at Ekron in the eleventh century. Harrison notes a decided Philistine presence along with Canaanites.[46] The Bible suggests Canaanites joined with Philistines in fighting King Saul. Philistines came down this valley for the final battle with Saul.

City VI, according to Harrison, was a remarkably heterogeneous community made up of people from widely varying social and cultural backgrounds. Any Israelite traces in this city would be explained by the fact that the Bible notes that a certain element in Israel joined with Philistines and Canaanites at this time.[47] It would not prove city VI was an Israelite city to be dated to Solomon's time.

The city following the terrible destruction of city VI was Israelite in culture. The pottery styles coincide with tenth-century BC sites nearby from the time of David. Harrison concludes that comparative stratigraphy, pottery and radiocarbon data point decisively to the tenth century. Therefore a down dating of this city to Ahab's time can no longer be seriously considered.[48]

Davidic bombshell

For years critics claimed that the absence of royal inscriptions and archives with David's name showed he either never existed or was a minor tribal chieftain. This assumption ignored the fact that in neighbouring cities of this time royal inscriptions were almost nonexistent. In the immediate area there are only three engravings mentioning kings in a town called Gebal (Byblos).[49] Even royal inscriptions in Assyria are mostly of local matters and do not deal with kings.

If Davidic inscriptions ever existed, they were defaced and destroyed by queen Athaliah (841–835 BC), who tried to wipe out the Davidic dynasty.[50] Any that did survive would now be under the Temple Mount in Jerusalem, where it is impossible to excavate due to the Islamic structures there. More likely, kings in Jerusalem did not put their name on inscriptions. In the time of Hezekiah (c. 700 BC), the famous Siloam inscription includes only the work crews for the massive tunnel. No civil leader is even mentioned.

No archival records will ever be found. Speaking of a similar lack of records for the Phoenicians, H.L. Katzenstein remarks that 'the scarcity of inscriptions cannot be used as evidence that the Phoenicians did not write; rather it is evidence that they wrote on perishable material'.[51] An Egyptian papyrus from a little before David's time describes a Phoenician king bringing out 'journal rolls' to check a commercial transaction.[52] Papyrus rolls in the Jerusalem archives would not have survived the burning of the city by the Babylonians in 586 BC. Absence of evidence is not evidence of absence. If that were true, then the lack of royal archives for the Phoenicians would prove that the people who invented writing could not write!

Nonetheless, in the past many negative critics have relegated David to the level of the mythical King Arthur. If he existed at all, they argued, he probably supervised a small clan on a single hilltop in Judah.

But then in 1993 a bomb was suddenly dropped. In Northern Tel Dan, where Jeroboam put one of his golden calves, an inscription with David's name on it was discovered. It was found in broad daylight by the excavating team with witnesses all around who could validate where it came from.

The inscription was buried in an area with ninth-century BC pottery. It had the type of letters and language a ninth-century Aramaic inscription should have. Evidence was that it had been there for centuries. A forger would have to know how to write ninth-century Aramaic and would have to know ninth-century pottery was in this area without ever having dug there.[53]

The monument was written by King Hazael of Aram (modern Syria), who invaded Israel in about 842 BC after Jehu's revolution.[54] The inscription celebrates the defeat of the kings of Judah and Israel. It refers to the northern monarch as the 'king of Israel', but the king of Judah is referred to as 'the house of David'. This shows that the Aramaean king was still smarting over his people's stinging, humiliating defeat under David and his general Joab nearly two hundred years before.[55]

This royal inscription comes from less than a century after the death of Solomon in 932 BC. Even 140 years after David's death, the Aramaeans still remembered him as the founder of the dynasty. Something very strong obviously made this indelible impression. This is a silent witness to the expansion of the Davidic empire into the north. David is suddenly elevated from an obscure clan chieftain to an international figure.

Space does not permit a full discussion of the debate on the 'house of David' inscription. It was claimed that this was a place name and not a personal name because there was no 'dot' word divider between 'house' and 'David'. Inscription specialist Gary Rendsburg has shown that this was not valid by comparing known styles of writing for this period. He noted that both personal names and place names often lacked this dot. Moreover, he has shown that it was characteristic of the Aramaeans to write 'house of' followed by the person who founded the house. 'House of Hazael' is an example.[56]

Efforts were also made to show that the name 'David' might be translated 'kettle', 'uncle' or 'beloved'. All of these suggestions require a vowel letter not present in Aramaic at this time.[57] In short, every early effort to avoid the import of the 'house of David' name have been thoroughly dismissed by inscription specialists.[58]

Supporting this is the work of eminent French epigrapher Andre Lemaire. Lemaire has taken a fresh look at the Moabite Stone.[59] This is one of the longest inscriptions written in a language related to Hebrew. It comes from the land of Moab, which was on the east side of the Jordan by the Dead Sea. Moab was a vassal of King Ahab and the northern kingdom of Israel. About the time of the Tel Dan inscription in the ninth century BC, Moab's King Mesha revolted and set up this inscription on a basalt rock to celebrate the nation's independence.

As in the Tel Dan inscription, the northern monarch, Ahab's son, is called the king of Israel. But Lemaire found the name of another enemy King Mesha of Moab was fighting. In line 31 he discovered the name 'house of David'. This too would have been about 140 years after David's death. Thus like King Hazael of Aram, the king of Moab remembered David's stinging defeat of his country.[60] This is another silent witness to the existence of David's empire.

Repentance of the Ninevites

About a century ago British liberal Old Testament scholar S.R. Driver complained that it was not easy to imagine 'a monarch of the type depicted in the inscriptions' as acting the way the king of Nineveh acted in Jonah 3.[61] He could not picture a great Assyrian king getting up from his throne, removing his royal robes and sitting in the dust. Liberal scholars were also 'rumpled' over the fact that the directions for this extreme repentance seem to come from nobles rather than the king himself. It was claimed this kind of administrative practice did not take place till the Persian period.

Hypercritic Andre Feuillet held that Jonah was an unhistorical novella composed like a papier mâché collage from phrases selected here and there from the rest of the Old Testament. He stated that he considered 'the sudden conversion of an immense and powerful pagan city to the monotheism of the small people of Israel' to be a much greater improbability than the episode about the big fish swallowing the prophet.[62]

No one, of course, could do any archaeology on the 'fish-tale' (even though there is a rumour that Muslims in Syria have the whale's skeleton on display in a mosque!). However, thanks to modern advancements in archaeology, the repentance story can be checked. It will be seen that the kind of behaviour of the Assyrian king that Jonah describes not only *could* take place but *did* take place in many different situations. In fact, in the light of the historical situation in Jonah's time (c. 763 BC), the biblical account may seem, if anything, understated. It will also be seen that directives about religious observances came, not from the king first, but from nobles who formed his omen department.[63]

Jonah's account of repentance in Nineveh

The book of Jonah tells the story of a Hebrew prophet sent by God in the eighth century BC to the Assyrian city of Nineveh, to 'cry out against it; for their wickedness has come up before me [God]'.[64] After his initial refusal (resulting in his run-in with the giant fish), Jonah goes and preaches: 'Forty days more, and Nineveh shall be overthrown!'[65] Here is how the Bible describes the response of the city's inhabitants:

And the people of Nineveh believed God; they proclaimed a fast, and everyone, great and small, put on sackcloth.

When the news reached the king of Nineveh, he rose from his throne, removed his robe, covered himself with sackcloth, and sat in ashes. Then he had a proclamation made in Nineveh: 'By the decree of the king and his nobles: No human being or animal, no herd or flock, shall taste anything. They shall not feed, nor shall they drink water. Human beings and animals shall be covered with sackcloth, and they shall cry mightily to God. All shall turn from their evil ways and from the violence that is in their hands.'[66]

It is important to note that the Ninevites do not become Jews keeping Sabbath and food laws. They do not accept Israel's covenant view of God and call him Yahweh. They pray, show humility, fast and change their ethical behaviour. They do not accept the Law of Moses or refer to any part of the Hebrew Bible.

Several questions have been raised about this behaviour of a great Assyrian monarch, each of them casting doubt on the veracity of the Bible's account. But in the 1990s light was thrown on this problem when Assyrian astrological reports and royal letters were published by the University of Helsinki. These records contain about 250 letters to the king from his omen experts. They show that the king did not interpret the signs of the times himself but relied on a large staff of specialists. They even told him when his son could go outside. From these examples it is easy to see that the author of Jonah had some 'inside information' about the religious practice of Assyrian kings.

From this evidence we can address several key questions.

1. *Would a powerful Assyrian monarch fast enough to worry his advisers?* It is clear from the records that the answer to this is 'yes'. The following quotation is taken from a letter entitled 'The king must give up fasting':

> The king, our lord, will pardon us. Is one day not enough for the king to mope and to eat nothing? . . . This is the third day (when) the king does not eat anything. The king is a beggar![67]

The writers of this letter note further that they are writing because they are worried and afraid the king is fasting too much.

According to Simo Parpola, the editor of the collection, a group of ancient scholars was attached to the Assyrian king as 'his spiritual guardians and advisers'. The king himself could not interpret divine signs.[68] Occasionally a king even protested the severity of conditions imposed by these experts. The mighty Ashurbanipal was required to fast from New Year's Day until the new moon. Impatiently he pleaded, 'Release me. Have I not waited (long enough?) . . . I want to eat. I want to drink wine.' The reply was to be patient; he would have to finish the fasting period whether he liked it or not.[69]

2. *Would such a powerful king take off his royal robes and wear plain clothes?* In one instance, omen experts had declared a time of mourning and instructed the king to wear special clothes in a previous letter. The king had then asked them, 'How many days should I wear them (that is, the special clothes)?' Their answer: 'The king should wear them on the 20th and the 21st . . . On the 22nd he can gird himself . . . then resume his normal activities.' These were days of fasting.[70]

In response to the king's question about this ritual, they replied: 'The king does not eat anything cooked; the king wears the clothes of a nurse.' After these directions are followed he is to go down to the river and perform further rituals.[71]

In both Babylonia and Assyria the same New Year rites were observed.[72] On the fifth day of this festival, every year, the king took off all his royal regalia and made vows to the deity after which his kingship was renewed.[73] Thereafter the destiny of the kingdom was set as the last act of the festival.

Thus the humiliating divesting of royal insignia was a very regular practice of Assyrian kings.[74] Every year the Assyrians did repent. Every year the king did get up off his throne, lay aside his regal insignia and undergo deep humiliation. The people hoped with a suspenseful 'perhaps' that the great god would forget about the violence and evil of the preceding year and not destroy them. In this New Year festival the people were confronted annually with the prospect of destruction.

3. *Would an Assyrian king leave his throne and palace to sit on the floor?* Again, the answer is 'yes'. A letter states:

> He (the king) sits seven days in a reed hut . . . He is treated as a sick person . . . hand-lifting prayers . . . are performed.[75]

In reply to the king's question about this he receives the answer: 'Your majesty should, as he usually does, act exactly as directed.' That is, spend seven days on the floor in a reed hut in prayer.[76]

After Henri Frankfort comments on the requirement of the king to sit in the reed hut, he states that in Assyrian times: 'The king consented when necessary to forego his role as steersman of state to become its rudder, manipulated by those who watch the portents in order that all might remain afloat upon the waves of the unknown.'[77] In times of extreme emergency, the king completely left his throne and a substitute king was installed. This leaving the throne and bringing up a substitute king was regularly performed during eclipses. It was a time of solemn fasting when the substitute king reigned until the danger passed.[78]

The eclipse of 763 BC

According to 2 Kings, Jonah prophesied during the reign of Jeroboam II (793–753 BC).[79] It is a matter of record that an eclipse of the sun occurred in Assyria on June 15, 763 BC. This is an anchor point in Assyrian dating. It is the earliest date in Assyrian history astronomically fixed.[80] According to A.T. Olmstead, a University of Chicago specialist in Assyrian history, this eclipse was the last manifestation needed by the Assyrians to prove the necessity of a change in dynasty. A revolt was raised in the city of Ashur. Other revolts occurred in the next four years in the provinces.[81]

In 1994 Alan Millard published the events that gave names to the various years in Assyria around the time of Jonah. Notice the central events two years before and four years after the eclipse:

765 BC	plague
764 BC	in the land (that is, no military expeditions)
763 BC	*revolt in the citadel, eclipse of the sun*
762 BC	revolt in the citadel
761 BC	revolt in Arrapha

760 BC   revolt in Arrapha
759 BC   revolt in Guzana, plague[82]

Assyrian chronicles usually named the year after some military campaign of the king, but during the years before and after the eclipse, years were noted by plagues, famine and earthquakes.

This 763 BC eclipse may be relevant to the question of Jonah. An Assyrian astrological report states that in certain kinds of eclipses, famine and locust plagues may occur. Supposed results of a particular kind of eclipse are described as follows:

> Rise of a rebel king; the throne will change ... there will be a rebellion ... a son will kill his father ... that king will die ... a devastating flood will occur.[83]

Eclipses of the moon might mean floods, the death of the king, the destruction of all lands, famine, the king's son seizing the throne, hostilities and the fall of the army.[84] D.J. Wiseman makes reference to a tablet from Guzanu during the reign of Ashurdan III (773–755 BC) that mentions an eclipse of the sun followed by flooding and famine.[85] Olmstead mentions that the bubonic plague raged so strongly during this period that military campaigns into Syria had to be abandoned.[86]

If Jonah had arrived in Nineveh around the time of this eclipse, when Assyria was about to fold up and collapse, it would not have taken much of a catalyst to start the kind of mass repentance described in Jonah 3. The behaviour of the Assyrian king described in the royal letters took place on certain days specified by experts during times when life was progressing as usual. How much more intense would it be in the New Year following this ominous event? One might even wonder if the book of Jonah understates the situation.

What about the common people?

The king of Assyria might have been capable of acting as Jonah describes, but would the people have acted like the men of Nineveh in Jonah 3? James Pritchard's *Ancient Near Eastern Texts* includes a prayer to 'Any God'. It comes from Ashurbanipal's library at Nineveh but was copied from an older original. It is thought to be much older than the seventh century BC. It states, 'May the god whom I know or do not know be quieted toward me.'[87]

Carl Keller footnotes two 100-day fasts in Assyria. One took place right before the time of Jonah in the days of King Adad-nirari III (810–783 BC).[88] Wiseman records a decree of an Assyrian king to a governor similar to Jonah 3 that states:

> Decree of the king. You and all the people of your land, your meadows, will mourn and pray for three days before the god Adad and repent. You will perform purification rites so there may be rest.[89]

This decree was to the governor of Gozan (c. 793 BC) and comes from the general period of the book of Jonah. Wiseman cites another decree which commanded, 'This mourning in the month of Siwan concerns all the people of the land.'

Alexander Heidel quotes a description of an Assyrian prince who had a dream about the netherworld. His reaction is as follows:

He uttered a loud lamentation, crying, 'Woe is my heart!' Like an arrow he flew into the street and scooped up into his mouth the dust on the street ... he continued to call with a loud voice, 'Woe! Alas!'[90]

Conclusion

The miracle God works in turning the hearts of men to repentance is far greater that his manipulation of a great but practically brainless fish. It has been shown that the Assyrian king and his people were fully capable of acting the way they are described in Jonah 3. S.R. Driver, one of the most capable Hebrew scholars who has ever lived, may have had difficulty understanding this, but now in the light of newly available Assyrian records it is difficult to see them acting in any other way. After seeing how archaeology has so greatly increased our knowledge of Assyrian repentances, it should be very easy for us to also accept the role of the great fish in this drama.

Wiseman has observed that many features in the narrative in Jonah 3 'exhibit an intimate and accurate knowledge of Assyria which could stem from a historical event' in the eighth century BC.[91] Perhaps when Jonah said, 'Forty more days and Nineveh will be destroyed', the unbelief might have been that the destruction was that far away. In view of all the famines, earthquakes, plagues, invasions, economic collapse cited by Wiseman, someone may have cried out, 'Forty days? Do we really have *that* much time?'

The reliability of archaeology versus reliability of the Bible

It is fashionable today in some circles to claim that 'the Bible, as history, flunks the new archaeological tests'.[92] The usual 'archaeological test' is 'we haven't found it so it must never have existed'.[93] Thus what people find or don't find in the ground becomes an ultimate authority by which the Bible is judged.

But is this really a proper, viable methodology?

Paul Bahn describes archaeology as a 'vast fiendish jigsaw puzzle invented by the devil as an instrument of tantalizing torture'.[94] He lists three reasons for this statement: (1) it will never be finished, (2) you don't know how many pieces are missing, and (3) most of these pieces are lost forever. In short, all the facts aren't in.

German archaeologist Manfred Bietak excavated for thirty-five double seasons on the ancient Hyksos capital in Egypt. This is equivalent to seventy years of excavation. Egyptologist James Hoffmeier asked Bietak what percentage of the material from the site he had uncovered. Bietak replied, 'About three percent'.[95]

A case of how this 'methodology' sometimes works is with the city of Jericho. Finkelstein and Silberman claim that because no early Bronze Age walls have been found there, the story in Joshua of walls coming down at the trumpet is 'simply a romantic mirage'.[96] In an article describing Egyptian Pharaoh Tuthmosis III's account of his conquest of Megiddo, Finkelstein also observes that no walls and no Egyptian camp from this period have been found there. But Finkelstein overdoes himself trying to defend the historical accuracy of this Pharaoh. He suggests earlier walls were reused. A belt of houses protected Megiddo (like Rahab's house at Jericho?). There weren't any fortifications. The Egyptians simply wanted to wait outside for the city to surrender. Finkelstein suggests that to solve the problem more archaeology needs to be done there![97] These are exactly the protests evangelicals have been making about Jericho. Their objections, of course, have been dismissed as special pleading for a weak or impossible premise.

An example of how something found very recently has shed light on the historical reliability of the Bible is at Beersheba. The book of Joshua seems to suggest there were two settlements in this area called Beersheba.[98] Does this sound like a desperate evangelical ploy to bolster an impossible position? The only city anyone knew of named Beersheba up till 2003 was seven miles away from the modern city. In the summer of 2003 workers were laying a new drainage ditch inside the modern city. Peter Fabian of the Israeli antiquity authority did a salvage operation there and found another settlement from biblical times. Fabian inferred this was the second settlement mentioned in Joshua.[99]

An Israeli archeologist at the University of Haifa has this to say about the subject:

> Although archaeology uses some modern technologies, many of its conclusions are drawn on the basis of intuition, rather than on objective measure. The quality of excavation, surveying and publication of results is very uneven . . . It is easy to be overwhelmed by the number of hypotheses, suppositions and presuppositions, supported or not by a mass of data.[100]

This chapter has shown some examples that support the fact that where all the facts are known, the Bible proves to be historically accurate. This does not mean that every archeological excavation will automatically 'prove' the Bible is true. In most cases, very few of the facts are in. As liberal archaeologist William Dever put it, 'Archaeology is not a science any more than history is a science'.[101] Sometimes archaeology resembles what Will Rogers said about the weather in Oklahoma: 'If you don't like it, wait awhile.' It certainly is not an omniscient, immutable crystal ball that can be used to convict the Bible of error and high crimes in the modern court of negative scholarship.

19

Archaeology and the Reliability of the New Testament

John McRay

We have more than the natural world to support our faith and enlighten us about the presence and purpose of God in the universe, and specifically in our lives. We have the Scriptures telling of God's preparation for the coming of Christ in the Old Testament and of his life and its impact on those who knew him in the New Testament.

Since none of the men who wrote the Bible are still alive, it is necessary to rely on the written record they left behind for what we can reliably know about Jesus of Nazareth. Faith is dependent upon what is contained in the Scriptures. John wrote in the fourth gospel: 'Jesus did many other signs in the presence of his disciples, which are not written in this book. But these are written so that you may come to believe that Jesus is the Messiah, the Son of God, and that through believing you may have life in his name.'[1]

Because our faith rests in the reliability of these documents, it is mandatory that we retain our confidence in their divine origin and their spiritual validity. The science of archaeology is constantly making significant contributions to our confidence in these texts in a number of ways. One of the most important contributions is the discovery of significant inscriptions relating to the biblical text.

Inscriptions containing people named in the Bible

The names of several people specifically mentioned in the Bible have been attested epigraphically in archaeological discoveries.

Caiaphas

The high priest before whom Jesus appeared just before his death was named Caiaphas.[2] Later both Simon Peter and John appeared before him in Jerusalem.[3]

In November 1990, in Jerusalem in the Peace Forest, south of the Gehenna Valley, near the Government House where the United Nations was located, a tomb was discovered that appears to be the burial cave of the family of Caiaphas. Although it is impossible to prove that the bones of the man in the ossuary are those of Caiaphas, it does seem highly probable that the ossuary once held his bones, and there is no real reason to doubt that we do indeed have the burial vault and the ancient stone casket of the high priest Caiaphas with his name inscribed on it.

Herod the Great

Herod the Great was the king of the Jewish people when Jesus was born in Bethlehem.[4] Among the approximately 7500 inscriptions that have been found in the market places of Athens in Greece, two little known inscriptions were discovered on the Acropolis that mention Herod the Great as a benefactor of Athens. They are very similar in content. The first reads: 'The people [erect this monument to] King Herod, Lover of Romans, because of his benefaction and goodwill.' The second reads: 'The people [erect this monument to] King Herod, Devout and a Lover of Caesar, because of his virtue and benefaction.'

Pontius Pilate

Pontius Pilate was prefect of Judaea from 26 to 36 AD. Jesus' appearance before him during the last days of his life is recorded by all four Gospels,[5] but he is known from extra-biblical records as well.

During the excavations of Caesarea Maritima, the city of Pilate's residence on the coast of Israel, an Italian expedition discovered a stone in the Roman theatre which contains the name 'Pontius Pilate, prefect of Judaea'. The stone was undoubtedly part of an important building, possibly a temple, dedicated in honour of the Emperor Tiberius and called a Tiberium in the inscription. It provides the first archaeological evidence of Pilate.

Erastus

Before 50 AD, an area 62 feet square was paved with stone at the north-east corner of the theatre in Corinth, Greece. Excavations there revealed part of a Latin inscription carved into the pavement which reads: 'Erastus in return for his aedileship laid [the pavement] at his own expense'.

The Erastus of this inscription is identified in the excavation publication as the Erastus mentioned by Paul in Romans, a letter written from Corinth, in which Erastus is referred to as 'the city treasurer'.[6] Three main points favouring the identification are set forth by the editor of the inscriptions: (1) the pavement was laid around 50 AD, the time when Erastus would likely have been converted; (2) the name 'Erastus', an uncommon cognomen, is not found in Corinth other than in this inscription; and (3) the particular Greek word used by Paul for 'treasurer' (*oikonomos*) is an appropriate term to describe the work of a Corinthian aedile or magistrate supervising public works.

Gallio

While the apostle Paul was in Corinth on his third missionary journey, he was brought before the Roman proconsul Gallio, who was visiting Corinth from his official residence in Delphi across the Corinthian Gulf.[7] Four fragments of an inscription carved in stone which had been mounted on the wall of a public building in Delphi have been excavated. They contain information about the accession of Gallio and help to determine the date of his tenure in office.

The fragments are from a copy of a letter sent from the Emporer Claudius to the city of Delphi, either to the people of Delphi or to the successor of Gallio, who had the letter carved into stone and attached to the wall of the building. It contains the name of 'Gallio proconsul of Asia', in addition to that of the Roman emperor Claudius, with dates for his reign. The letter is dated to 52 AD.

Since proconsuls normally held office for one year, and since these provincial governors were required to leave Rome for their posts not later than the middle of April, Gallio probably began his term of office in May of 51. And since Paul had arrived in Corinth eighteen months earlier than his appearance before Gallio,[8] he would have entered Corinth in the winter of 49/50—perhaps in January of 50. This would coincide well with Luke's statement that when Paul arrived in Corinth on his second journey, he found Aquila and Priscilla, Jews who had 'recently come from Italy . . . because Claudius had ordered all Jews to leave Rome'.[9] This expulsion is also referred to in other ancient sources and can be dated to 49 AD. Suetonius, chief secretary to the Emperor Hadrian (117–38 AD), wrote a biographical account of the Roman emperors entitled *The Twelve Caesars*, in which he said: 'Because the Jews at Rome caused continuous disturbances at the instigation of Christ, he [Claudius] expelled them from the City.'

Thus the accuracy of Luke's account in Acts is confirmed and illustrated.

Inscriptions reproducing New Testament texts

Another category of inscriptions includes those that reproduce portions of the text of the New Testament. Although these are not abundant, some were made at the same time as most of the oldest extant copies of our New Testament in Greek, the language in which it was originally written.

Inscription in Caesarea Maritima

Two mosaic inscriptions containing the Greek text of Romans 13:3 have been found in Caesarea Maritima on the coast of Israel. We found the longer one when I was excavating there in 1972. The shorter one was previously found by Abraham Negev in 1960.

The two texts, dating back to at least the fifth century AD, are part of a mosaic floor of a large public building (perhaps a praetorium or archives building) and are identical to that passage in the Greek New Testament. These are as old as some of our oldest manuscripts of the New Testament and are obviously important sources of comparison with them.

1 Corinthians 15:52–53 inscription

In the Judaean desert in southern Israel, excavations have produced a mosaic inscription of 1 Corinthians 15:52 and 53. It was part of the floor in front of a burial cave near the Khirbet ed Deir monastery east of Hebron.

The inscription, speaking of the resurrection of the body, was situated so that it could be read as one stood facing the tomb. It reads: 'This perishable nature must put on the imperishable, and the mortal nature must put on immortality. For the trumpet will sound, and the dead will be raised.' Portions of each verse were used and verse 53 was placed before verse 52. The mosaic dates from the fifth or sixth century AD.

New Testament passages contextualised in history

The politarchs in Thessalonica

Archaeology has made a positive contribution to a problem that has centred around the city of Thessalonica in Northern Greece for many years. Critics of the New Testament asserted that Luke was mistaken in his use of the term 'politarchs' for the officials before whom Christians associated with the apostle Paul were taken in this city, because no such office existed at that time.[10]

However, a fragment of an inscription containing this term has been found in Thessalonica and is now beautifully displayed in the British Museum. The inscription, which was attached to a first-century arch on Egnatia Street, begins: 'In the time of the politarchs ...' Thirty-five inscriptions have now been discovered which contain this term; nineteen of them come from Thessalonica, and at least three date to the first century AD.

These inscriptions prove that the office known as politarch existed in Macedonia at least as early as the reign of King Perseus in the second century BC. It is significant that scholarly discussion has now shifted from the question of whether the office of politarch actually existed to the question of how long before the first century it originated! It is now incontrovertible that politarchs existed in Macedonia both before and during the time of Paul.

A riot in Ephesus

Archaeology and history come together in dramatic confirmation of the accuracy of the literary context of Acts 19:23–38. In this passage, a demonstration of hostility to Paul's teaching in Ephesus is held in the theatre adjacent to the forum (the mall of an ancient city), and a number of specific references by Luke, the author of Acts, to persons, places and things have been shown to be accurate by archaeological investigation.

The demonstration against Paul was prompted by his teaching against idolatry in Ephesus and motivated in particular by the silversmiths, who made idols of the pagan deity whose Greek name was Artemis and whose Roman name was Diana. A number of references in a newly published volume of inscriptions on Ephesus mention silversmiths. Two well-preserved statues of Artemis, which are on display in the Seljuk museum near Ephesus, were excavated in the office of the Town Clerk, which was adjacent to the south side of the theatre. This clerk went to placate the mob gathered in the theatre because Roman law forbade such demonstrations. Several inscriptions found in the city refer to this civic office.

It is significant that Paul was encouraged not to enter the theatre during this riot by Roman officials named 'asiarchs', political figures of wealth and power, who were friends of his.[11] Numerous inscriptions mentioning asiarchs have been found in Ephesus, some of them dating to within fifty years of the events described in Acts. The recent monumental publication of the inscriptions from Ephesus, containing 3500 previously known and new inscriptions, contains references to 106 asiarchs in Ephesus, including both men and women.

An inscription in Ephesus touches on another trade in the city mentioned by Paul, who wrote to Timothy, his young assistant, about an individual called 'Alexander the coppersmith', a man who opposed him while he taught in that city.[12] Timothy was probably in Ephesus when he received the letter. One of the inscriptions refers to the '(work place) of Diogenes the coppersmith'.

The arrest of Paul

A recent archaeological discovery has illuminated the story Luke records of Paul's arrest in Jerusalem at the end of his third missionary journey.[13] While he was in the temple in Jerusalem for a purification ceremony, some of the Jews accused him of taking Trophimus the Ephesian, a Gentile from Asia, into an inner area of the temple forbidden to non-Jews, and immediately tried to kill him. The temple building and its courts were accessible only to Jewish worshippers. The entire temple court was separated from the outer Court of the Gentiles by a small partition wall called the balustrade.

Josephus, the first-century Jewish historian, described this arrangement. The Jewish portion was surrounded by a stone balustrade with inscriptions in Latin and Greek on it at regular intervals, prohibiting the entrance of a foreigner under threat of the death penalty. The Jews were allowed by the Romans to put to death any non-Jew who crossed this boundary wall, even Roman citizens. Two of these stone slabs containing the inscriptions described by Josephus have been found and published. The inscription reads: 'No foreigner is to enter within the forecourt and the balustrade around the sanctuary. Whoever is caught will have himself to blame for his subsequent death.'

The accuracy of geographical references

Archaeological excavations have identified a number of geographical sites referred to in the New Testament and provided confirmation of the accuracy of the references to these sites.

The Capernaum synagogue and the house of Peter

When Jesus began his public ministry in Nazareth of Galilee at about thirty years of age and was rejected in the synagogue there,[14] he went to Capernaum on the north shore of the Sea of Galilee, where he apparently lived in the home of Simon Peter, one of his disciples[15] and taught in the synagogue.[16] Archaeological excavations conducted in Capernaum have discovered this synagogue under the fourth/fifth-century limestone structure still standing there. Portions of the floor and walls of the first-century synagogue were found beneath the floor of this building.

Immediately south of the synagogue, excavators found remains of a small residence that may be the house of Peter. One room in the first-century house was set apart in the middle of that century as a public area and venerated from that time until the fourth century as the house of Peter.

The pool of Bethesda in Jerusalem

In the city of Jerusalem, the Pool of Bethesda is named in the Gospel of John as the place where Jesus healed a crippled man.[17] The pool has been found. Evidence of the existence of a pool called Bethesda appears in a list of places in Jerusalem in the Copper Scroll, one of the Dead Sea Scrolls found at Qumran, which was written between 25 and 68 AD.

About one hundred yards north of the Temple Mount's northern wall and about the same distance west of the Lion Gate (Stephen's Gate), excavations were conducted near the Church of Saint Anne. Two large pools were identified, just a few metres apart, cut into the rock and plastered. Many fragments of column bases, capitals and drums, which probably belonged to the five porches (that is, porticoes or colonnaded walkways) of the pool that John mentions, were discovered. This two-pool complex is referred to as the Sheep Pool by Eusebius and the 'twin pools' by the Bordeaux Pilgrim, both in the early fourth century. It is located near the ancient Sheep Gate, as we would expect by its name.

The Pool of Siloam in Southern Jerusalem

Another pool in Jerusalem was immortalised by Jesus when he healed a blind man and told him to wash his eyes in its water.[18] It is called the Pool of Siloam in the Gospel of John and was built by Hezekiah, a king of Judah in the eighth century BC, at the southern end of a 1749-foot long tunnel that he cut through solid rock to bring water from the Gihon Spring outside the city walls to the pool inside the walls.

Here, in 1897, excavators uncovered a court about 75 feet square, in the centre of which was the pool. It was probably surrounded by a colonnaded walkway, which must have been the one described by the Pilgrim of Bordeaux in 333 AD as a 'quadriporticus' (fourfold porch). After the excavations, the people of the village of Silwan (the modern spelling of Siloam) built a mosque on the north-west corner of the pool with a minaret, which still stands above the pool.

The pinnacle of the temple in Jerusalem

The New Testament says that after Jesus was baptised by John the Immerser, he was taken by Satan to the city of Jerusalem, where he stood on 'the pinnacle of the temple' and was tempted to throw himself down, so that an angel could prevent his being harmed and prove he was really the Son of God.[19] This pinnacle was the south-east corner of the southern extension of the city wall built by Herod the Great, which looked down into the Kidron Valley. Huge vaults were built in this south-eastern corner of the temple platform to support a large stone pavement built above it and the Royal Porch, which ran east and west along the southern wall. This platform and the porch on it were probably the pinnacle referred to in the gospels because of their height and prominent position.

The pavement covered an area of 1800 square feet, supported by eighty-eight pillars which rested on huge Herodian blocks and formed twelve rows of galleries. The original city wall was destroyed by the Romans in 70 AD; however, the foundation blocks of these Herodian pillars may still be seen today below the present pavement in the vaulted area erroneously called 'Solomons Stables'.

Rolling stones at tombs

In the Gospel of Matthew it is stated that an angel descended from heaven to the tomb of Jesus, 'rolled back the stone and sat on it'.[20] Many tombs from the time of Christ have been discovered in Jerusalem, and some of them still have these rolling stones by the entrance to the tomb.

North of the Damascus Gate, a tomb from the time of Jesus was built for the burial of Queen Helena of Adiabene and has the stone still in place. Another, better preserved rolling stone still stands beside the entrance to the family tomb of Herod the Great, south of the King David Hotel. More than sixty rolling stone tombs have been found and studied in Israel and Jordan in recent years.

The lecture hall of Tyrannus in Ephesus

Luke writes that while Paul was in Ephesus he taught daily for two years in the lecture hall of Tyrannus, 'so that all the residents of Asia, both Jews and Greeks, heard the word of the Lord'.[21] Archaeological evidence of this hall has been found in a first-century AD inscription, which refers to it.

The hall was situated adjacent to the Mazaeus and Mithridates gates, which provided entrance to the commercial forum of Ephesus from the south. According to some of the ancient texts of the book of Acts, Paul used the hall 'from the fifth hour to the tenth'—that is, from 11.00 a.m. to 4.00 p.m.—when most people in the Mediterranean world, then as now, closed their businesses, ate lunch and rested. The students of Tyrannus may have also gone home during these hours, giving Paul free use of the building.

The tribunal in Corinth

At Corinth, one of the most important discoveries relating to the New Testament is the tribunal (Greek *bema*), or speaker's platform, from which official proclamations were read and at which citizens appeared before appropriate officials. It still stands in the heart of the forum. The stone structure was identified by several pieces of an inscription found nearby and dated to the period between 25 and 50 AD, just prior to Paul's arrival in the city.

Paul spent eighteen months in Corinth on his second

missionary journey. At the end of that time, the Jews took advantage of the inauguration of Gallio as proconsul of Achaia in May or June of 51 to bring Paul before him on a charge of violating their law.[22] Gallio found no violation of Roman law by Paul, no 'wrongdoing or vicious' crime, and, refusing to be a judge of Jewish law, drove Paul's accusers from this 'tribunal', where he was seated.[23]

Gallio was the brother of Seneca, a Greek Stoic philosopher who later became an adviser to the Emperor Nero. Seneca perhaps informed the emperor of the fact that Paul had already been acquitted before Gallio in Corinth and thus influenced the favourable outcome of Paul's first arrest in Rome as implied in the last verses of Acts.

The demoniac at Gergesa

Variant readings in ancient manuscripts of the gospels give three slightly different names—Gerasa, Gadara and Gergesa—for the place where Jesus cast the demons out of a man and into a herd of swine, which ran over a cliff into the Sea of Galilee and drowned.[24] Archaeology helps us decide which is correct.

Two of these towns are located in Jordan—Gerasa (modern Jerash) and Gadara (modern Um Qeis). The other, Gergesa (modern El Kursi), is in Israel on the eastern shore of the Sea of Galilee. Excavations at Gerasa and Gadara have revealed remains of these cities from the Roman period, the time of Jesus. However, both are miles from the Sea of Galilee and cannot fit the story because the swine obviously could not have run off a cliff into the sea in these distant locations.

However, in 1970, when a road was being constructed on

the east side of the sea, the foundations of a Byzantine church building were found at El Kursi, built over an earlier burial vault. This indicated a long-standing settlement. Since the Byzantines customarily built churches over sites that marked significant events in Christian tradition, El Kursi is most likely ancient Gergesa. It is the only place on the entire east coast of the sea where a cliff extends out to the sea and would thus be the appropriate setting for that story. Furthermore, tombs have been found in the adjacent hills, consistent with the story that the man from whom demons were cast out was said to have lived among the tombs.

The healing of blind Bartimaeus at Jericho

Archaeological illumination has been provided for the enigmatic statements in the gospels that Jesus healed a blind man named Bartimaeus when he was going *into* Jericho and when he was coming *out* of Jericho.[25]

Excavations have shown that settlements in the various periods of Jericho's history, like those at other sites such as Beersheba, had town centres that were several miles apart. The palace complex of Herod the Great, for example, was south of the Old Testament site of Jericho, and the New Testament city is yet to be found, though it probably lies just east of the excavated Herodian palace.

Thus Jesus was apparently going out of one section of Jericho into another when he healed Bartimaeus, and both gospels are correct.

* * *

These examples of archaeological findings that confirm the fine detail of the New Testament give us confidence to believe that the Bible is indeed a true and accurate record of events that actually occurred.[26]

20

What about the Bible's Scientific Reliability?

Timothy G. Standish

Imagine what it must have been like to live around the year 1000 AD. In Western Europe, the church represented the principal repository of knowledge, and much of that knowledge came from earlier civilisations. Science as we know it was not practised; instead, those wanting information used as their primary source the writings of earlier thinkers.

In addition to this, the church had a different view of the Bible from that which most Protestant Christians would embrace today. This view, called *prima Scriptura*, stated that the Scriptures were an important source of knowledge but had to be interpreted through other sources, church tradition being very important among these. This view of Scripture is quite different from the Reformation idea of *sola Scriptura*—the Bible alone—in which the Bible was to be, as far as possible, its own interpreter.

Not surprisingly, making Scripture subject to other authorities like church tradition led to some ideas that seem unusual to us today. Suppose you were to visit a monastery (where most scholars resided at the time) and asked a monk, 'Where do clams come from?' The monk would most likely have referred you to the writings of Aristotle, not the Bible or some other religious source. Reading Aristotle you would learn that clams and their relatives come about spontaneously from various non-living sources. In his book *The History of Animals* (c. 350 BC), Aristotle wrote that organisms with shells are generated by mud and other inorganic substances:

> As a general rule, then, all testaceans [shell-bearing animals] grow by spontaneous generation in mud, differing from one another according to the differences of the material; oysters growing in slime, and cockles and the other testaceans above mentioned on sandy bottoms.[1]

The idea that living creatures could somehow come into being from mud seems preposterous to us now, and yet that was the best science available in 1000 AD.

Stranger still, the idea that living creatures come about spontaneously from various combinations of the elements had incredible staying power. As late as the early 1600s, the brilliant Belgian scientist Jan Baptista van Helmont wrote out the following recipe for making mice: 'Place a dirty shirt or some rags in an open pot or barrel containing a few grains of wheat or some wheat bran, and in twenty-one days, mice will appear. There will be adult males and females present, and they will be capable of mating and reproducing more mice.'

It was not until the work of a number of scientists, starting with Francesco Redi in the mid-1600s, that science started to question the theory that life could be spontaneously generated from lifeless things. Only after experiments by Abbe Spallanzani in the late 1700s, Theodor Schwann and Franz Schulze in 1854, and finally Louis Pasteur in 1861 was the idea of the spontaneous generation of life ultimately laid to rest.

We now read about the experiments of Pasteur, Redi, Spallanzani, Schwann and Schulze in most high school biology textbooks. Ironically, most of these textbooks continue to promote the idea of spontaneous generation. For example, the textbook that I used when teaching high school biology students in the United States discusses the experiments of Redi, Spallanzani and Pasteur in the same chapter that talks about chemical evolution—the evolution of the first cells from a mixture of naturally occurring chemicals.[2] It may be that we who live in the twenty-first century are not as different as we would like to think from our ancestors who lived over a thousand years ago.

This is because many modern scientists, like the mediaeval scholars, embrace certain Greek philosophies. In the case of the spontaneous generation of life, the philosophy of materialism dictates that it must be true. In the philosophical sense, materialism means that the material world is all that exists; there is no supernatural God or gods. If there is nothing supernatural to make life, then living things *must* have come about by some natural process, presumably from the matter available on earth.

When it comes to spontaneous generation of life, the Bible is very clear. It states: 'You are worthy, our Lord and God, to receive glory and honour and power, for you created all things, and by your will they were created and have their being.'[3] It was not the mud that spontaneously created the clams; it was God who created clams.

The first few chapters of the Bible tell us how he did this and detail how new clams, worms, birds and all other living things continue to be made. After God created the living things using the same materials from which the earth is composed,[4] he told them to multiply.[5] In other words, when creating the initial kinds of living things, God endowed them with the ability to reproduce. The clear claim of Scripture is that God made the living things using materials on the earth, and the living things that we see today are descendants of these first creatures. The earth is not capable of making living things itself; it is God who intervenes to create life.

Even though the Bible's meaning seems plain, what it says about where living things come from can be contorted. Twisting the meaning seems to be possible with almost anything that is written. For example, I remember once attending a performance of *Hamlet* in which the Elizabethan characters had been adapted into the setting of a communist eastern European country in the 1930s. Clearly that is not what William Shakespeare had in mind when writing a play about a Danish prince and the royal court.

The principles of *prima Scriptura* and *sola Scriptura* appear in direct opposition when it comes to figuring out what the Bible says. While both affirm the importance of Scripture, *sola Scriptura* asks, 'What is the most obvious reading of the Bible?' Where there are apparent ambiguities, the Scriptures themselves are searched for any information that may shed light on the meaning. *Prima Scriptura*, on the other hand, allows

Scripture to be interpreted according to whatever the current understanding of reality is. In the case of the mediaeval church, tradition was the filter through which Scripture was to be interpreted. Since the traditional understanding of nature was the one proposed by Greek philosophers and those who followed them, it doesn't seem so unreasonable that the Greek philosopher Aristotle was considered to be *the* authority on the origin of clams and other organisms. It was simply the way things were done at the time.

Unfortunately, because the church conflated the science of the day with beliefs based on Scripture, it is easy to get the impression that the Scriptures themselves make claims that now seem bizarre. Scripture did not teach that clams came about by spontaneous generation from mud, just as it did not teach that the earth could not revolve around the sun or any number of other strange ideas that are attributed to the mediaeval church as a result of belief in Scripture. In reality, most of the ideas held by mediaeval scholars can be traced to their willingness to believe Greek philosophers rather than look at the Bible and see what it actually says.

Perhaps even worse than the actual misconceptions attributed to our mediaeval antecedents are fictitious ideas that are attributed to them—for example, the myth that the mediaeval church believed the earth was flat. Ironically, the Greeks, on whom mediaeval scholars relied so much, knew that the earth was round, just as every educated person since about 300 BC has known. People only began to believe that the mediaeval church thought the earth was flat after an imaginary account of Christopher Columbus' voyage to the New World written by

Washington Irving appeared around 1830.[6] In Irving's fanciful account, Columbus had to convince a very unwilling church that the earth might not in fact be flat.

What can we learn from imagining life in the year 1000 AD? We can learn that our preconceptions about where mediaeval people got their knowledge may be wrong. The church did not get its ideas from the Bible, especially those ideas that dealt with the world we study in science. The example of spontaneous generation of life is one of many possible examples where, if the Bible had been taken at face value, it would have suggested an understanding of nature that is closer to the truth than the Greek philosophy used in mediaeval times. Ironically, the same is true today. When we embrace the philosophy of materialism, we automatically revert back to the idea of life's spontaneous generation, in which the earth somehow makes the living things.

If we are going to embrace the same ideas as the mediaeval church, we should be slow to criticise it. But the alternative seems far more attractive. That alternative is to reject the flawed mediaeval understanding of Scripture and look at what the Bible actually says.

21

Amazing Biblical Prophecies That Came to Pass

William Shea

A knowledge of the future could be very advantageous to us in a number of ways, especially financial. This is the principle on which inside traders on the stock market operate. By possessing knowledge of what will happen to the stock of a particular company before it has been announced or become generally known, inside traders can buy or sell stock in that company and make large profits. Of course, governments consider this kind of 'future' knowledge to be unfair trading, so it is illegal.

However, the Bible contains a number of accounts of knowledge of future events that enabled the original readers to have 'inside information' from the creator God, who sees the 'end from the beginning'.[1] This inside information—or prophecy, as it is generally known—strengthened the faith of believers or provided direct warnings so they could survive troubled and difficult times. One of the books of the Bible that is particularly devoted to prophecy and that provided its readers with 'inside information' about the future is the book of Daniel.

This book presents its hero as a youth from Judah who was taken away into exile in Babylon during the early days of Nebuchadnezzar's conquests in the land of Judah. According to the internal testimony of the book itself, Daniel lived in Babylon through the entire period of the Neo-Babylonian kingdom, from 605 to 539 BC, and into the earliest portion of the Persian period, from 539 BC to at least 536 BC. The first half of the book records historical narratives from the times of these kings while the second half consists of prophecies given during those times. Some critics think the book was not written in the era that its setting presupposes (the sixth century BC). But there is evidence from both the historical chapters and the prophetic chapters that its writing does belong to this time. We do not have space to consider the historical evidence for this in detail, but some of this evidence will come out as we evaluate prophecies from this book.

The prophecy of the fall of King Belshazzar

Chapter five of Daniel narrates what happened in the palace of Babylon on the night the city fell to the Medes and Persians. The king, who is identified as 'Belshazzar', summoned his nobles and officials to a great banquet. He felt quite secure behind his thick double walls. He undoubtedly felt that the Persians besieging Babylon from the outside had no chance of conquering the city in view of its extraordinarily strong fortifications.

In the course of the feast, some writing supernaturally

appeared on the wall of the palace audience chamber where the banquet was held. The four words written there were sufficiently mysterious that none of the wise men of Babylon could interpret them. Daniel, who was remembered from a previous episode of interpretation, was summoned. He was able to read the writing and tell the king that it meant that he had been weighed in the balances of divine judgment and had been found wanting. His kingdom was to be taken from him and given to the Medes and Persians. (Incidentally, this is where we get the well-known expression 'the writing is on the wall'.)

Belshazzar was shocked by this but kept his head enough to carry out his promise to appoint the person who could interpret the writing to high public office. Daniel was, therefore, accorded the appropriate honours to make him the third ruler in the kingdom. This did not do much good for Daniel and even less good for Belshazzar, because the Medes and Persians entered the city that very night by the strategy of diverting the Euphrates River. Babylon was conquered without a battle. So Belshazzar was slain and his kingdom passed into the hands of the Medes and Persians.

At first one might think that there is no way, through historical sources, to check on the fulfilment of this prophecy. While it is true that it would be very difficult to demonstrate it was given the very night it was fulfilled, there are indirect methods through which we can evaluate the setting.

First, we know there was a Belshazzar. Until the late nineteenth century AD his existence and position were questioned. In the king lists and the classical sources of Greek and Roman historians, it is recorded that Nabonidus was the last official king of Babylon. A knowledge of Belshazzar had been lost from all other historical sources outside of the Bible.

Then, beginning in 1861, the name of Belshazzar as crown prince began to appear in cuneiform tablets that were being translated at that time. These references continued to accumulate until a tablet known as 'The Verse Account of Nabonidus' was published in 1929. This important tablet indicated that Nabonidus 'entrusted the kingship' to Belshazzar when he went off to Tema in Arabia for a prolonged period (it turned out to be ten years). So evidence for Belshazzar as a kind of co-king, known as a co-regent, became available.

But the episode described in Daniel 5 is even more specific. It indicates that when Daniel came into the throne room to read the writing on the wall, the king who was there was Belshazzar, not Nabonidus. It would have been more likely for Nabonidus to be conducting this banquet, but he is not even mentioned in the narrative. The direct implication is that Nabonidus was not in the palace that night. If he was not in the palace, where was he?

A Babylonian text known as 'The Nabonidus Chronicle' tells us. When it looked like the Persian threat to Babylonia from the east was unavoidable, Nabonidus returned from Arabia to Babylon to manage its defence. This he did at least by the end of 540 BC because some of the actions he took in that regard are dated in this text to early in 539 BC. The text also tells us that Babylon was taken without a battle on 16 Tishri of Nabonidus' seventeenth and last year of reign. This can be equated to October 12, 539 BC. At this time, the text says, Nabonidus was out in the field with a division of the Babylonian army, fighting Cyrus and the Persians at the site of a city named Opis on the Tigris River. Cyrus defeated Nabonidus, who fled from the scene of the battle but did not return to Babylon until after it had fallen to the other division of the Persian army. Thus there was no possible way he could have been in Babylon the night it fell.

This would have been an easy place in which to catch the writer of Daniel in a mistake, if he had put Nabonidus in the banquet hall on that night. But the writer knew which king was there—Belshazzar, the junior co-regent—and he knew which king was not—Nabonidus, the senior co-regent, who was out in the field with the Babylonian army.

How could the writer of this chapter have had such accurate knowledge about who was in the city and who was not on that very night? Why did he pick an obscure ruler whose name was lost to later history instead of the more common and well-known last official ruler of the city and country? The answer is, because he was an eyewitness in the palace that night. That is how he knew who was there and who was not there. If his knowledge about this central fact was so accurate, then I believe we can trust his record as to the prophecy of what was to happen on that very night.

A prophecy about the fate of Babylon

A longer time prophecy about the fate of Babylon is found in another book of the Bible, the book of Jeremiah (chapters 50–51). When the Persians conquered Babylon in 539 BC they did not destroy the city. They wanted to make use of its riches and opportunities. Thus Babylon did not immediately fall into destruction and obscurity. It became one of the provinces of the Persian Empire and one of the three major capitals of the Persian kings.

But a prophecy at the end of Jeremiah, written prior to the capture of Jerusalem by Nebuchadnezzar about seventy years earlier, said that Babylon would fall into desolation and be abandoned. It would become the haunt only for wild birds and wild animals. It would be abandoned and desolate, a heap of ruins.[2]

For a time it looked as if this prophecy would not come to pass. But the first great step downward came when Babylon revolted against the Persian King Xerxes in the fourth year of his reign (482 BC). As a punishment for this revolt, Xerxes destroyed the city's walls and temples and melted down the golden statue of Marduk that was housed in its main temple. The next stage in the city's decline came when a new capital of the country was built under the Greek kings who ruled it after the death of Alexander the Great. The new capital was named Seleucia and was located on the Tigris River. Much of the remaining population of Babylon was transferred there in 275 BC. After that, Babylon faded away. By the end of the first century AD, it was no longer inhabited.

Babylon remained in this ruined and abandoned state, but knowledge of where it had been located was not lost, as was the case with Nineveh (discussed below). The modern study of Babylon began with some survey work and superficial excavations carried out there by C.J. Rich in 1811. The major excavations of the city came under the direction of the German Robert Koldewey from 1895 to 1917. Restoration work has been carried out sporadically since that time so that today the site is like a great museum piece. When I visited it in 1966, the only inhabitant was the night watchman. Some have said that Saddam Hussein intended to rebuild Babylon as a modern city. This is incorrect. The Iraq Department of Antiquities has carried out some restoration work there in recent times, but this is not for the purpose of modern occupation. It is to demonstrate the greatness of the city in antiquity.

The extensive ruins of ancient Babylon, which cover more than 1400 acres, stand as mute testimony to the truth and accuracy of the prophecy of Jeremiah. It should also be recalled that Jeremiah gave this prophecy while Babylon was still at the height of its power. He was in Judah on each of the three occasions when Nebuchadnezzar conquered it (605, 597 and 587 BC). Because he saw this as having been permitted by God, Jeremiah advised the people not to rebel against Nebuchadnezzar. He counselled the people to surrender, submit to the Babylonian king and become loyal vassals to him. For this he was considered a traitor and suffered various kinds of punishments and imprisonments. Having had to suffer through all this, and yet see his people go down to destruction in spite of his counsel that could have saved them, he has been characterised as the 'weeping prophet'.

An international prophecy: the four kingdoms of Daniel

Daniel chapters 2 and 7 provide parallel prophecies about four world—or perhaps more accurately, Mediterranean and Near Eastern—empires. Daniel 2 gives the account of events that occurred around a dream given to King Nebuchadnezzar that the wise men of Babylon could not describe or interpret. However, Daniel was able to successfully describe the dream and interpret it. The king had dreamed about a great statue made of different metals, gold for the head, silver for the chest and arms, bronze for the stomach and thighs, and iron for the legs. Each metal represented a future political empire. In the feet, the iron was mixed with clay to show that the fourth kingdom would not be succeeded by a fifth but would end up as a mixture representing the parts into which the fourth kingdom was divided. In the dream a stone struck the feet and smashed the whole image into pieces. These were blown away, and the stone became a mountain that filled the whole earth.

Daniel 7 gives the same outline but uses different prophetic symbols. In this case four animals were seen by the prophet in a symbolic vision given directly to him. The first was a lion, the second was a bear, the third was a leopard and only some parts of the fourth beast were described. One feature of the fourth beast was that it had ten horns representing the divisions that came out of the fourth kingdom, just as the iron was mixed with clay to represent those divisions in Daniel 2. Another feature of the fourth beast was that it had great iron teeth. These iron teeth connected it with the kingdom of iron in the statue of Nebuchadnezzar's dream.

So four great kingdoms were to appear on the scene from the time of Daniel in the sixth century BC until the fourth kingdom was divided up into various parts. The question is, which were these four kingdoms that Daniel saw in the future?

The first was clearly identified as Babylon itself when Daniel interpreted the king's dream to him: 'You, O king . . . are that head of gold.'[3] Nebuchadnezzar ruled the Neo-Babylonian kingdom for forty-three of its seventy years and built the famous Hanging Gardens of Babylon, one of the seven wonders of the ancient world. Thus it was appropriate to identify Nebuchadnezzar as the one great king of this kingdom.

The silver kingdom that followed Babylon's gold arose during Daniel's own time, as is evidenced by the story of the fall of Babylon to the Medes and Persians in Daniel 5. Of this kingdom Daniel told Nebuchadnezzar, 'After you, another kingdom will rise, inferior to yours.'[4] The silver kingdom is paralleled by the bear in Daniel 7, which reveals more characteristics of this kingdom: it was 'raised up on one of its sides', which means the other side was lower.[5] This represented the Medes and Persians. The Medes were the stronger power earlier on, but when the Persians conquered them to make up the combined kingdom of Medo-Persia, the Persian element became stronger. This same feature is shown in Daniel 8, where yet another visionary prophecy describes a ram with one horn higher than the other and specifically identifies this combined kingdom as that of the Medes and the Persians.[6]

In the vision in Daniel 8 the Medo-Persian ram was succeeded by a goat that had four horns on its head and flew over the ground. The interpretation of this symbol is given in Daniel 8:21 as Greece. This parallels the leopard in Daniel 7 that had four heads and wings on its back with which it could fly.

Thus far there are names in the book of Daniel itself for the first three of these four great kingdoms. The gold and the lion represented Babylon. The silver and the bear represented Medo-Persia. The bronze and the leopard represented Greece. The fourth kingdom is represented by the iron and the nondescript beast. It is not named in the book. It must be identified by a conclusion drawn from history. What kingdom succeeded Greece on the Mediterranean world scene? The answer is obvious: Rome.

It was Rome that picked up the four pieces into which the empire of Alexander the Great divided when his generals took over after his death. Rome defeated Greece at the battle of Pydna in 168 BC. The kingdom of Pergamum in Asia Minor was willed to Rome in 133 BC when the king died there without a male heir. The Roman general Pompey conquered Syria and Judaea some sixty years before the birth of Christ, and Julius Caesar wrested Egypt from the line of Greek kings there known as the Ptolemies in 31 BC. So Rome took over the pieces of the Greek empire of Alexander and was the historical successor to Greece.

In this way the four kingdoms of Daniel 2 and 7 have been identified as Babylon, Medo-Persia, Greece and Rome. This means that in the dream of Nebuchadnezzar described in chapter 2 and the vision given to Daniel himself in chapter 7, Daniel was shown what was to follow in history for at least the next four centuries after his own time.

History as 'prophecy'?

But there are those who do not like this much direct evidence for divine foreknowledge in prophecy, and they have argued against this view. They say that the author of Daniel did not live in the sixth century BC when this prophecy was given. Rather he is supposed to have lived in the second century BC and to have used the pen name of Daniel to write up history he had already seen take place. Thus these critics claim that Daniel is actually history written up as if it were prophecy.

This view began with a Platonic philosopher named Porphyry, who wrote against Christianity in the fifth century AD. His writings have, for the most part, been lost. Most of them did not bother the church fathers of the time, but his criticism of Daniel did. For that reason this part of his attack has been preserved in the works of theologians such as Jerome, who wrote against it. Unfortunately, Porphyry's view was taken up by later scholars who wrote against divine foreknowledge in Daniel.

The argument can be evaluated, however, to see how well it fits the data. We do not have space to repeat the many historical evidences from Babylon which indicate that Daniel was written in the sixth century BC. We will deal only with the issue of prophecy. This can be done by asking the question: If the writer of Daniel wrote in the second century BC, what would he write about the things that would follow in the future? After all, the stone that became a mountain in Daniel 2 represented God's kingdom that would be set up at the end of human history;[7] thus there had to be something between that terminal point and the second century BC. The same point is evident from Daniel 7, which states that the kingdom of heaven will be given to the saints of the Most High and they will reign for ever and ever.[8]

If the author of Daniel was writing in the second century BC and he was only a historian, not a true prophet, what kind of future history would he guess at? There are two main possibilities. First, he could have said that the fourth kingdom, Rome, which was stronger than all the previous kingdoms, would stand forever. That was probably the most common view of the future as seen from the second century BC, by which time Rome had come to pre-eminence. (It actually was the view of Flavius Josephus, the first-century Jewish historian, when he treated this portion of the book of Daniel. He did not mention the divisions or the stone kingdom to follow.) Alternatively the writer could have reasoned that, if there had been four great world kingdoms, there should be a fifth, a sixth, a seventh and so on. In other words, the sequence should just keep going. After Rome another great world power should come, and then another, and another.

These then would have been the two main alternatives for a historian writing in the second century BC without information from divine foreknowledge: either that Rome would stand forever or that other great world powers would follow it.

But the writer of Daniel did not follow either of these two logical possibilities. In contrast to the idea that there would be further world powers, he said 'no'—the fourth power would be broken into pieces and those pieces would continue and contend with each other until God set up his kingdom. In contrast to the idea that Rome would stand forever, he also said 'no'—this fourth kingdom would be broken up. And indeed that is just what happened with the barbarian invasions of Rome.

Jerome, who wrote his commentary on Daniel from Bethlehem in Judaea during the fifth century AD, right when the barbarian invasions were taking place, observed that he was living in the time when the legs of iron were disintegrating into the feet of iron and clay. This was seven centuries after the supposed date for the writings of the pseudo-Daniel in the second century BC. It was eleven centuries after the time of the real historical Daniel in Babylon in the sixth century BC.

How was it then that the author of Daniel knew several centuries in advance that Rome would break up into pieces, that it would not stand forever or be replaced by another great world kingdom? How was it that he chose the least likely possibility for the future from the standpoint of ordinary human logic? The point is that he did not rely on ordinary human logic; he relied on the foreknowledge that God gave to him.

All historians agree that history has followed the pattern outlined in the prophecy. Although that pattern has been followed so precisely, some commentators do not want to admit that it came from a divine source, so they turn the book of Daniel on its head and claim it is history written up as prophecy. In fact, however, this is a book of true prophecy that was given to a real Daniel who lived in Babylon as an exile from Judah in the sixth century BC. While there, he received dreams and visions from God of what would come to pass after his time. These predictions were fulfilled. The standard and well-documented sequence of Babylon, Medo-Persia, Greece and Rome, followed by the division of Rome, stands as eloquent testimony to the fact that this book contains true prophecy.

The fall of Nineveh

Long before Daniel, the nation of Assyria, located in the northern part of Mesopotamia, dominated the ancient Near East from the early ninth century to the late seventh century BC. Its empire ultimately stretched all the way to southern Egypt. It was also known as the most bloodthirsty and ruthless nation of the ancient world. For this reason it was very much feared and hated by the captive peoples under its control. It has been said that the foreign policy of Assyria was based on only one principle, that of creating terror in the hearts of the peoples it conquered. Any failure to pay the heavy annual tribute resulted in progressively more severe penalties that culminated in destruction, death and deportation. In the Bible it is evident that the northern kingdom of Israel fell under Assyria's sway by the middle of the ninth century BC and was destroyed by it in the late eighth century BC. Then the Assyrians moved farther south into Judah and eventually into Egypt where they even burned Thebes, 400 miles south of the Nile Delta.

Given Assyria's power, it is natural that God turned the attention of his prophets to these developments. Two prophets late in the seventh century were directed to pronounce prophecies concerning the fate of Nineveh, the capital of Assyria through the last half of this period. Enclosing an area of about 1600 acres within its city walls, Nineveh was even larger than Babylon. One of the prophets who was given a prophecy about Nineveh was Nahum, and his writings contain a vivid description of the final battle for that city and the defeat of its defenders:

> He summons his picked troops,
>   yet they stumble on their way.
> They dash to the city wall;
>   the protective shield is put in place.

> The river gates are thrown open
>   and the palace collapses.
> It is decreed that the city
>   be exiled and carried away.
> Its slave girls moan like doves
>   and beat upon their breasts.
> Nineveh is like a pool,
>   and its water is draining away.
> 'Stop! Stop!' they cry,
>   but no one turns back.
> Plunder the silver!
>   Plunder the gold!
> The supply is endless,
>   the wealth from all its treasures!
> She is pillaged, plundered, stripped!
>   Hearts melt, knees give way,
> bodies tremble, every face grows pale.[9]

The description continues with more of the same vivid language depicting this great city in its death throes. Its defenders are unavailing and the city goes down to defeat.

The same event is recorded in an extra-biblical source, the Babylonian Chronicles for the years 614 and 612 BC written under Nabopolassar, the father of Nebuchadnezzar. Because Assyria was so hated in the ancient world, a coalition was formed between the Medes in the north-east and the Babylonians in the south, and together they attacked Nineveh during the campaigns of those two years. In the first year, 614, they conquered another Assyrian city and then attacked

Nineveh itself, but they were not able to take it. For that reason they came back two years later, in 612, and finished off the job. Nineveh fell and was destroyed. Only a small group of soldiers escaped and retreated west to the city of Carchemish on the Upper Euphrates River. They survived there only a few years and then the last vestige of the once great Assyrian empire was gone.

The details of this event as described in the book of Nahum have been starkly confirmed by the most recent excavations at Nineveh, those conducted by the University of California just before the 1991 Gulf War broke out. While digging along the southern wall, the archaeologists found skeletons of Assyrian soldiers who died trying to defend their city. They were buried under the fallen walls of the city, and the various positions in which they were found shows the agony in which they died.

Unlike the city of Babylon, which took several centuries to erode and blow away, Nineveh was destroyed as soon as it was conquered and was never reoccupied. The ruined and abandoned condition of Nineveh over the next 2500 years is vividly predicted by another biblical prophet, Zephaniah:

> He [God] will stretch out his hand against the north
>   and destroy Assyria,
> leaving Nineveh utterly desolate
>   and dry as the desert.
> Flocks and herds will lie down there,
>   creatures of every kind.
> The desert owl and the screech owl
>   will roost on her columns.

> Their calls will echo through the windows,
>> rubble will be in the doorways,
>> the beams of cedar will be exposed.
> This is the carefree city
>> that lived in safety.
> She said to herself,
>> 'I am, and there is none besides me.'
> What a ruin she has become,
>> a lair for wild beasts!
> All who pass by her scoff
>> and shake their fists.[10]

If anything, this description of Nineveh's desolation goes beyond even what was written about Babylon in Jeremiah 50 and 51.

The sudden destruction of Nineveh is probably the main reason that knowledge of its location was lost. In the early days of Mesopotamian archaeology in the nineteenth century, one of the major quests was to find Nineveh. The story of its first excavators illustrates just how lost the knowledge of this place was. Excavations were begun in 1842 by a French physician, Paul Emile Botta, who chose a site across the Tigris River from the city of Mosul, where he served at the court of the Pasha. He dug there for a short time but found nothing of significance. Then a man came from a town 15 miles to the east who told him that the ancient statues and other objects he was seeking were sticking out of the ground nearby. So Botta moved his excavations to Khorsabad where he dug up what turned out to be Dur-Sharrukin, the ancient capital of the Assyrian king

Sargon II (mentioned in Isaiah 20:1). Not realising the site's true identity, Botta announced in April 1843 that he had found Nineveh at Khorsabad.

Shortly after that, in 1845, an Englishman by the name of Austen Henry Layard came to Mosul to search for Nineveh. He began his excavations first at modern Nimrud, ancient Calah, south of Mosul, but then he turned his attention to the site of the ruins directly across the river from Mosul where Botta had begun. There he found the true Nineveh. All of this early archaeological confusion about where Nineveh was located illustrates the point that a clear knowledge of where that great city was located had disappeared.

From the two prophecies quoted above about Nineveh's destruction, it can be seen that Nahum speaks, in vivid terms, about the final battle for the city. From our perspective today it is difficult to tell whether this prophecy was given before, during or after that famous battle. We do not have a specific way of determining how much is prophecy and how much history in this short biblical book.

The situation is different with Zephaniah. This prophecy does not speak about the final battle for Nineveh. Rather it tells what will happen to the city *after* the final battle. It will be destroyed, abandoned and ruined. It will become only a place for birds and animals, not people, to dwell. The fulfilment of this prophecy we can evaluate directly. The 2500 years when the site of this city lay abandoned and forgotten demonstrates directly how clearly the words of the ancient prophet came to pass. Even if those words were written after its destruction, they still contain a long-term prediction that has been fulfilled.

Since the city was so important down to the time of its destruction, one might have expected that it would be reoccupied and rebuilt, as happened to many cities in the ancient world. But Nineveh was so hated that it never again became the site of human habitation. Zephaniah was correct.

A prophecy about Alexander the Great

The prophecy in Daniel chapter 8 begins with a depiction of what the Medo-Persian kingdom would accomplish, using the symbol of a rampaging ram. This ram is identified as Medo-Persia.[11] It is followed symbolically by a goat, representing Greece.[12] This goat starts out with one prominent horn, like a unicorn. That great horn represents its first king as it started out on its war with the Persian ram.

Historically we know this 'horn' to be Alexander the Great, who mustered his army and invaded the Near East, defeating the Persians and conquering all their territory in a lightning-like campaign that lasted only three years. Then, on returning to Babylon after his campaign to the Indus River, Alexander died at the young age of thirty-three. His kingdom was divided among his generals, a development symbolised in the prophecy by four horns that came up after the one chief horn was broken off.[13]

Once again, critics of the book have argued that this was not prophecy, but history later written up as if it were prophecy. However, there is an interesting story in the writings of Josephus that indicates this prophecy was already known by the fourth century BC—well before the critics claim it was written in the second century BC.

The story is about Alexander as he campaigned down the coast of Syria and Palestine. On his way to Egypt he decided to turn off the main road south and go up to Jerusalem. When he came to the city, one of the priests took the scroll of Daniel to him and showed him where he was located in this prophecy, as the Greek that would overthrow the empire of the Persians. Impressed by this prophetic reference to himself, Alexander asked the Jewish leaders what he could do for them. They asked him for relief from taxes during their sabbatical years when they let their fields lie fallow and did not harvest their crops. Alexander is said to have granted their request. The passage in Josephus runs as follows:

> And when the book of Daniel was showed him, wherein Daniel declared that one of the Greeks should destroy the empire of the Persians, he supposed that himself was the person intended; and as he was then glad, he dismissed the multitude for the present, but the next day he called them to him, and bade them ask what favours they pleased of him: whereupon the high priest desired that they might enjoy the laws of their forefathers, and might pay no tribute on the seventh year. He granted all they desired: and when they entreated him that he would permit the Jews in Babylon and Media to enjoy their own laws also, he willingly promised to do hereafter what they desired.[14]

If this story in Josephus is accurate, then the prophecy of Daniel 8, including the element of the great horn of Greece which was Alexander, was already in existence by the fourth century BC. Not only does this give evidence of the early date for the composition of Daniel, but it also shows how one element of this prophecy met its fulfilment and was recognised at the time when it did so.

Needless to say, once again critics of the predictive nature of Daniel dismiss this story as unhistorical because it contradicts their theory about history being written as if it were prophecy. There is some evidence in the story itself, however, which testifies to the historical nature of the meeting of Alexander and the priests at Jerusalem. That evidence comes from the reference to a sabbatical year in this context.

About a dozen references to sabbatical years have been found in extra-biblical sources. These texts and inscriptions give the equivalent to those sabbatical years in terms of other calendars. Thus a table of sabbatical years can be filled out. The year in which this interview with Alexander occurred was 331 BC. According to the table of sabbatical years, 331 was indeed a sabbatical year. Now that Judaea had been taken over by the Macedonian king, the Jewish leaders could see the problem that would confront them when they had to pay taxes to him. They would have no harvest with which to pay the tax. Thus the urgency of their request.

This minor feature, the request based on the sabbatical year, gives evidence that the episode really did happen, and that the historical transition that then took place actually was prophesied by Daniel before it happened.[15]

The Roman destruction of Jerusalem

All the prophecies studied so far have been taken from the Old Testament. This is natural since there is a much larger body of prophetic literature in the Old Testament, which stretches over about 1000 years between Moses and Ezra.

We would be remiss, however, if we did not consider at least one prophecy from the New Testament to illustrate that it partakes of the same prophetic spirit. And who better to look to for a prophetic word than Jesus?

Probably the best-known prophecy of Jesus is that which is found in his sermon from the Mount of Olives recorded in the books of Matthew (chapter 24), Mark (chapter 13) and Luke (chapter 21). His teaching here concerned the signs that would precede his return to earth at the Second Coming. In giving this prophecy, however, Jesus blended two events: the end of Jerusalem brought about by the Romans, and the end of the world.

While there are elements here that deal with the experience of the early and later church, there are two specific elements that deal with Jerusalem itself. The first is found in the first two verses of Matthew 24, which serve as a prologue to the rest of the sermon. This part of the narrative occurred while Jesus and his disciples were passing out of the temple area on their way to the Mount of Olives:

> Jesus left the temple and was walking away when his disciples came up to him to call his attention to its buildings. 'Do you see all these things?' he asked. 'I tell you the truth, not one stone here will be left on another; every one will be thrown down.'

It is very easy to check the fulfilment of this prophecy because the site where the second temple (otherwise known as Herod's temple) stood is well known. It was somewhere within an area of about 20 acres enclosed by retaining walls and known today as the Temple Mount. The eastern retaining wall facing the Kidron Valley has been open to view through the centuries. The western wall, now also known as the Wailing Wall, has also been open to view, and it has been cleared more in recent years. The southern wall too has been cleared by archaeologists so that the portals that served as entryways to the temple area can now be seen.

But that still leaves open the question as to where the temple itself stood within those 20 acres. Today the area is occupied by the silver-domed al-Aqsa mosque on the southern edge of the platform and the golden Dome of the Rock near the centre. The Dome of the Rock is not a mosque but a memorial building built over the rock scarp protruding inside the building that is taken traditionally as the place from which Mohammed took his night ride to heaven. But where was the Jewish temple that was destroyed by the Romans located? This is unknown as there are no archaeological remnants left from it.

Archaeologists have excavated about fifty buildings in Syria and Israel that they classify as temples. These identifications can be made on the basis of the stumps of walls that are left in their ruins. From these remains, temples can generally be classified as one-room, two-room or three-room temples. They can also be divided into three architectural types: broad room temples, long-axis temples and square temples. From the description of Solomon's temple in the Bible we know that it was a long-axis building, and the second temple was built to the same pattern. But where? No remnants of its walls have been found, so archaeologists do not know where it was located on the Temple Mount.

This absence has led to several theories. The most common is that the rock under the Dome of the Rock was the base for the altar of burnt offering. If this is correct, it would mean that the temple building proper was built west of where the Dome of the Rock stands today. Other archaeologists think it was built farther east on the temple platform. Still others think it was built farther north. Theories abound like this because we do not have any archaeological evidence. The temple was so thoroughly destroyed that not even remnants remain: not one block upon another. The prophecy of Jesus has been literally fulfilled.

The second part of Jesus' prophecy dealing with Jerusalem is a warning that when the people of Jerusalem and Judaea saw the banners of the Roman legions standing in the holy place of the temple, they were to flee.[16] This refers to the fact that there was a brief lull at the beginning of the Roman war that presented an opportunity for the inhabitants to flee.

The Roman attack on Jerusalem began in the autumn of 66 AD, when Cestius Gallus led his troops to Gibeon about five or six miles north of Jerusalem. At the time the Jews were celebrating the Feast of Tabernacles, which generally falls in our month of September or early October. The Roman army fought its way right up to the northern wall of the temple area. Many Jews in the city thought that their case was hopeless. But then for some unexplained reason the Roman commander withdrew his troops from that forward position. Encouraged by this withdrawal, the Jews counterattacked and drove the Romans all the way down to the coast by way of the Beth Horon pass.[17]

This temporary Roman retreat gave Jerusalem and Judaea a brief breathing space through the winter of 66–67 AD. Unfortunately for the inhabitants there, the Roman army returned in the spring of 67 AD and continued in unrelenting pursuit of its enemies until all of Judaea was conquered and Jerusalem was taken and destroyed in 70 AD. Today there are two museums of burnt houses from the first century in the Jewish Quarter of the city that give mute evidence to the ferocity of that destruction.

One could say that this prophetic warning to flee was written up after the war was over, but there is evidence that this was not the case. The evidence comes from the fact that the Christians, warned by Jesus' prophecy, did flee. We know this from the writings of Eusebius, the Christian bishop of Caesarea in the early fourth century AD. He says that during the interval when the Roman troops withdrew, the Christians fled down to the east to the Jordan River and then crossed the river and went up its east side to a town named Pella, where they sat out the war and were saved from the death and destruction that ensued with the Roman victory.

Thus the prophetic warning of Jesus was not just an academic question of whether history met prophecy or not; it was a matter of life and death for the Christians of that time and history indicates that they heeded the warning.

Biblical prophecies

There are many prophecies in the Bible that the Bible itself says were fulfilled. But the records of these fulfilments are found only in the Bible. In these instances there is no external evidence confirming the fulfilment. However, there are also many biblical prophecies for which there is external evidence which indicates they were fulfilled. The cases assembled here demonstrate that point.

Biblical prophecies operate on various levels. Some were addressed to individuals, some to towns or cities and some to kingdoms or nations. The same is true in terms of time range. Some prophecies concerned immediate circumstances and others events in the relatively near future, while still others can be classified as long range predictions that extend over centuries. The cases described in this chapter cover most of these possibilities.

The one common factor in all these cases is that there is external evidence to demonstrate the accuracy of the predictions. This provides evidence that the prophecies were written on the basis of more than just educated human guesses. They testify to the God who gave inside information to his servants the prophets. And this is one more good reason to believe that the biblical God exists.[18]

22

The Messianic Prophecies Fulfilled in Jesus

Arnold G. Fruchtenbaum

The paradox
The Old Testament of the Bible was written hundreds of years before the birth of Jesus, yet it contains accurate prophecies about his life. In this chapter we will take a closer look at some of these prophecies.

Anyone who sets himself the task of seeking to know what the Old Testament has to say about the coming of the Messiah soon finds himself involved with a seeming paradox. At times one even seems to be faced with an outright contradiction. On the one hand, the inquirer will find numerous predictions regarding the Messiah which portray him as one who is going to suffer humiliation, physical harm and finally death in a violent manner. This death was stated by the prophets to be a 'substitutionary' death for the sins of the Jewish people. On the other hand, the prophets also spoke of the Messiah coming as a 'conquering king' who would destroy the enemies of Israel and set up a Messianic Kingdom of peace and prosperity. Thus the Jewish prophets gave a two-fold picture of the Messiah who was to come.

For centuries past, during the formulation of the Talmud (rabbinic commentaries), our rabbis made serious studies of messianic prophecies. They came up with this conclusion: the prophets spoke of two different Messiahs. The Messiah who was to come, suffer and die they termed *Mashiach ben Yosef*, 'Messiah, the Son of Joseph'. The second Messiah who would then follow the first they termed *Mashiach ben David*, 'Messiah, the Son of David'. This one would raise the first Messiah back to life and establish the Messianic Kingdom of peace on earth.

That the Old Testament presents these two lines of messianic prophecy was something that all the early rabbis recognised. The Old Testament never clearly states that there will be two Messiahs. In fact, many of the paradoxical descriptions are found side by side in the same passages, in which it seems that only one person is meant. But to the early rabbis, the two Messiahs theory seemed to be the best answer.

For centuries, Orthodox Judaism held the concept of two Messiahs. In Jewish history since the Talmudic period, however, Mashiach ben David alone has been played up in the imaginations of Jewish hearts and minds. The other messianic figure, Mashiach ben Yosef—the 'suffering' one—has been ignored. He has been there in Jewish theology when needed to explain the Suffering Messiah passages contained in the Old Testament, for his existence provided an escape clause when thorny questions were raised. Otherwise, this messianic figure has been largely neglected. Few Jews today have heard of

Mashiach ben Yosef or know of his existence in Jewish theology of the past. The one that Jews today know of is the one who is to conquer, Mashiach ben David.

The source of the paradox

One of the major sources from which the rabbis developed their concept of the Suffering Messiah, the Son of Joseph, was Isaiah 53. The present-day bone of contention regarding what the Old Testament says about the Messiah centres on this chapter. The passage speaks of a Servant, the Servant of Jehovah. This Servant undergoes a great deal of suffering, ending in death. The chapter goes on to state that this suffering is a vicarious suffering, that the death is a substitutionary death for sin—he is suffering and dying for the sins of others. Finally, the passage indicates that this Servant is resurrected.

The bone of contention is not so much over what the passage says, but of whom it speaks. Did Isaiah prophesy concerning the Messiah here? Rabbis say that this is the 'Christian interpretation', not the Jewish one. The 'Jewish interpretation', they say, is that Isaiah is speaking about the people of Israel suffering in the Gentile world. According to the rabbis, Isaiah 53 does not speak of the Messiah at all.

But to make the passage speak of the collective body of Israel seems almost to force an interpretation. Taken by itself, the passage seems to have only one individual in mind.

In a book I wrote several years ago, *Jesus Was a Jew*,[1] I quoted source after source showing that the historical Jewish interpretation of Isaiah 53 is that it speaks of the Messiah, not of the nation. In fact, the first rabbi ever to claim that Isaiah 53 speaks of the nation and not of an individual was Rashi, about the year 1100 AD. I might add that he was opposed in this interpretation by the majority of the rabbis of his day, and the rabbis continued to oppose his interpretation for centuries after him. Historically speaking, it was not until the 1800s that the 'national interpretation' of Isaiah 53 became the dominant rabbinical view.

To interpret Isaiah 53 as speaking of the Messiah is not un-Jewish. In fact, if we speak of the traditional Jewish interpretation, it would be that this passage speaks of the Messiah. Rashi's position was contrary to all rabbinical teaching of his day and of the preceding one thousand years. Today his view has become dominant but it is not the traditional Jewish view. The traditional view is that Isaiah 53 speaks of the Messiah, not of the nation.

The text of Isaiah 52:13 to 53:12

The text itself should be able to help us determine whether the Suffering Servant is the individual Messiah or the nation of Israel. Before dealing with some specific details, it might prove helpful to quote the entire Isaiah passage, and then make a summary of it.[2]

> 52:13Behold, my servant shall deal wisely, he shall be exalted and lifted up, and shall be very high. 14Like as many were astonished at thee (his visage was so marred more than any man, and his form more than the sons of men), 15so shall he sprinkle many nations; kings shall shut their mouths at him: for that which had not been told them shall they see; and that which they had not heard shall they understand.

⁵³:¹Who hath believed our message? and to whom hath the arm of Jehovah been revealed? ²For he grew up before him as a tender plant, and as a root out of a dry ground: he hath no form nor comeliness; and when we see him, there is no beauty that we should desire him. ³He was despised, and rejected of men; a man of sorrows, and acquainted with grief: and as one from whom men hide their face he was despised; and we esteemed him not. ⁴Surely he hath borne our griefs, and carried our sorrows; yet we did esteem him stricken, smitten of God, and afflicted. ⁵But he was wounded for our transgressions, he was bruised for our iniquities; the chastisement of our peace was upon him; and with his stripes we are healed. ⁶All we like sheep have gone astray; we have turned every one to his own way; and Jehovah hath laid on him the iniquity of us all.

⁷He was oppressed, yet when he was afflicted he opened not his mouth; as a lamb that is led to the slaughter, and as a sheep that before its shearers is dumb, so he opened not his mouth. ⁸By oppression and judgment he was taken away; and as for his generation, who among them considered that he was cut off out of the land of the living for the transgression of my people to whom the stroke was due? ⁹And they made his grave with the wicked, and with a rich man in his death; although he had done no violence, neither was any deceit in his mouth.

¹⁰Yet it pleased Jehovah to bruise him; he hath put him to grief: when thou shalt make his soul an offering for sin, he shall see his seed, he shall prolong his days, and the pleasure of Jehovah shall prosper in his hand. ¹¹He shall see of the travail of his soul, and shall be satisfied: by the knowledge of himself shall my righteous servant justify many; and he shall bear their iniquities.

¹²Therefore will I divide him a portion with the great, and he shall divide the spoil with the strong; because he poured out his soul unto death, and was numbered with the transgressors: yet he bare the sin of many, and made intercession for the transgressors.

52:13–15

In 52:13–15, God is doing the speaking. He is calling the attention of all to the Suffering Servant. God declares that his Servant will act 'wisely', and his actions will gain him a position of glory. God further states that his Servant will suffer, but this suffering will eventually gain the silent attention of world rulers when they begin to understand the purpose of his suffering. The Servant will be terribly disfigured but will, in the end, save many.

53:1–9

After God has thus drawn the attention of the people to his Servant, the people now respond in 53:1–9.

In verses 1–3 they confess their non-recognition of the Servant in his person and calling. In verse 2 they claim to be surprised at what they have just learned from the three preceding verses. They note that at the time the Servant was with them, there did not seem to be anything special about him. His childhood and growth were no different from that of others. There was nothing particularly charismatic in his personality that would attract men to him. His outward features were hardly unique. On the contrary, the opposite was true. Instead of drawing people to him, he was 'despised and rejected of men' (in general), and he was 'a man of sorrows, acquainted with grief' (personally). His rejection was not merely passive; it was active, and the people did their best to avoid him.

In verses 4–6 the people confess that at the time of his suffering, they considered it to be the punishment of God for his own sins. Now, however, they acknowledge that the Servant's suffering was vicarious: he suffered for the sins of the people and not for his own sins. The people confess that it was they who went astray; they each one had gone their own selfish ways, and the punishment of their sins was laid upon this Servant of Jehovah. This passage, then, is a confession of a change of attitude on the part of the people towards the Servant as they recognise the true nature of his sufferings. The severe judgment that the Servant had suffered led the people to form an opinion of him, since his suffering seemed to mark him as a special victim of Jehovah's anger. But now they reverse this opinion, marking the beginning of repentance.

In verse 5 the people confess that the vicarious suffering of the Servant of Jehovah resulted in reconciliation and spiritual healing. This verse penetrates more deeply into the meaning of the Servant's sufferings, seeing the connection between his passion and their sins. The connection is two-fold: the chastisement for our sins (suffering was the penalty due to the people's transgression) and the means of reconciliation (it was the remedy by which the people were restored to spiritual health). It was for the sins of the people that he was suffering and not for his own sins.

In verse 6 the people confess why the sufferings spoken of in the preceding verses were necessary. It was because the people were so wholly estranged from God that substitution was required for reconciliation. They had strayed and selfishly sought their own way; yet Jehovah laid their sins on the Servant. Thus the people confess with penitence that they have long-mistaken him whom God has sent to them for their good, even when they have gone astray to their own ruin.

In verses 7–9 the prophet Isaiah appears to be doing the speaking as he describes and details the sufferings of the Servant which lead to his death.

In verse 7 the Servant is pictured as humbly submitting himself to unjust treatment. He does not speak a word in his own defence. He suffers quietly, never crying out against the injustice done to him.

In verse 8 we find the death of the Servant of Jehovah. Here we are told that after a judicial trial and judgment, he was taken away for execution. The Servant of Jehovah was being executed for the sins of the prophet's own people, the ones who deserved the judgment of judicial execution. But no one seemed to realise the holy purpose of God in this event. Verse 8 is a key verse to the entire passage, in that we learn that this was a sentence of death pronounced in a court of law and then executed. This verse clearly states that he did not deserve death. Those for whom he was dying never realised the true reason for his death. But, as verses 4–6 have related, they assumed he was dying for his own sins.

In verse 9 the burial of the Servant is described. After his death, those who executed him assigned a criminal's grave for him along with other criminals. A criminal is what they considered him to be, and that is the way he was executed. Yet he would be buried in a rich man's tomb! This is true poetic justice since, in actuality, the Servant had done nothing wrong, nor was there anything wrong in his character.

53:10–12

In verses 10–12 we have the results of the sufferings and death of the Servant of Jehovah. These results, in the end, are very beneficial.

Verse 10 records how God was pleased to allow the Servant to suffer and die. This was the means by which God was going to make atonement for the people. The death of the Servant was an offering for the sins of the people. The ones who had 'gone astray' and sinned would now be forgiven on the basis of the death of the Servant, for by his substitutionary death he provided atonement for the people. God punished the Servant in place of the people, and thus the sins of the people were atoned for. This verse further states that the Servant will see his posterity, and his days will be prolonged. How can that be if the Servant is killed? The only way this would be possible is by means of resurrection. So the 'good pleasure' of the Lord, the verse concludes, will continue to 'prosper in his hand', for he will live again because of his resurrection.

Verse 11 declares that God will 'be satisfied' with the work of the Servant. The Servant of Jehovah dies a substitutionary death for the sins of the people. The question now is, will God accept this substitution? And the answer is yes. For God will see the sufferings and death of the Servant and God's justice will be satisfied. Therefore, God can make the next statement, that because of the Servant's vicarious suffering and death, the righteous Servant will 'justify many'. To *justify* means to 'declare righteous'. So the Servant who suffered and died, and is now resurrected, will be able to make many righteous. The people who were sinners and could do nothing because of their separation from God will be able to be made righteous by the Servant. This verse concludes by telling us how this is possible —the Servant bears their sins. Their sins are put on the Servant's account, and the account is considered 'paid in full' by the Servant's blood.

Verse 12 records that in the end the Servant will be tremendously and greatly blessed by God above all others. The reasons for this are given in the verse. First of all, he willingly and voluntarily suffered and died. Secondly, he was humble enough to allow others to consider him a sinner and to consider him as suffering and dying for his own sins. Thirdly, however, he actually 'bore the sin of many'. For the 'many' who are justified and made righteous are made so only because he has put their sins on his account. Fourth and finally, the Servant 'intercedes' and pleads with God on behalf of the sinners.

This, essentially, summarises the content of the passage. If the Servant is Israel, the people are the Gentiles. If the Servant is the Messiah, the people are Israel, the Jewish people. Until Rashi, all Jewish theology taught that the passage refers to the Messiah. Since the 1800s, most rabbinical theology has taught that it refers to Israel. But if the passage is taken literally and read simply, it speaks of a single individual.

Clues to interpretation

The text itself provides a number of clues as to whether it refers to an individual Messiah or to the collective body of Israel.

The consistent usage of pronouns

An important clue is the consistent usage of pronouns. A distinction is maintained between *we*, *us* and *our* over against *he*, *him* and *his*. *We*, *us* and *our* must refer to Isaiah the prophet and the people to whom Isaiah is speaking. *He*, *him* and *his* must refer to the Suffering Servant. Now Isaiah was a Jew, as were also the people to whom he was speaking. It will be good to requote a portion of this passage to bring out the emphasis of the various pronouns in order to get a clearer understanding of the point being made. Here is 53:4–9:

> [4]Surely *he* hath borne *our* griefs, and carried *our* sorrows; yet *we* did esteem *him* stricken, smitten of God, and afflicted. [5]But *he* was wounded for *our* transgressions, *he* was bruised for *our* iniquities; the chastisement of *our* peace was upon *him*; and with *his* stripes *we* are healed. [6]All *we* like sheep have gone astray; *we* have turned every one to his own way; and Jehovah hath laid on *him* the iniquity of *us* all.
>
> [7]*He* was oppressed, yet when *he* was afflicted *he* opened not *his* mouth; as a lamb that is led to the slaughter, and as a sheep that before its shearers is dumb, so *he* opened not *his* mouth. [8]By oppression and judgment *he* was taken away; and as for *his* generation, who among them considered that *he* was cut off out of the land of the living for the transgression of *my* people to whom the stroke was due? [9]And they made *his* grave with the wicked, and with a rich man in *his* death; although *he* had done no violence, neither was any deceit in *his* mouth.

Obviously, the *we*, *us* and *our* are the Jews. Isaiah is speaking to the nation of Israel, the Jewish people as a whole. He is including himself with the collective body of Israel. The Suffering Servant is in a different category: *he, him* and *his*. *He* is the one who is suffering for *us*; *he* is the one upon whom God is laying *our* sins; *he* is the one who is going to die for *our* sins so that *we* can have salvation through *him*.

The constant and consistent use of pronouns and the identification of the pronouns exclude the Suffering Servant from being interpreted as the nation of Israel. Rather, the Suffering Servant is the Messiah himself.

Isaiah's view of the Servant's death

The second clue to the Servant's identity is in the closing sentence of verse 8, which also serves to exclude Israel from being the Servant. It reads: 'he was cut off out of the land of the living for the transgression of my people to whom the stroke was due . . .'

As Isaiah views the death of the Suffering Servant, he discloses that his death is for the sins of 'my people'. Who are Isaiah's people? No one questions that Isaiah was a Jew. Thus Isaiah's people must be the Jews; they must be the people of Israel as well. And if 'my people' are Israel, they cannot also be the Suffering Servant. Hence, the Suffering Servant must refer to the individual Messiah.

A single human personality portrayed

A third clue lies in the fact that throughout the passage, the Suffering Servant is portrayed as a singular human personality. There is no hint of allegory or any suggestion that the Suffering Servant is to be taken symbolically as referring to Israel. The

Servant goes through all the functions that an individual personality goes through. Israel is kept distinct from the Suffering Servant. The Messiah is viewed as a future, historical person who would accomplish the prophecy of Isaiah. Israel is the people looking on while this is happening. There is no personification of Israel at all in the passage.

An innocent sufferer

The fourth clue lies in the fact that the Suffering Servant is presented in the passage as an innocent sufferer (verses 4-6, 8b, 9b). It is easy to see how this can be true of the Messiah, but it is impossible of Israel. Moses and the prophets never told Israel, 'You will suffer for being innocent.' Rather they said, 'You will suffer for your sins unless you repent and conform to the revealed will of God.' God punished Israel many times and in various ways, and it was always because of sins. According to the prophets, both the Babylonian exile and the present-day dispersion were the result of disobedience on the part of Israel to the revealed will of God. This is in sharp contrast with the Suffering Servant, who is portrayed as an innocent sufferer.

A voluntary, willing and silent sufferer

The fifth clue is the fact that the Servant is further portrayed as a voluntary, willing and silent sufferer (verse 7). He willingly submits to the suffering he undergoes and voices no complaint about the injustice done him. Furthermore, as he undergoes these sufferings that lead to his death, he is silent.

In Israel's history, the Jews have been oppressed and have gone into captivity, exile and finally present-day dispersion.

But none of these occurred voluntarily on Israel's part. Israel has generally fought back. These things fell on Israel only because she was defeated, and Israel was never defeated willingly. But the Messiah would be a willing sufferer.

Furthermore, reading through the literature of Jewish history, it can hardly be said that Israel has been a silent sufferer. Rather, during her sufferings Israel has always cried out against the inhumanity of those who were afflicting her. Israel has produced a large body of literature cataloguing her sufferings and complaints. The activities of the Jewish Defense League show that there is sometimes violence directed against anti-Semites and a desire to see them destroyed. This, too, rules out interpreting the Suffering Servant as the personification of Israel and again points to him being the Messiah.

The Servant's vicarious and substitutionary death

The sixth clue is that in this passage the Suffering Servant suffers a vicarious and substitutionary death (vervses 4–6, 8, 10, 12). He suffers for the sake of others so that they need not suffer for their own sins. Nowhere in the Scriptures or in Jewish history do we ever see Israel suffering for the Gentiles. Israel often suffers *because* of the Gentiles, but never *for* the Gentiles. Israel suffers, but always for her own sins. There is no substitution where Israel is concerned, only where the Messiah is concerned.

The Servant's justification and spiritual healing

The seventh clue is that the sufferings of the Servant bring justification and spiritual healing to those who accept them (verses 5b, 11b). The sufferings of Israel have failed to bring justification and spiritual healing to the Gentiles. After three thousand years of Jewish suffering, the Gentiles are hardly justified and are still spiritually sick, as became obvious by the way the Gentile nations were involved in the Holocaust. But the Messiah's suffering was to bring this justification and spiritual healing to Jewish lives.

The death of the Servant

The eighth clue is a crucial one. The Suffering Servant dies (verses 8, 12). The sufferings of the Servant lead to and end in death. This especially makes the personification of Israel in this passage impossible. The Jewish people are alive and well in spite of many attempts to destroy them by anti-Semites through the centuries. This again forces one to the conclusion that the Suffering Servant cannot be Israel personified, but rather an individual Messiah.

The resurrection of the Servant

The ninth and final clue naturally follows: the Suffering Servant is resurrected (verses 10–11). The one who died for sins does not stay dead but is resurrected, and we can see the results of his suffering in that he brings justification and spiritual healing to many. Since Israel never died, there is no need for a resurrection. But if a person like the Messiah dies, God will certainly resurrect him to life again.

This, then, is the conflict over Isaiah 53. If one simply reads the chapter as one would read any chapter of another book, no other conclusion can be reached than that Isaiah speaks of an individual person suffering for the sins of the Jewish people. And for centuries, this was the only conclusion that Judaism ever had—they labelled the Suffering Servant as the Messiah, the Son of Joseph. Later rabbinical interpretation that made the Suffering Servant the personification of Israel seemed more like an attempt to explain away the passage rather than actually explain it.

Isaiah 53 must be read without prejudice and taken simply for what it is saying. It must not be interpreted in any way that is merely a defence against Christian polemics. The traditional Jewish viewpoint is most in harmony with the simple statements of the text itself, speaking of the sufferings of the Messiah for the sins of Israel.

The uniqueness of his birth

If the Old Testament only spoke of the Messiah in terms of his suffering, it would hardly give us enough to go on. But there is much more to the Old Testament picture of the Messiah than Isaiah 53.

Genesis 3:15

Following the account of the creation, the Old Testament continues with the story of Adam and Eve. In the guise of a serpent, Satan deceives Eve and causes her to break the one commandment of God. Adam follows suit. The result is that sin enters human experience. Human beings now stand under the righteous judgment of God.

Nevertheless, at the time of the Fall, God provides for future redemption. As he addresses Satan, God says in Genesis 3:15:

> And I will put enmity between you and the woman, and between your seed and her seed; he shall bruise you on the head, and you shall bruise him on the heel.

The key to this verse is the term 'her seed'. In and of itself this statement may not seem remarkable, but in the context of biblical teaching it is most unusual. Throughout the Hebrew Scriptures, lineage was always reckoned after the man, never the woman. In all the genealogies we have in the Old Testament record, women are virtually ignored because they were considered unimportant in determining genealogy. Yet the future person who would bruise Satan's head would not be reckoned after a man, but after a woman.

In spite of the normal biblical pattern, here we have a clear statement that the future Redeemer will come from the 'seed' of the woman. His birth will take into account his mother only. For a reason that is not explained here, the father will not be taken into account at all. Yet this goes totally contrary to the whole Hebrew view regarding genealogies.

That this verse was taken to be messianic is clear from the 'Targums of Jonathan' and the 'Jerusalem Targums' (Aramaic paraphrases of the Hebrew text). Furthermore, the Talmudic expression 'Heels of the Messiah' seems to have been taken from this verse. But Genesis itself does not explain how or why this Redeemer can be labelled 'seed of the woman' when it goes contrary to the biblical pattern.

Isaiah 7:14

Centuries later, Israel had a great prophet in the person of Isaiah. It was left to this prophet to explain the meaning and reason why the Messiah would be reckoned only after the seed of the woman. Isaiah writes in Isaiah 7:14 (NIV):

> Therefore the Lord himself will give you a sign: The virgin will be with child and will give birth to a son, and will call him Immanuel.

The very fact that the birth of the Person spoken of in this passage is described as 'a sign' points to some unusual circumstance regarding the birth. In other words, the birth could not be normal, for that would not fulfil the requirement of the word 'sign'. It had to be unusual in some way—perhaps miraculous, or at least attention-getting.

The very existence of the Jewish people is derived from the sign of a birth. The Scriptures make it clear that both Abraham and Sarah were beyond the point of being able to bear children; Abraham was ninety-nine years old, and Sarah eighty-nine. She had, of course, already undergone menopause when God promised that she would have a son within one year![3] This would be the sign that God would keep his covenant with Abraham and would make a great nation from him. A year later this sign took place with the birth of Isaac, through whom the Jewish people came. It was the sign needed to authenticate the covenant. This was a miraculous birth.

The birth of the 'son' in Isaiah 7:14 was also to be a sign—to be unusual in some way. But this time, the unusual nature of the birth was not going to be due to the great age of the mother. It would be a sign by virtue of the fact that this son would be born to 'a virgin'.

At this point, another conflict often ensues. Rabbis today

claim that the Hebrew word *almah* does not mean 'virgin' but 'young woman'. But what they fail to explain is how this would be a sign. A young woman giving birth to a baby is hardly unusual.

In other passages where *almah* is used, it clearly means 'virgin'. It occurs in six other places in the Old Testament outside Isaiah 7:14, and in all six no one disagrees that it means 'virgin'. If it means 'virgin' in those six other passages, there is no way it could mean 'non-virgin' in Isaiah 7:14.

This was certainly the view of the seventy Jewish rabbis who translated the Old Testament into Greek around 250 BC (the translation known as the Septuagint). These rabbis all translated *almah* with *parthenos*, the simple Greek word for 'virgin'.

Even if *almah* is allowed to mean 'young woman', it still must be admitted that the word can refer to a 'virginal young woman'. But it must not be ignored that this birth was to be a sign—an unusual birth. This is best seen if taken to mean a 'virgin birth'.

This, then, is the explanation of the mystery of Genesis 3:15. The Messiah would be reckoned after the seed of a woman because He would *not* have a human father. Because of a virgin birth, his lineage could be traced only through his mother and not his father.

The place of his birth

Not only was the means of the Messiah's birth prophesied, but also the place. This was done by the prophet Micah, a contemporary of Isaiah. In chapter 5 of his book, we read in verse 2:

> But as for you, Bethlehem Ephrathah,
> Too little to be among the clans of Judah,
> From you One will go forth for Me to be ruler in Israel.
> His goings forth are from long ago,
> From the days of eternity.

Concerning this verse there is far less disagreement among Orthodox rabbis, since they generally take it to mean that the Messiah will originate from Bethlehem. This is the view taken by 'The Soncino Books of the Bible', which is the Orthodox Jewish commentary on the Old Testament and which takes as its source some earlier Jewish commentaries.

The lineage of the Messiah

Another point that is uncontested is that the Messiah would be a descendant of King David. From this comes the rabbinical ascription 'Mashiach ben David'.

Of the numerous passages that might be cited, we will limit ourselves to the following two, both from Isaiah. The first is Isaiah 11:1:

> Then a shoot will spring from the stem of Jesse,
> And a branch from his roots will bear fruit.

The second is Isaiah 11:10:

> Then it will come about in that day
> That the nations will resort to the root of Jesse,
> Who will stand as a signal for the peoples;
> And His resting place will be glorious.

Jesse was the father of David; thus, these passages show that the Messiah will come from the House of David. With this all Orthodox Judaism agrees. Other passages regarding this same point will be cited later in a different context.

The sufferings of the Messiah

Isaiah 53

As we have seen, that the Messiah will suffer and die was something upon which all early rabbis agreed. The central passage which supports this view is Isaiah 53, but there are others.

Psalm 22:1–21

This is another important passage dealing with the suffering of the Messiah.

> ^{1}My God, my God, why hast Thou forsaken me?
> Far from my deliverance are the words of my groaning.
> ^{2}O my God, I cry by day, but Thou dost not answer;
> And by night, but I have no rest.
> ^{3}Yet Thou are holy,
> O Thou who are enthroned upon the praises of Israel.
> ^{4}In Thee our fathers trusted;
> They trusted, and Thou didst deliver them.
> ^{5}To Thee they cried out, and were delivered;
> In Thee they trusted, and were not disappointed.
> ^{6}But I am a worm, and not a man,
> A reproach of men, and despised by the people.
> ^{7}All who see me sneer at me;
> They separate with the lip, they wag the head, saying,

⁸'Commit yourself to the LORD; let Him deliver him;
Let Him rescue him, because He delights in him.'
⁹Yet Thou art He who didst bring me forth from the womb;
Thou didst make me trust when upon my mother's breasts.
¹⁰Upon Thee I was cast from birth;
Thou hast been my God from my mother's womb.
¹¹Be not far from me, for trouble is near;
For there is none to help.
¹²Many bulls have surrounded me;
Strong bulls of Bashan have encircled me.
¹³They open wide their mouth at me,
As a ravening and a roaring lion.
¹⁴I am poured out like water,
And all my bones are out of joint;
My heart is like wax;
It is melted within me.
¹⁵My strength is dried up like a potsherd,
And my tongue cleaves to my jaws;
And Thou dost lay me in the dust of death.
¹⁶For dogs have surrounded me;
A band of evildoers has encompassed me;
They pierced my hands and my feet.
¹⁷I can count all my bones.
They look, they stare at me.
¹⁸They divide my garments among them,
And for my clothing they cast lots.
¹⁹But Thou, O LORD, be not far off;
O Thou my help, hasten to my assistance.

[20]Deliver my soul from the sword,
My only life from the power of the dog.
[21]Save me from the lion's mouth;
And from the horns of the wild oxen Thou dost answer me.

To summarise this passage, we find that the Messiah is forsaken by God and ridiculed and tormented by the people. His clothes are gambled away by his tormentors. He suffers such agony that all his bones come 'out of joint'. His heart is 'like wax', a Hebrew phrase meaning 'a ruptured heart', evidenced by the pouring out of blood and water. His hands and feet are 'pierced'.

In many ways, this Psalm is similar to Isaiah 53, providing even more detail as to the type of suffering and agony that the Messiah must undergo. The rabbis in the 'Yalkut' (another Aramaic paraphrase) also understood this passage to refer to Maschiach ben Yosef.

The kingship of the Messiah

In all the passages discussed so far, the Messiah has been portrayed as a man, but as 'a man of sorrows'; he is to suffer and die. The earlier rabbis all recognised that these passages spoke of the Messiah and called him Mashiach ben Yosef. For as Joseph the patriarch suffered at the hands of his brethren, the Messiah would also suffer.

But other Old Testament passages speak of another kind of Messiah: not a sufferer but a conqueror; not a dying Messiah but a reigning one: Mashiach ben David. Most of what is said about the Messiah in Moses and the prophets revolves around the Messiah's coming to bring peace and to establish the Messianic Kingdom in Israel.

There are far too many such passages to even begin to list them here, but one passage will be quoted in full to illustrate the point. It should be noted how differently this Messiah is portrayed compared with all the previous passages thus far discussed. It is little wonder that the early rabbis were confused and so devised the theory of the two Messiahs, with each Messiah coming only once.

Isaiah 11:1–10

> [1]Then a shoot will spring from the stem of Jesse,
> And a branch from his roots will bear fruit.
> [2]And the Spirit of the LORD will rest on Him,
> The spirit of wisdom and understanding,
> The spirit of counsel and strength,
> The spirit of knowledge and the fear of the LORD.
> [3]And He will delight in the fear of the LORD,
> And He will not judge by what His eyes see,
> Nor make a decision by what His ears hear;
> [4]But with righteousness He will judge the poor,
> And decide with fairness for the afflicted of the earth;
> And He will strike the earth with the rod of His mouth,
> And with the breath of His lips will He slay the wicked.
> [5]Also righteousness will be the belt about His loins,
> And faithfulness the belt about His waist.
> [6]And the wolf will dwell with the lamb,
> And the leopard will lie down with the kid,
> And the calf and the young lion and the fatling together;
> And a little boy will lead them.

> ⁷Also the cow and the bear will graze;
> Their young will lie down together;
> And the lion will eat straw like the ox.
> ⁸And the nursing child will play by the hole of the cobra,
> And the weaned child will put his hand on the viper's den.
> ⁹They will not hurt or destroy in all My holy mountain,
> For the earth will be full of the knowledge of the LORD,
> As the waters cover the sea.
> ¹⁰Then it will come about in that day
> That the nations will resort to the root of Jesse,
> Who will stand as a signal for the peoples;
> And His resting place will be glorious.

Both the ancient and modern rabbis agree that this passage speaks of the Messiah and the Messianic Age. But unlike the previous passages, there is no picture of a dying Messiah being rebuked and despised by his people. The picture we get here is of a reigning Messiah who brings peace and prosperity to the entire world. This peace extends even down to the animal kingdom, and the wicked are removed in judgment.

The God-Man concept and the Messiah

Another aspect of the kingship of the Messiah is the strange God-Man concept that emerges in some passages. These passages add a whole new dimension to the person of the Messiah, making him a man—and yet more than a man.

Isaiah 9:6–7

> ⁶For a child will be born to us, a son will be given to us;
> And the government will rest on His shoulders;

> And His name will be called Wonderful Counsellor, Mighty God, Eternal Father, Prince of Peace.
> ⁷There will be no end to the increase of His government or of peace,
> On the throne of David and over his kingdom,
> To establish it and to uphold it with justice and with righteousness
> From then on and forevermore.
> The zeal of the LORD of hosts will accomplish this.

Verse 6 declares that 'a son' will be born into the Jewish world who will eventually control the reins of 'government'. Verse 7 identifies him as the Messianic descendant of David; it gives a dramatic description of his reign, which will be characterised by peace and justice. But in verse 6, he is given names that can only be true of God himself: 'Mighty God' and 'Everlasting Father' ('Wonderful Counsellor' and 'Prince of Peace' can be true of a man).

This new dimension is that the Messiah had to be a man, a descendant of David, but he was to be God as well. It further explains what Isaiah says in Isaiah 7:14, which we discussed earlier:

> Therefore the Lord himself will give you a sign: The virgin will be with child and will give birth to a son, and will call him Immanuel.

Here Isaiah declares that the son who is going to be born of a virgin is to be named 'Immanuel'. In the Bible, when parents name a child, it shows the thinking of the parents. However, when God gives a person a name, it actually represents the person's very character, which only God can foresee. So when this child is named Immanuel by God, the name portrays the actual character of the child.

What does Immanuel mean? It means: 'With us, God' or 'God is among us'. So here we have a child that is born of a virgin and who is 'With us, God'.

Jeremiah 23:5–6

Isaiah is not alone in presenting this picture of the Messiah as the God-Man. The thought is echoed in Jeremiah 23:5–6:

> ⁵'Behold, the days are coming,' declares the LORD,
> 'When I shall raise up for David a righteous Branch;
> And He will reign as king and act wisely
> And do justice and righteousness in the land.
> ⁶'In His days Judah will be saved,
> And Israel will dwell securely;
> And this is His name by which He will be called,
> "The LORD our righteousness." '

Here, too, a descendant of David reigns upon the throne of David, and the character of his reign is described as one of peace and security for Israel. Yet he is given the very name of God, which can only belong to God himself—*Adonai Tzidkenu*, 'Jehovah our righteousness'. This is the *YHVH*, the very name that God revealed to Moses as being his own personal name: 'I AM'. So once again, the future King Messiah of Israel is seen as a man on the one hand, but as God on the other. As with the Sonship concept, the God-Man concept is related to the Messiah's kingship.

Micah 5:2

Look again at this passage we studied earlier when we talked about the place of the Messiah's birth (Bethlehem):

> But as for you, Bethlehem Ephrathah,
> Too little to be among the clans of Judah,
> From you One will go forth for Me to be ruler in Israel.
> His goings forth are from long ago,
> From the days of eternity.

According to Micah, the Messiah's human origin would be Bethlehem. But the prophet states even further that 'His goings forth are from long ago, from the days of eternity'. This individual, who is to be born in Bethlehem, has his origins from eternity. Only one person is eternal from eternity past, and that is God himself. As to his human origin, this person was born in Bethlehem; as to his divine origin, he is from eternity, which means he is both God and man at the same time.

We can see from these passages the picture of the Messiah given in the Old Testament. On the one hand, he is a suffering and dying Messiah. On the other, he is a conquering and reigning Messiah called God and the Son of God.

The New Testament never claims that Yeshua—Jesus—was a man who became God. This is heresy, and it is contrary to Judaism of any form, biblical, rabbinical or otherwise. It is also contrary to faith in the Messiah. Neither the New Testament nor Yeshua himself ever taught that there was a man who became God. The New Testament claims the reverse: it was God who became a man in the person of Jesus of Nazareth.

23

The Evidence for Jesus' Resurrection

Michael R. Licona

This book has looked at various data pointing to the existence of God. We have observed that this God is eternal and extremely intelligent. But can we know anything else about him/her/it? Has God ever revealed himself to humans in a manner where we can know who it is who created us? Can we know if God is good or evil?

Most religions attempt to answer these questions with the god they present. But is there any way of determining which of these religions is true—or whether any of them are? Several religions provide a truth-test. For example, the Book of Mormon claims that if you read it with an open mind and a sincere heart while asking God to reveal its truth to you, he will.[1] In a similar manner, the Qur'an challenges anyone wanting to know if it is from God simply to try to produce a sura (chapter) like one in it. In fact, it invites the doubter to enlist any number of wise people to assist him. After a frustrating attempt at reproducing the same beauty, the doubter will know that only Allah could have written it.²

These truth-tests are quite subjective. Mormon missionaries are surprised when I inform them I have read a large portion of the Book of Mormon while sincerely praying with an open mind and have received no revelation from God of its truth. In fact, while seeking I believe God showed me it was not from him. The data from archaeology and huge problems relating to another Mormon scripture called the Book of Abraham pose serious challenges to the validity of the Mormon faith that may very well be insurmountable.³

Likewise, readers may find it helpful to compare the first sura in the Qur'an with Psalm 19 in the Christian Bible and judge for themselves which one is superior. Both contain similar content. While many may hold Psalm 19 to be more beautiful and pregnant with meaning, a Muslim might reply that the Qur'an must be read in Arabic in order to appreciate its full beauty, since the Arabic employed has a poetic rhythm. This may be true, but the Psalms were songs and likewise have a poetic rhythm when sung in Hebrew. Thus it seems that the Qur'an's truth-test comes down to one's opinion concerning which language sounds more beautiful. Such truth-tests are not very helpful since their answers appear to be based on feelings and subjective judgments.

Jesus offered a truth-test, too. But his was not subjective. Rather, it was linked to a historical event: his resurrection. While on earth Jesus made some radical claims. He claimed to be the uniquely divine Son of God[4] and that our eternal destiny depends on what we do with him.[5] When someone makes these sorts of claims, he can expect his critics to ask him for evidence supporting them. Jesus' critics were no different. He said he would provide them one key piece of evidence: his resurrection from the dead.[6] Therefore, if Jesus did not rise from the dead, we can know Christianity is false. On the other hand, if Jesus rose from the dead, then it seems he did this as a confirmation of his claims.

Jesus' truth-test is not very helpful if we cannot determine with a reasonable amount of confidence whether the event occurred. Can we put the resurrection of Jesus to the test in order to make a historical judgment? What if we were to look at this as a professional historian would, considering only the available data without any theological considerations? How would we go about this?

One of the challenges before us is that not everyone agrees on the data in which we can have confidence. For example, many critics are sceptical about the reports of Jesus' resurrection found in the gospels—they hold that the gospels were not written by the traditional authors, that they were written forty to seventy years after the event, that they contain legend and biased reporting, and/or that they were not intended to be understood literally. Each of these charges can be answered.[7] However, what if we were to avoid such a discussion altogether and consider data *earlier* than the gospels—data from sources held to be either the alleged eyewitnesses or those close to them? And what if we discovered that corroborating data exists from sources outside the New Testament, even from non-Christian sources?

In this chapter, our approach to the evidence for Jesus' resurrection will be to consider data so strongly evidenced that it is granted by the large majority of scholars who study the subject—even the rather sceptical ones. This is not to say that a consensus determines truth; however, it is important to note that the majority of even sceptical scholars are persuaded by the strength of this data.[8] We will focus on the reasons for accepting it.

To do this we will first select the data we will consider. Imagine a wealth of data dropped into a funnel that filters out any data that, while possibly true, does not have much supporting evidence. The data that remains is then dropped into a second funnel, which filters out data not granted by the large majority of scholars, including critics. The data that emerges after passing the screening of both filters is what we will call 'facts'. Even then, space prevents us considering all these facts. So we will look at the three most important. Finally, we will pass these three facts through a third funnel—one that, as much as possible, will leave intact only non-biblical sources that support them.

Fact 1: Jesus' death by crucifixion

In addition to what we read in the four gospels, Jesus' crucifixion is attested by even non-Christian ancient sources. Josephus (a first-century Jewish historian), Tacitus (an early second-century Roman historian), Lucian (a mid-second-century Greek satirist and historian) and Mara bar Serapion (a second to third-century prisoner writing to his son) all report the event.

That Jesus was killed by the process is quite clear. Crucifixion and the torture that usually preceded it was the worst way in which to die in antiquity. In the first century BC,

Cicero wrote that it was the most horrendous torture. Several ancient authors reported that the scourging that took place prior to crucifying the victim would many times expose his inner parts. Then the victim was usually nailed to the cross. Sometimes the victim's tongue was cut off and/or his eyes burned out while on the cross. In the first century, Seneca described a man hanging on the cross as being 'sickly', 'deformed' and swollen with 'ugly welts on shoulders and chest'.

In addition to the extreme physical pain there was psychological agony. The victim was humiliated by being hung on the cross naked. Occasionally a taller cross was used as a sign of greater humiliation. It was typical for the person to be mocked while on the cross. In the first century BC, the Jewish king Janaeus crucified a man and then held a banquet in front of him. While flute players played, he danced in front of the man suffering on the cross.

It was typical for a person to remain alive on the cross for days, with insects and severe muscle cramps adding to his already excruciating torment. Some debate exists among medical experts regarding the actual cause of death; however, most agree that death came as a result of asphyxiation. This is a condition of not enough oxygen. It is believed that on the cross, the victim would want to relieve the severe pain caused by his weight pressing down on his pierced feet. As a means of temporary relief, the victim would slump on the cross, allowing his entire weight to be supported by the nails in his hands or wrists.[9] In this slumped position, the victim would experience shallow breathing and find it difficult to exhale. In order to get some relief, he would push up on his pierced feet, exhale and return to the slumped position. In this setting, breaking the victim's legs was ironically considered a severe act of mercy to prevent the person pushing up. Their torment would be ended since they would die shortly after of asphyxiation.

Given this process, it is difficult to propose that Jesus somehow survived the cross. The Roman guards would be certain when a person was dead: he wouldn't be pushing up any longer. Moreover, given the scourging that took place prior to his crucifixion, even if Jesus had somehow survived the cross, there is no way in his pathetic and mutilated state that he could have convinced his disciples that he was the risen Lord.

Fact 2: The appearances

We can establish historically that shortly after Jesus' crucifixion, a number of people believed he had risen bodily and had appeared to them. Reports of Jesus' resurrection begin very early after his crucifixion, are multiply attested and come from eyewitnesses, one of which had been an enemy. From a historian's point of view this is quite strong material. At this point, we are not attempting to establish what it was these people experienced, only that they experienced something which they interpreted to have been the resurrected Jesus.

We will divide the experiences into two categories: the appearances to friends and the appearances to foes. For the former, let's consider Clement of Rome and Josephus.

Appearances to friends

Clement held a position of authority in the church in Rome and is believed to have been a disciple of Peter. In a letter he wrote to the church in Corinth around 95 AD, he reports how the disciples had been 'fully assured by the resurrection of our Lord Jesus Christ' and as a result preached the good news of God's kingdom.[10] This is pretty strong evidence, since it comes from someone who knew at least one of the disciples, and one of the three most prominent ones at that. Josephus, a non-Christian historian in the first century, likewise writes that the disciples 'reported that he had appeared to them three days after his crucifixion and that he was alive'.[11]

Many other sources exist that establish that Jesus' disciples claimed he had risen.[12] However, these two have been cited since they are outside the New Testament.

We can go one step further than establishing that this was the disciples' claim. No fewer than seven ancient sources attest to the disciples' willingness to suffer and even die for their belief in what they were claiming. Luke, Clement of Rome, Polycarp, Ignatius, Dionysius, Tertullian and Origen all speak of this fact. While it is certainly true that adherents of other worldviews and religions have been willing to die for their beliefs, this only confirms the point that the disciples sincerely regarded their beliefs as being true. Muslim terrorists willingly give up their lives for a cause they believe is true and just. While this speaks nothing regarding the truth of Islam, it certainly confirms the sincere beliefs of the terrorists that it is. Few people are willing to suffer and die for a known lie. Even less likely is it that all the members of a group would be willing to die for an event they all knew did not occur and for a cause they all knew offered no real hope. Thus, while the sincere beliefs of the disciples do not establish the truth of their claim, they do indicate that they were not intentionally lying about what they thought they had experienced: a post-resurrection appearance of Jesus.

Before we move on, it is helpful to note one further difference between the sincere beliefs of the disciples and those of today's adherents of Christianity, Islam, Hinduism and other religions who are willing to suffer and even die for their convictions. Today's adherents willingly suffer and die for what they *believe* is true. They trust the reports of earlier sources. On the other hand, the disciples suffered and died for what they *knew* was either true or false. Their belief had nothing to do with earlier sources since they themselves were the primary sources. Thus, if the disciples did not have the experiences of which they spoke, they suffered and died for a known lie. That all of the disciples suffered willingly to the point of death for a known lie stretches credulity. This is why there is a nearly universal consensus today, even among critical scholars, that Jesus' disciples were not lying. Rather, they sincerely believed he had risen from the dead and had appeared to them.

Appearances to foes

Now let's consider the second category: the appearances to foes. Not only did Jesus' friends believe he had risen and appeared to them; we know of at least two sceptics who likewise held that Jesus had appeared to them after his resurrection: Paul and James.

Paul is of special interest. In three letters known to have been written by him, he testified that he had been a persecutor of the church and had given up his status as a budding Jewish leader because the risen Jesus had appeared personally to him.[13] Luke described Paul's experience in Acts. There also appears to have been an early oral tradition circulating around Galatia noting that the one who once persecuted the church was now preaching the faith he once sought to destroy.[14] Thus Paul's conversion is documented by eyewitness, early and multiple sources. He was so convinced by this experience that he immediately became a Christian and was transformed from being a persecutor of the church to one of its most aggressive advocates. Paul's willingness to suffer and die for his conviction that he had seen the risen Jesus is testified to by no less than seven ancient sources: Paul, Luke, Clement of Rome, Polycarp, Dionysius, Tertullian and Origen.

The place of Paul's experience in our case for Jesus' resurrection may be more fully appreciated if we imaginatively place it in today's world. As this chapter is being written, it has been a little over two years since Osama bin Ladin and al-Qaida bombed the World Trade Center and Pentagon in the United States, and bin Ladin has still not been captured. Interestingly, bin Ladin and the Paul known to the Christian church prior to meeting Jesus have a few things in common. Bin Ladin hates the United States because he believes Americans are corrupting Islam's holy land with their military presence in Saudi Arabia. Paul hated Christians because he believed they were presenting a false Messiah and false teachings that were corrupting Judaism. Bin Ladin has responded by terrorist activities and killings. Paul responded by consenting to the execution of one Christian, pursuing, arresting and threatening to kill other Christians, and seeking to destroy the church.[15]

The Evidence for Jesus' Resurrection

Now suppose one night you are watching the evening news and the top story shows bin Ladin speaking to a group of Pakistani Muslims who have supported his efforts. He is speaking in Arabic, so the journalist provides a translation and summary of what he is saying. While in his hideout a week ago, bin Ladin relates, Jesus appeared to him and told him that Christianity was true and that his terrorist activities were actually persecutions against Jesus. In concluding his speech he says, 'Brothers, hear me! We must repent of our wickedness and rebellion against God. He will forgive us because of what he accomplished through the death and resurrection of his Son, the Messiah!' At this he is met with harsh threats and attempts by some in the crowd to kill him. During the next three months, while eluding capture, bin Ladin is seen proclaiming the same message to a large number of crowds. Finally he is killed by some angry Muslims.

This type of transformation is very close to what we have with Paul. The strength of this data is that it comes not from a friend of Jesus but from someone who had hated him.

We then come to James, the brother of Jesus. Not as much information is available on James as on Paul. There are two reports in the gospels that Jesus' brothers did not believe in him prior to Jesus' resurrection.[16] Most scholars hold that we can trust these reports concerning his brothers because this would have been a very embarrassing thing for the early church; someone inventing bits of information about Jesus in the gospels would not have invented this piece of potentially damaging material. Moreover, these reports by Mark and John are different and appear to be independent of one another. So if it is very unlikely that one author would invent such a story, it is even more improbable that two would commit the same unlikely blunder. John also reports that at the cross Jesus entrusted the care of his mother to one of his disciples.[17] Why do this when that task would naturally fall to the next oldest brother? The best answer is that at that point none of Jesus' brothers were believers and he wanted the care of his mother to be in the hands of a spiritual son.

What we know about James after Jesus' crucifixion is very interesting. Paul and Luke both report that he became a prominent leader in the Jerusalem church.[18] Paul reports this among comments that he and James had a difference of opinion on some matters.[19] Thus it seems very unlikely that James' position of leadership in the church was an invention. Josephus reports the martyrdom of James by the Jewish leadership on the conviction that he was a breaker of the Jewish Law,[20] something his brother Jesus and his disciples were constantly accused of doing.[21] From hints found in the New Testament as well as accounts outside it, James appears to have been very pious in his Jewish then Christian beliefs.[22] As historians we must ask what it was that changed the mind of James from that of a sceptic who would have regarded his brother as a false prophet accursed by God to someone who became a prominent Christian leader who died for his faith.

What would it take for you to believe your brother was the Lord? The best answer appears to be found in the very early Christian creed formed shortly after Jesus' crucifixion: 'Then he appeared to James.'[23]

Fact 3: The Empty Tomb

Is there any evidence that the tomb of Jesus was empty? Surprisingly, there is quite a bit supporting the New Testament records that it was. Let's consider three evidences for the empty tomb.

The first is the *Jerusalem factor.* Had the disciples left Jerusalem and travelled to the Far East proclaiming the empty tomb, we might not think too much of it, since not many people would have been likely to make the long trip to Jerusalem in order to verify their claims. However, given the proclamation of Jesus' resurrection and the empty tomb, it would have been impossible for the Christian movement to get off the ground in Jerusalem if the body had still been in the tomb. All the Jewish or Roman authorities would have had to do was exhume the body and publicly expose the Christians' claims as a hoax. And yet, we know from Tacitus that after Jesus' crucifixion Christianity exploded, igniting in Judaea where Jerusalem is located.[24]

Apparently the authorities were unable to produce the body of Jesus. Not only is there no record of such an event, it is fascinating to observe what the Jewish leaders *did* say had happened: the disciples stole his body. Justin (150 AD) and Tertullian (200 AD) both testify that the Jewish authorities were still claiming this in their day, and this corroborates Matthew's testimony that these authorities told the guards, 'You are to say, "His disciples came during the night and stole him away while we were asleep."'[25]

This seems to be an attempt to account for the empty tomb. If my nine-year-old son told his teacher that our dog ate his homework, this would appear to be an attempt to explain why he could not produce it. So not only does it seem highly unlikely that the corpse of Jesus was ever exposed by the authorities, despite a compelling motive for them to do so; the response from these same authorities implied an empty tomb. Ironically, then, our second evidence for the empty tomb comes from *enemy testimony*. Not only were Jesus' friends reporting that his tomb was empty; we even have the testimony of his enemies strongly implying it.

The third piece of evidence for the empty tomb is that *women were the first and primary witnesses* to it. They are reported as witnesses in all four gospels, whereas the male disciples are only reported as witnesses in two. This is significant because in Jesus' day the testimony of a woman was worth very little. In the first century Josephus wrote, 'Do not let the testimony of women be admitted, because of the changeability and imprudence of their gender.'[26] Many similar statements in other ancient writings reveal the low view of women in antiquity.[27] As historians we must ask why, if the empty tomb story was an invention, would its author, who wanted us to believe this hardly believable event, list unbelievable women as its primary witnesses? Most historians conclude that an inventor would not have done so. Rather, it is more likely that the authors of the gospels were attempting, in spite of the ridicule they would be expected to receive from others, to report the event accurately.[28]

Together, these three facts present a strong case for the empty tomb.

Accounting for the facts

Let's see where we have come in these last few pages. We observed that the truth of Christianity stands or falls on Jesus' resurrection. If Jesus did not rise, Christianity should be abandoned. If he in fact rose, Christianity is true and eternal life is ours if we embrace Jesus as Lord.

We decided that we would look at the evidence as a historian would. We did not assume that the Bible is anything more than an ancient library of Jewish and Christian writings regarded as authoritative by the early church. We considered only evidence that passed two tests: (1) it was strongly-evidenced and (2) it was granted by an impressive majority of scholars who study the subject, including agnostic and atheist scholars. From the evidence that emerged, we took the three most important facts and dropped them through a final funnel that, as much as possible, left only non-biblical sources to support them. The three facts we worked with were (1) Jesus' death by crucifixion, (2) the beliefs of a number of people, both friend and foe, that Jesus had been resurrected and had appeared to them and (3) the empty tomb.

As historians, we must now account for these three facts. What happened subsequent to Jesus' violent death that caused the tomb to be empty and a number of people, both friend and foe, to conclude he had risen from the dead and had appeared to them? Certainly, the idea of resurrection perfectly accounts for all three. But are there any natural explanations that can account for them? Can we explain the three key facts given what we know from science, psychology, history or philosophy? If a natural explanation exists that is equally plausible to the resurrection explanation, we might prefer it.

Alternate theories

Ever since shortly after Jesus' crucifixion, his opponents have sought to provide explanations from natural causes in order to account for what happened. As we begin to look at these, we first note that certain natural theories can be ruled out immediately.

For example, let's take the embellishment form of legend theory. This theory states that the testimonies of Jesus' appearances and empty tomb were the results of legendary embellishments that were tagged onto the original story, now unknown to us. This idea, however, can be dismissed immediately since we can establish that the original disciples had claimed Jesus had risen and had appeared to them. Remember the reports by Josephus and Clement. Also recall that the empty tomb was attested by Jesus' enemies and was unlikely to have been the result of an inventor.

The theory that the disciples lied about the appearances and possibly stole the body can also be dismissed based on our three evidences. We have established that the disciples sincerely believed the risen Jesus had appeared to them. Furthermore, the sceptics Paul and James would have been the first to suspect fraud on the part of the disciples. Yet they left their scepticism and became Christians because they too believed the risen Jesus had appeared to them.

What about grief hallucinations on the part of the disciples? They had lived with Jesus for several years with the hopes of reigning with him. Then the devastating events of his arrest and brutal execution occurred. Disoriented, hopeless and longing for the days that had just passed, perhaps they began experiencing hallucinations of Jesus risen from the dead. Even if this were the case with the disciples, however, the sceptics Paul and James would not have been in this frame of mind. Paul especially was out to finish the job the Jewish and Roman leadership had started and rid the world of Christianity. He was far from grieving. Moreover, if the disciples had experienced hallucinations of the risen Jesus, wouldn't his decomposing corpse still be in the tomb?

Thus we can see that the known facts themselves refute several of the major natural theories that attempt to account for what happened. The late and highly respected New Testament scholar Raymond Brown wrote that twentieth-century critical scholars had rejected opposing theories. He added that today's scholars both ignore them and even treat them as unrespectable.[29] Space prohibits us from going further into the area of natural theories.[30]

Natural and supernatural causes

Since the majority of scholars reject natural theories proposed as alternative explanations to the resurrection of Jesus, on what basis do unbelieving scholars maintain their unbelief that the event occurred? Reasons differ for each person. However, scepticism toward God's existence and the miraculous governs the minds of many.

A surgeon friend once asked how he as a physician could believe that a man rose from the dead when he knew medically and from professional experience that once a person dies, he or she stays dead. Although this was an honest question, it showed he had failed to grasp the claim being made about Jesus. What his medical knowledge and experience revealed was that a person was not going to return from the dead *by natural causes*. That is quite clear and I doubt anyone would question it. However, the early Christians did not claim that Jesus was resurrected by natural causes. They claimed that *God raised Jesus*. If God exists and wants to raise someone from the dead, a history of millions of people dying and staying dead does nothing to address this claim.

We may want to ask why God would have wanted to raise Jesus when he did not raise others. That question is easily answered when we consider that Jesus claimed to be God's unique Son who had a special relationship with him. If Jesus was telling the truth, there is plenty of reason to believe God would have desired to raise him.

Scepticism is articulated in a number of other ways.[31] Let's consider just one. The famous eighteenth-century Scottish sceptic David Hume said that if someone told him he had seen a dead man alive again, he would consider three options: (1) the one proclaiming the news was a liar, (2) the one proclaiming the news was somehow deceived or (3) the event really occurred. In judging which of these options was correct, Hume applied two criteria: (1) reject the greater miracle and (2) only accept the miracle claim when the falsehood of the testimony concerning the miracle would be more miraculous than the miracle itself.[32] In other words, the evidence for the miracle has to be so strong that it would be more of a miracle for it *not* to have occurred.

Near the same time, the American sceptic Thomas Paine asked a similar question whether it was more probable that nature would alter its course or that a man would lie. He added that we have never witnessed nature altering its course but we have witnessed people lying and have good reasons for believing millions to have done so. Thus the odds are millions to one that a lie has been told when it comes to a miracle claim.[33]

While Hume and Paine are certainly justified in rejecting the large majority of miracle claims, their scepticism goes too far. All miracle claims are not equal. The large majority can be dismissed immediately from our serious consideration since no credible evidence exists for them or since they can easily be accounted for by plausible natural explanations. But as we have seen, Jesus' resurrection does not fit into this category. Good evidence exists for it and there are no plausible natural theories that can account for our three facts.

We might reply to Hume and Paine that if God does not exist, then of course it is more probable that men would lie than that nature would alter its course. However, if God exists, we would need to ask at least three other questions: (1) Is there good evidence that the event in question occurred? (2) Is there reasonable evidence that those making the claim lied or were lied to? (3) Does a context exist in which we might think it reasonable for God to act miraculously? If good evidence exists that the event occurred, no good evidence exists for the involvement of a lie, and a context exists where we might expect God to act miraculously, then there is no reason to believe that a lie is more probable than that the miracle under consideration occurred.

We might address two further questions to Paine. First, who says we have never witnessed nature altering its course? History is full of claims that God has interacted with our world. Many

Christians can testify to answers to their prayers that are of such a nature that the answers go far beyond coincidence. While false miracle claims abound, counterfeit miracle claims no more undermine authentic ones than counterfeit money nullifies the real thing.

Second, we might also ask Paine a question pertaining to his probability statement. While it may be that the number of lies told by people over thousands of years significantly outweighs the number of credible miracle claims, does the equation he proposes make much sense—that is, that the option that has the greater probability of occurring should be accepted? Given this logic, we would have to conclude that no one ever wins the lottery, since the number of times known where people have lost significantly outweighs the number of times we have a credible claim for a winner. It is certainly much more probable that a person will lose when playing the lottery than that they will win. Paine's line of reasoning would likewise keep us from believing in the occurrence of singular events such as the big bang. Although scepticism is valuable in that it puts a check on the uncritical acceptance of miracle claims, it is far from effective in providing a reason for rejecting Jesus' resurrection as an event that occurred in history.

One final question may be asked, and it is directed toward Hume's claim that the probability that a miracle occurred must be so great that its falsification would be even more miraculous. Is a burden of proof to this degree merited? Why must evidence to this degree be provided before we are justified in concluding that a miracle has occurred? In spite of Hume's reluctance, it is nothing short of remarkable that the sharpest sceptical minds over two thousand years have not been able to present a plausible theory that can account for all of the known data concerning Jesus' resurrection.

Conclusion

We have seen that no plausible alternative theories exist that can explain the known historical facts of Jesus' death by crucifixion, the empty tomb and the beliefs of a number of Jesus' friends and foes that he had been resurrected and had appeared to them. These three facts are pieces of a puzzle that, no matter how you arrange them, only look like the resurrection of Jesus.

Why should we conclude that Jesus' resurrection occurred rather than hold that we have an irregularity in nature for which no explanation presently exists? It is helpful here to look at the context in which these things took place. Suppose we read an ancient account of an average man who died and for no apparent reason rose from the dead. The ancient account goes on to say that he only said 'hello' to a few people and then disappeared. We wouldn't give much more thought to it. If the evidence for it was very strong, we might throw our arms in the air and conclude that this is highly irregular and no plausible explanation currently exists for what had occurred.

Now suppose that we read another ancient account of a man who claimed to be divine and to have a close relationship with his Father, who happens to be God. Let us also suppose that this man performed amazing deeds that many interpreted as miracles. Furthermore, this man predicted that he would be killed but would rise from the dead. The man was then killed.

The account goes on to say that he rose from the dead and that several eyewitnesses claimed that he had appeared to them. In this case we may not be so quick to throw our arms in the air and claim we have no idea of what occurred. The difference from the first scenario, of course, is that in the second case we have a context that is charged with religious significance.[34] Thus, if the details in the second scenario can be established, forming a religious context, if good evidence exists that the event took place, and if there are no plausible explanations that can account for the known historical facts surrounding the event, we have good reasons for concluding that the event occurred.

This is precisely what we have when we come to Jesus' resurrection. And this should greatly encourage us. Jesus' resurrection not only provides evidence for the existence of God, it also shows us that God has interacted with our world and has actually revealed himself to humankind in Jesus, who taught that it is actually possible for us to have a relationship with God and that this is, in fact, God's desire. His resurrection was God telling the world that what his Son was saying is true.

Conclusion

24

The Absurdity of Life Without God

Phil Fernandes

There are many rational arguments that provide strong evidence that the theistic God exists. Many of these have been discussed in this book. However, clues to God's existence can also be found in human experience. Many philosophers and theologians have reasoned that if we are to make sense of our lives—if our lives are to have true significance—then a personal God must exist.

In this chapter we will demonstrate the truth of this. In order to make sense of our moral/rational experience and of human existence, God must exist. If God does not exist, human life is absurd. We will attempt to show that humans need God and that, since what humans need exists, therefore God exists. We will also examine evidence for God from religious experience and changed lives due to religious conversions.

Is life meaningful?

Many Christian thinkers have concluded that without God life is meaningless. In fact, many atheist philosophers also acknowledge that if God does not exist, then life is without meaning. The nineteenth-century German philosopher and militant atheist Friedrich Nietzsche reasoned that since God was dead (that is, non-existent), then truth, morality and meaning in life were mere illusions. We will argue that without God there is no way to make sense of our moral or rational experiences, and that, if God does not exist, human existence is absurd.

Making sense of our moral experience

Do the lives we live really make sense? Are our actions significant? Is there really a distinction between the lives of Adolph Hitler and Mother Teresa? A million years from now, will it really matter whether we lived our lives like Hitler (who slaughtered the innocent) or Mother Teresa (who cared for the helpless)? If God does not exist and we cease to exist after death, then it seems that both Hitler and Mother Teresa (as well as everyone they influenced) will reach the same destination: extinction. Why should we refrain from evil and do good if there will be no lasting, eternal significance to our actions? Many philosophers believe that, if there is no God, we cannot make sense of our moral experience.[1]

We all make moral value judgments when we call the actions of another person wrong. When we do this we appeal to a moral law, a standard of what is right and wrong. Moral relativists argue that this moral law originates with each individual—each person determines their own list of rights and wrongs. However, this cannot be the case, for everyone makes moral value judgments, even moral relativists. Moral relativists argue that it is *wrong* to call the actions of another person *wrong* because there is no such thing as right and *wrong*. Obviously, this is a self-refuting view. In short, the moral law could not originate with each individual, for then we could not condemn the actions of another individual such as Adolph Hitler. Since even moral relativists condemn the actions of Adolph Hitler, the moral standard must be above individuals.

At this point, the moral relativist may respond by saying that the moral law is produced by each society. However, this view fairs no better, for then one society could not condemn the actions of another society. The Allies could not justifiably condemn the actions of Nazi Germany; the United States could not rightly judge as evil the actions of the Taliban. Therefore, if we wish to condemn the actions of the Nazis or the Taliban, we must appeal to a moral law above all societies.

The moral law also does not come from a world consensus, for world consensus is often mistaken. The world once thought the sun revolved around the earth, slavery was morally permissible and women were the property of their husbands. If the world consensus was wrong in all these instances, it is not a good candidate for the source of the moral law. Even atheists and moral relativists protest to try to make the world a *better* place to live. But this makes no sense if the world consensus is the creator of the moral law.

Appealing to society or world consensus will never give us an adequate explanation for our moral experience as humans.

CONCLUSION

We all make moral value judgments. Appealing to society or world consensus only quantitatively adds men and women. What we need is a moral law *qualitatively* above all people. And if we wish to condemn the actions of the past (slavery, the Holocaust and so on), this moral standard must be eternal and unchanging. Therefore, there must exist an eternal, unchanging moral law that stands above all individuals, societies and any world consensus if we are to make sense of our moral experiences as human beings. And if there exists an eternal, unchanging moral law above all individuals, societies and any world consensus, there must exist as its cause an eternal, unchanging moral Lawgiver above all individuals, societies and any world consensus.

Making sense of our rational experience

Human beings not only have moral experiences; we also have rational experiences. We think and seek the truth. When a person rejects reality and lives in a world of lies, we say they are mentally unstable. However, this presupposes that there is a reality and there is truth. If truth does not exist, there is no distinction between reason and non-reason.

The theist claims to know something (that God exists), and the atheist claims to know something (that God does not exist). Even the agnostic claims to know something (that the supposed evidence for God's existence is insufficient). The theist, atheist and agnostic all believe their positions are true. But this assumes the existence of truth and man's ability to know truth.

What guarantee does the atheist or agnostic have that we can know reality as it is? How do the atheist and agnostic know that we can know reality as it is, not just reality as it appears to us? Atheism and agnosticism offer no good reason why we should assume that the gap between reality and appearance can be bridged.

However, theism entails the doctrine that a rational God created both man in his own image (as a rational being) and a coherent universe, so that through reason man could find out about the universe in which he lives. Remove the rational God from the equation and the basis for human knowledge crumbles.

Scepticism fails on this point, for the sceptic believes that man cannot know truth. However, the claim that 'man cannot know truth' is itself a claim to know the truth that 'man cannot know truth'. Therefore, it is possible for man to know truth. In this way, too, the denial of the existence of absolute truth is self-refuting. For the statement 'there is no absolute truth' to be true, absolute truth would have to exist. Therefore the statement 'there is no absolute truth' must be false. Absolute truth exists, and, as stated above, man can know truth.

Some truths are universal, unchanging and eternal. For example, we do not invent mathematical truths such as $1 + 1 = 2$; rather, we discover them. This is also true of the laws of logic (the law of non-contradiction, the law of the excluded middle, the law of identity and so on). These laws stand above human minds and judge them. For instance, if we add $1 + 1$ and we conclude with 3, the eternal truth that $1 + 1 = 2$ will declare us wrong.

Augustine argued that human minds, since they are finite and fallible, could not possibly be the source of these eternal truths. He reasoned that only an eternal, unchanging Mind—God—could be the source of these truths.[2]

Atheism has no basis for eternal, unchanging truths. If atheism is true then there may have been a time when 1 + 1 equalled 3. There may also have been a time when torturing innocent babies was good. The existence of eternal, unchanging truths and humanity's ability to know these truths are strong clues to the existence of God, the Truth-giver.[3]

Making sense of human existence

The seventeenth-century French thinker Blaise Pascal graphically described the human situation apart from God in the following words:

> Imagine a number of men in chains, all under the sentence of death, some of whom are each day butchered in the sight of others; those remaining see their own condition in that of their fellows, and looking at each other with grief and despair await their turn. This is an image of the human condition.[4]

In response to our inevitable death, Pascal stated: 'Being unable to cure death, wretchedness and ignorance, men have decided, in order to be happy, not to think about such things.'[5] Human beings look to temporary pleasures to divert their attention from the fact that they know they will someday die. Rather than despair, they choose to ignore their sad state.

All people face their own inevitable death. As they go through life, they seek to hide this dreadful fact from themselves through temporary, empty pleasures. But, as far as Pascal is concerned, this is meaningless existence. We can only find genuine meaning in life if we find the God of the Bible. Apart from God, life is absurd.[6] Death is real, and atheism offers no real solution—no real hope—that this enemy will be defeated.

Several Christian thinkers have argued that man's psychological needs can only be met by the God of the Bible.[7] All people long for loving acceptance, forgiveness, peace and significance. And since we contemplate eternity, we need to be loved eternally and unconditionally. We need to be forgiven for eternity. We need to experience eternal peace and significance. It offers no genuine consolation to people to be loved for seventy years if they know they are going to cease to exist for the eternity that follows. We desperately need forgiveness to remove our guilt and hope to obliterate our despair. But without Christ's atoning death on Calvary, there is no forgiveness. And without Christ's resurrection, there can be no lasting hope that death will be defeated. Only in Christianity can humanity's deepest psychological needs be met. We can only be eternally significant if God exists; we can only be eternally loved if God is real. Ultimately, human anxiety will not be healed if there is no God and no life after death. Only God can meet our greatest needs.

All men acknowledge the existence of evil (at least in practice if not in their beliefs). But nothing less than the God of the Bible can guarantee the ultimate defeat of evil. Life is hopeless and without meaning if God does not exist.[8]

Even the atheist philosopher Bertrand Russell acknowledged the absurdity of life without God. His rejection of God led him to refer to the universe as 'purposeless' and 'void of meaning'. He wrote:

> That man is the product of causes which had no prevision of the end they were achieving; that his origin, his growth, his hopes and fears, his loves and his beliefs, are but the outcome of accidental collocations of atoms . . .[9]

In this rare moment of honesty, Russell was right. He goes on to point out that 'no fire, no heroism, no intensity of thought and feeling, can preserve an individual life beyond the grave'. He rightly notes that 'all the labours of the ages, all the devotion, all the inspiration, all the noonday brightness of human genius, are destined to extinction in the vast death of the solar system'.

If the God of the Bible does not exist, man is damned to a life of meaningless existence. To hide from this fact, a person can focus his attention on Pascalian diversions (or maybe choose a Kierkegaardian leap into the non-rational realm). But for those who have the courage to deal with reality head on, a choice must be made between despair and the God of the Bible. As Pascal said, the wise man will wager on God.

Divine longing—humanity's thirst for God

Though many people currently deny or ignore the existence of the God of the Bible, their lives display a vacuum only he can fill. Christian philosophers Norman Geisler and Winfried Corduan argue that all people 'sense a basic need for God'.[10] Geisler and Corduan give several examples to make their case.[11]

The fact that atheist psychologist Sigmund Freud attempted to explain away this phenomenon shows that even he recognised this need for God in himself and others. Freud admitted that man feels powerless and insignificant in the face of the vast universe in which he finds himself. According to Freud, man invents God through his imagination to calm his fears.

Friedrich Schleiermacher taught that all people have a feeling of absolute dependence, even though they do not all explain it in the same way. Believers and non-believers alike recognise their absolute dependence on something that transcends their earthly experience.

Martin Heidegger viewed man as a 'being-unto-death'. Man finds himself thrown into the world and has no say about his being here. He does not know why he is here; he only knows he is destined for nothingness. Man has no control over his birth or his death. He must die. He finds himself thrust into this world en route to extinction. He is without a ground for his being.

Paul Tillich recognised that man is limited and dependent. Man needs a ground for his being, something to anchor his existence. Tillich spoke of this need as man's 'ultimate concern'.

Jean-Paul Sartre, a French atheist, admitted his need for God. Sartre taught that man needs God to give his existence definition and meaning. But, since Sartre rejected God's existence, he felt that the entire project was absurd. According to Sartre, man has a need for God, but there is no God who can meet this need.

Walter Kaufmann referred to man as the 'God-intoxicated ape'. The late German philosopher Friedrich Nietzsche considered his own atheistic views to be so unbearable that he wished he could be convinced he was wrong. He felt a strong thirst for God but rejected the possibility of God's existence.[12]

After surveying the above list of non-Christian thinkers, Geisler and Corduan conclude that 'people generally, if not universally, manifest a need for the Transcendent' (God). They add that 'the sense of contingency, the feeling of cosmic dependence, the need to believe in some sort of Transcendent is apparently present in all men'.[13]

In God just wishful thinking?

The evidence indicates that all people sense a need for the God of the Bible. Even atheists and other non-Christians have expressed this need. Still, it must be shown that this need points to the actual existence of God.

Though both atheists and Christians alike recognise the universal thirst for God, atheists deny that God actually exists. Instead, they speculate as to why so many people believe he exists. An example of this kind of speculation is found in the thought of Sigmund Freud.

Freud was convinced that God did not exist. But if atheism was true, then why did so many people believe in God? Freud tried to answer this question. He suggested that 'primitive' man felt extremely threatened by nature due to natural disasters such as storms, floods, diseases and death. 'Primitive' man had no control over nature; he was totally helpless in this regard. There was nowhere he could turn for help. Freud theorised that 'primitive' men therefore decided to personalise nature. In this way, man could attempt to plead with or appease nature. Imagining nature to be a personal being enabled people to offer sacrifices to nature in the hope that nature would be kind to them in return.[14]

Freud's speculation did not stop there. He assumed that originally mankind banded together in small groups consisting of a male, his several wives and their offspring. Freud believed that 'primitive' male children desired to have sexual relations with their mothers and therefore became extremely jealous of their father. Though they loved their father since he was their protector, they began to hate him due to their jealousy. Eventually they banded together and murdered their father. After the murder, they ate their father's flesh in a ritualistic meal. But soon they were overcome with feelings of guilt. As a result, they deified the father image and began offering sacrifices to him as a god.[15]

Therefore, Freud taught that God was nothing but a product of human imagination. God did not create man; man created God. In this way, reasoned Freud, the belief in the Father-God originated in man's wishful thinking.

This highly speculative theory does not do justice to humanity's universal thirst for God. This theory—or myth—appears to be 'wishful thinking' on the part of Freud. Whatever the case, his proposed explanation deserves a response. We will now examine how several Christian thinkers have answered Freud's unverified suppositions.

Christian theologian R.C. Sproul is quick to point out that Freud's line of reasoning does not disprove God's existence; instead, it presupposes his non-existence. In other words, Freud was not trying to answer the question, 'Does God exist?' Rather, he was attempting to answer the question, 'Since we already know God does not exist, why do so many people believe that he does?'[16]

CONCLUSION

Hence Freud's wild speculation should not be viewed as a disproof of God's existence. It is simply a desperate attempt to explain away strong evidence for God's existence—the fact that possibly everyone senses a need for God. Freud's endeavour focuses on the question, 'If atheism is true (and Freud believed that it was), why are there so few atheists?' He answers this by accusing all who disagree with him of being deluded.

Sproul points out that Freud's speculation explains how men use their imagination to invent idols (false gods that do not exist). But it does not explain the God of the Bible, for the God of the Bible is far too demanding. No one would wish for the existence of a Being who requires the submission and obedience demanded by the Christian God. The gods of other religions are attractive candidates for projection, but the Holy God of the Scriptures is the type of Being from whom men run. No one would invent him through wishful thinking.[17]

According to Christian philosopher J.P. Moreland (citing psychologist Paul Vitz), 'atheism is a result of the desire to kill the father figure (in Freudian language) because one wishes to be autonomous'.[18] In my opinion, man's two greatest drives are his thirst for God and his desire to be autonomous. Man has a void that can only be filled by God, yet still he wants to be his own king. The Christian chooses God over autonomy while the atheist chooses autonomy. Using the Freudian belief that humans have an unconscious desire to kill the father figure—an unconscious desire to free themselves from the main authority figure in their life—we have an explanation for the atheism of the small percentage of people who reject God's existence.

On the other hand, Christians and other theists make up such a diverse collection of people that there is no way to place them into one psychological group. Vitz shows that the world's leading atheists have had either no relationship or a bad relationship with their fathers.[19] No such generalisation can be made of Christians.

Moreland adds that even if Freud was right, his argument would still be guilty of committing what philosophers call the genetic fallacy.[20] The genetic fallacy claims that a belief can be shown to be false just by showing that its origin is unreasonable. But this is not the case. Even if humanity, due to fear and guilt, originated the idea of God, this does not disprove the existence of God. God might still exist even if believers arrived at this conclusion through faulty reasoning.

Philosophers Geisler and Corduan argue that what people really need actually exists.[21] Humans need food and water; food and water exist. Even if a person dies of hunger or thirst, the fact is that food and water do exist. It is just that the person did not find them. Geisler and Corduan reason that all people really need God; the thirst for God is universal. God is not something people merely desire; he is something people need. And since whatever else man needs exists, we are justified in concluding that God exists. This is true even if a person does not find God (just as some people do not find food or water).

This argument is not meant to be an air-tight proof, but it does seem to have a high degree of probability since everything else man genuinely needs does in fact exist.

If all people have a void that only God can fill, then one would expect the Bible to address this issue. And this, of course, is the case. Jesus said, 'Man does not live on bread alone, but on every word that comes from the mouth of God.'[22] Humans are more than physical beings with physical needs. They are also spiritual beings with spiritual needs. Not only do they need physical nourishment, they also need spiritual nourishment from the true God. Jesus proclaimed, 'I am the bread of life. He who comes to me will never go hungry, and he who believes in me will never be thirsty.'[23] Only Jesus can fully quench our thirst for God.

Unfortunately, most people attempt to quench this thirst with unworthy substitutes that can never meet their ultimate needs. The Bible declares, 'My people have committed two sins: They have forsaken me, the spring of living water, and have dug their own cisterns, broken cisterns that cannot hold water.'[24] Rather than turn to Christ, most people look elsewhere to find fulfilment in life. For some, their 'broken cistern' is a false religion. For others, it is material wealth. Many people try sexual immorality, drugs or alcohol. But there is no worthy substitute for the true living water. Only Jesus can meet humanity's deepest needs. As Augustine prayed, 'You have made us for yourself and our hearts find no peace until they rest in you.'[25]

Evidence for God from religious experience

Moreland tells us that there are different types of religious experience. The *monistic* type involves 'a union between the subject and the One'. The *theistic* type consists of a non-sensory awareness of a personal Being who is separate from the subject.[26]

Theistic religious experiences deal with encounters with a transcendent personal presence. These experiences can be triggered by somewhat 'ordinary' or 'natural' events such as a beautiful sunset, feelings of guilt, feelings of being watched when alone, singing religious songs, and prayer during times of stress (even many atheists have admitted praying during difficult times). In theistic religious experiences, there is an encounter with a personal transcendent Being, a personal 'Other'.

Many people have reported having the theistic sort of religious experience. Is there any way to find objective evidence for these subjective experiences? Many theistic thinkers think there is.

Moreland uses what he calls the 'God hypothesis' to explain the transformation that has occurred in the lives of many believers.[27] He reasons that the strength of the 'God hypothesis' to account for theistic religious experiences is its ability to explain the data more adequately than rival hypotheses. History has recorded a vast list of names of believers (Paul, Peter, James, the apostles, Augustine, John Newton, C.S. Lewis, Chuck Colson and many more) whose lives have been transformed for the good by a religious experience with the God of the Bible. These transformed individuals display a zeal for holiness and self-sacrifice not previously evidenced in their lives before their conversion experiences. Mere psychological factors cannot account for the data as adequately as the 'God hypothesis'—their lives were changed by a real encounter with the personal God.

Moreland adds that the explanatory power of the 'God

hypothesis' is strengthened if one can show that a great diversity exists among those experiencing such transformation. The more diverse the group of people experiencing the transformation, the more difficult it is to explain their transformation merely on the basis of common psychological or sociological causes. In an experiential group as diverse as Christianity, the only common factor shared by believers is their faith in God. Since real effects evidence real causes, God is the real cause for the real effect of the transformed lives of believers.

Moreland argues that the case for the Christian type of theistic religious experience is strengthened since it is tied to objective events like Christ's bodily resurrection from the dead.[28] Many Christian thinkers have provided solid historical evidences for Christ's bodily resurrection from the dead.[29] The case for the resurrection offers objective, historical confirmation of the validity of the religious experiences of Christians. Philosophical and scientific evidences would also corroborate the 'God hypothesis'.

Denying the reality of a person's theistic religious experience is not as easy as it may first appear. In some respects, denying the existence of God would be similar to denying the existence of my brother Mark, for I have had personal encounters with both. If others had never met my brother Mark, it would not lessen my belief in his existence, for I have personally encountered him; I have a personal relationship with him. The same can be said for God. Just because Christians have never seen God or experienced him through the five senses does not mean their experiences of him are less than genuine. My experience of God is non-sensory (though it may be triggered by sensory data such as a sunset). However, my experiences of love, moral absolutes and the laws of logic are also non-sensory; yet that does not make love, moral absolutes and the laws of logic less than real.

An illustration may help. Suppose there is a room filled with millions of white ping-pong balls. While searching the room I find one red ping-pong ball. Others who have been in the room but never found the red ping-pong ball might deny its existence. Nevertheless, the argument that the red ping-pong ball does not exist because they have never experienced it is fallacious. The same can be said for the argument that God does not exist because the atheist claims he or she has not experienced him.

Moreland points out that it may be the case that a condition for experiencing the true God is the desire to find him. It may be that only those who seek God experience him.[30] It is possible that those who do not desire God may never find him simply because they refuse to look for him.

Conclusion

Besides rational, historical and scientific evidences for the existence of God, a case for God's existence can be made based on human experience. Our moral experiences, our thought life (rational experiences) and our sense of divine longing (our thirst for God) are all clues to God's existence. Without God, human existence would be meaningless; if we do not posit the existence of God we can make no sense of the world in which we live.

Finally, theistic religious experience and the transformed

CONCLUSION

lives that flow from it provide strong evidence for the existence of God. With the apostle Paul we can proclaim, 'in him we live and move and have our being'.[31]

The Contributors

Jerry Bergman, PhD (evaluation and research, Wayne State University), PhD (biology, Columbia Pacific University), is a faculty member at Northwest State College where he has served as chairman of the academic affairs committee and as faculty adviser for degree programs.

Stephen Caesar, MA (anthropology and archaeology, Harvard University), is a staff member at Associates for Biblical Research and serves as adjunct professor of literature at Newbury College, Massachusetts.

David Catchpoole, PhD (plant physiology, University of New England), is a research scientist with *Answers in Genesis*, Australia.

Steven B. Cowan, PhD (philosophy, University of Arkansas), is the Associate Director of the Apologetics Resource Center in Birmingham, Alabama.

David K. Down is editor of *Archaeological Diggings*, Australia's leading ancient history and archaeology magazine.

Danny R. Faulkner, PhD (astronomy, Indiana University), is Professor of Astronomy and Physics at the University of South Carolina, Lancaster.

Paul Ferguson, PhD (theology, Chicago Theological Seminary), is Professor of Old Testament at Christian Life College, Illinois.

Phil Fernandes, PhD (philosophy of religion, Greenwich University), is the president of the Institute of Biblical Defense.

Arnold G. Fruchtenbaum, PhD (systematic theology, New York University), is the founder and director of Ariel Ministries.

Werner Gitt, DEng (engineering, Technical University of Aachen), was Director and Professor at the German Federal Institute of Physics, Braunschweig, from 1978 to 2002.

Kenneth E. Himma, JD (University of Washington School of Law), PhD (philosophy of law, University of Washington), teaches in the Department of Philosophy, the Information School and the Law School at the University of Washington.

George Javor, PhD (biochemistry, Columbia University), is Professor of Biochemistry in the School of Medicine, Loma Linda University, California.

Michael R. Licona, MA (religious studies, Liberty University), is the co-author with Gary R. Habermas of *The Case for the Resurrection of Jesus*, Kregel Publications, 2004.

John McRay, PhD (New Testament and early Christian literature, University of Chicago), is Professor Emeritus of New Testament and Archaeology at Wheaton College Graduate School, Wheaton, Illinois.

Jon Paulien, PhD (theology, Andrews University), is Professor of New Testament Interpretation at Andrews University.

Ariel A. Roth, PhD (biology, University of Michigan), is a former director of the Geoscience Research Institute, and a former professor and chairman of biology at Loma Linda University.

William Shea, MD (Loma Linda University), PhD (biblical and oriental studies, University of Michigan), was research associate at the Biblical Research Institute in Washington, DC from 1987 to 1999.

Frank J. Sherwin, MA (zoology, University of Northern Colorado), is a research scientist at the Institute for Creation Research in San Diego.

Andrew A. Snelling, PhD (geology, University of Sydney), is Professor of Geology at the Institute for Creation Research in San Diego.

Timothy G. Standish, PhD (biology and public policy, George Mason University), is a research scientist at the Geoscience Research Institute, Loma Linda University.

Eric Svendsen, PhD (New Testament theology, Greenwich School of Theology/North-West University, Potchefstroom, South Africa), is Founder and Director of New Testament Research Ministries.

Charles Taliaferro, PhD (philosophy, Brown University), is Professor of Philosophy at St Olaf College, Minnesota.

Steven Thompson, PhD (biblical studies, University of St Andrews), is senior lecturer in biblical studies, Faculty of Theology, Avondale College, Australia.

Barry L. Whitney, PhD (Christian theology and philosophy of religion, McMaster University, Canada) is Professor of Philosophy of Religion (Christian Apologetics) at the University of Windsor, Ontario.

The Editors

Michael J. Westacott is a freelance editor living in Cairns, Australia.

John F. Ashton, PhD (epistemology, University of Newcastle), is a Fellow of the Royal Australian Chemical Institute, and Honorary Associate in the School of Molecular and Microbial Biosciences at the University of Sydney. He is the editor of *The God Factor: 50 Scientists and Academics Explain Why They Believe in God* (HarperCollins, 2001) and *In Six Days: Why 50 Scientists Choose to Believe in Creation* (Strand, 2003).

Notes

Introduction

1. For Dr Carson's own more detailed account of what happened, see B. Carson & G. Lewis, *The Big Picture: Getting Perspective on What's Really Important in Life*, Zondervan, Grand Rapids, MI, 1999, pp. 259–263.

Chapter 1 Has Science Disproved God?

1. J. Ashton (ed.), *In Six Days: Why 50 Scientists Choose to Believe in Creation*, Master Books, Green Forest, 2001 (originally published by New Holland, Sydney, 1999); J. Ashton (ed.), *On the Seventh Day: 40 Scientists and Academics Explain Why They Believe in God*, Master Books, Green Forest, 2002 (originally published as *The God Factor: 50 Scientists and Academics Explain Why They Believe in God*, HarperCollins, Sydney, 2001).

2. Ayer ought to have considered the protest of theologian John Hick, who challenged empirical verification with his theory of 'eschatological verification'. Hick rightly held that we simply are not in the proper position to determine God's existence or

non-existence by empiricism. Just as we need to be in a proper position to verify whether there are mountains on the dark side of the planet Pluto, for example, we would need to be in the proper position to verify God's existence. That proper position would be an afterlife, the *eschaton*. (See both Hick's and Ayer's essays in J. Hick (ed.), *The Existence of God*, Macmillan, New York, 1964.)

3. C. Sagan, *Cosmos*, Random House, New York, 1980, p. 1.
4. Postmodernists have discredited the long-standing popular myth that scientists use the empirical test in an objective, neutral and unbiased manner. Philosophers of science like Thomas Kuhn and others have shown there is clearly a significant subjective bias in the scientists (as in all people), who bring to their analysis of empirical data the assumptions and presuppositions of the current scientific paradigm.
5. 1 Corinthians 13:9–12. Biblical quotations in this chapter are from the New American Standard Version and the New King James Bible.
6. Acts 8:30–31; 2 Peter 3:15–16.
7. Psalm 22:1–2; 10:1; 30:7; 44:23–24; 88:13–14; 89:46; Isaiah 45:15. For a systematic, comprehensive analysis of responses to the problem of evil, see B. Whitney, *Theodicy: An Annotated Bibliography on the Problem of Evil, 1960–1991*, Philosophy Documentation Center, Bowling Green, OH, 1998.
8. Ephesians 2:8; Romans 5:1–2.
9. Romans 2:1.
10. Romans 10:17.
11. Proverbs 1:29; 2:6; 3:5–7; 8:11; 9:10.
12. For lack of space and in light of the overall design of this chapter, it is important to call attention to what has *not* been discussed: the detailed theological justification for the views put

forth, especially on the relationship between divine grace and human freedom, and more especially, the central role of Christ not only in his atonement of human sin and as the means of salvation, but insofar as he is the object of faith, the one who has made repentance possible, available to all and acceptable for salvation.
13. Galatians 5:22–23.
14. Ephesians 3:12; 4:13; 6:16; Colossians 2:6; Galatians 3:11.
15. John 3:3; 1 Peter 1:23.
16. Matthew 23:37.
17. Psalm 19:1–4.
18. Pascal avoided both faithism and intellectualism (rationalism) in his understanding of the rationality of belief. His work has been outlined provocatively by philosopher P. Kreeft in *Christianity for Modern Pagans*, Ignatius, San Francisco, 1993.
19. John 3:19 (cf. Romans 1:18–32).
20. Matthew 13:58.
21. Romans 1:24–25.
22. Proverbs 1:7; 3:5; 9:10; Job 28:12, 28; Ecclesiastes 7:24.
23. Acts 7:51–53; Hebrews 3:7–8; Psalm 95:7–11.
24. Isaiah 7:13; Ephesians 4:30.
25. John 3:16; Romans 4:16; 2 Corinthians 5:19.
26. Historically there have been impressive rational defences of belief in the God of Christianity, culminating in the scholasticism of Aquinas (thirteenth century). Since then, in response to the increasing scepticism and secular focus in Western culture, there has been a corresponding withdrawal by theologians from the perceived necessity to defend the rational basis of religious beliefs. Outside the Roman Catholic Church, the majority of Christians have regarded belief in God as true but irrational (or better, non-rational, in the sense that it is not able to meet the

strict standards of truth determined by rationality and the empirical method). This has been a serious mistake, but it is encouraging that we are currently witnessing a revival of 'apologetics', the rational defence of beliefs. The common believer needs to be informed about the arsenal of evidence which is available.

Chapter 2 Fingerprints of the Divine Around Us
1. E. Knowles, *The Oxford Dictionary of Quotations*, 5th edn, Oxford University Press, Oxford, 1999, p. 290.
2. P. Johnson, *Reason in the Balance: The Case Against Naturalism in Science, Law, and Education*, InterVarsity Press, Downers Grove, IL, 1995.
3. C. Sagan, *Cosmos*, Random House, New York, 1980, p. 1.
4. J.D. Barrow & F.J. Tipler, *The Anthropic Cosmological Principle*, Clarendon Press, Oxford, 1986.
5. See, for example, H. Arp, *Seeing Red: Redshifts, Cosmology and Academic Science*, Aperion Press, Montreal, 1998.
6. W.J. Cocke & W.G. Tifft, 'Statistical procedure and the significance of periodicities in double-galaxy redshifts', *Astrophysical Journal*, vol. 368, no. 2 (1991), pp. 383–389.

Chapter 3 Where Did the Universe Come From?
1. This chapter draws on the work of William Lane Craig, in particular his essay 'The existence of God and the beginning of the universe', *Truth: A Journal of Modern Thought*, vol. 3 (1991), pp. 85–96. Also available from <http://www.leaderu.com/truth/3truth11.html>. All quotations from Lane are from this article.
2. Hilbert's Hotel was first described by the German mathematician David Hilbert and is discussed extensively by Craig (ibid).
3. For further reading refer to the following:

W.L. Craig, *The Kalam Cosmological Argument*, Macmillan, New York, 1979.

—— *Reasonable Faith,* Crossway Books, Wheaton, IL, 1994.

—— & Q. Smith, *Theism, Atheism, and Big Bang Cosmology*, Clarendon Press, Oxford, 1993.

J.P. Moreland, *Scaling the Secular City: A Defense of Christianity*, Baker Books, Grand Rapids, MI, 1987 (especially Chapter 1).

W. Rowe, *The Cosmological Argument*, Princeton University Press, Princeton, 1975.

Chapter 4 Design by Information

1. I wish to especially thank my friend Dr Carl Wieland (Answers in Genesis, Australia) for his assistance in preparing this chapter. He not only did an excellent job of translating it from German to English, he also played an appreciable role in the fundamental scientific deliberations. His technical insights and discussions led to significant improvements, especially in the vitally important conclusions.
2. See also W. Gitt, *In the Beginning Was Information*, 3rd English edn, Christliche Literatur-Verbreitung, Bielefeld, 2001; W. Gitt, *Am Anfang war die Information*, 3. überarbeitete und erweiterte Auflage 2002, Hänssler Verlag, Holzgerlingen.
3. *In the Beginning Was Information*, pp. 47–49.
4. ibid., pp. 128–131.
5. Energy can exist in various forms (for example, mechanical, electrical, magnetic, thermal). These are, however, equivalent to each other and are thus expressible in the same units (for example, joules).
6. See *In the Beginning Was Information*, pp. 170–180.
7. See *Am Anfang war die Information*, pp. 131–150.
8. By contrast, the triplet code carried on DNA can easily be shown

to meet the criterion of being freely chosen in the sense of being arbitrary. In other words, there is no physical/chemical reason why the biomachinery of cells has to assign to the triplet CUG, for instance, the meaning of the amino acid 'leucine'. In fact, in some yeast species it is translated as 'serine'. This underscores the point—since the code is not the inevitable outcome of the physics and chemistry of the system, it was at some prior time freely chosen.

9. W. Gitt, 'Information: The Third Fundamental Quantity', *Siemens Review*, Vol. 56, Nov/Dec. 1989, pp. 2–7; W. Gitt, 'Neues Maß zum Vergleich hoher Speicherdichten', *Jahresbericht 1997 der Physikalisch-Technischen Bundesanstalt in Braunschweig*, März, 1997.
10. See Chapter 3 of this book.
11. Isaiah 44:6.
12. 'For since the creation of the world God's invisible qualities—his eternal power and divine nature—have been clearly seen, being understood from what has been made, so that men are without excuse' (Romans 1:20).
13. Psalm 90:2; Isaiah 40:28; Daniel 6:27.
14. This does not even consider the question of the creation of the individual atoms, merely their arrangement into DNA, which does not spontaneously happen from physics and chemistry. In living things, it happens via programmed machinery. To synthesise them in the laboratory requires the application of energy plus intelligence.
15. Revelation 1:8; Luke 1:37. That which we cannot conclude from any natural law is revealed to us by God himself. The almighty one who made and created all this is Jesus Christ, the Son of God. Colossians 1:16–17 says of him: 'For by him all things were created: things in heaven and on earth, visible and invisible . . .

all things were created by him and for him. He is before all things, and in him all things hold together.'
16. John 4:24.
17. 1 Thessalonians 5:23.
18. Exodus 20:11.
19. B.-O. Küppers, *Leben = Physik + Chemie?* Piper-Verlag, München, Zürich, 2. Auflage 1990, p. 19.
20. This can be viewed as an imperfect analogy of reproduction. One organism 'programs' matter to generate another program-bearing creature capable of doing the same. No intelligent intervention is involved. But here, too, one cannot trace it back indefinitely. In reproduction we see no information created, but we see information frequently lost through genetic mistakes (mutation). Hence the original information had to arise through the application of intelligence (mind). This is consistent with God programming the first creatures of any particular kind, including building in the capacity to vary within limits.
21. Genesis 1:24–25.
22. D. Batten, K. Ham, J. Sarfati & C. Wieland, *The (Revised, Expanded) Answers Book*, 5th edn, Answers in Genesis, 2003, pp. 133–134. Batten and Safarti are PhD scientists, Ham and Wieland have scientific and/or medical qualifications. All are actively involved with writing and speaking on Bible/science topics. See also <http://www.answersingenesis.org>.

Chapter 5 The Human Body: Evidence for Intelligent Design

1. S.C. Meyer, 'The origin of life and the death of materialism', *The Intercollegiate Review*, Spring 1996, p. 24.
2. F. Crick, *What Mad Pursuit: A Personal View of Scientific Discovery*, Sloan Foundation Science, Penguin Books, London, 1988 (reprint), p. 138.

3. M. Schliwa & G. Woehlke, 'Molecular motors', *Nature*, 422 (2003), p. 759.
4. J.E. Molloy & C. Veigel, 'Myosin motors walk the walk', *Science*, 300 (2003), pp. 2045–2046.
5. J.W. Thornton & R. DeSalle, 'Genomics meets phylogenetics', *Annual Review of Genomics and Human Genetics* 2000, p. 64.
6. J. Gerhard & M. Kirschner, *Cells, Embryos and Evolution*, Blackwell Science, Oxford, 1997, p. 161.
7. C. Janeway, P. Travers, M. Walport & M. Shlomchik, *Immunobiology*, Garland Publishing, New York, 2001, pp. 598–599.
8. ibid.
9. ibid., p. 602.
10. E.R. Nobel, G.A. Nobel, G.A. Schad & A.J. McInnis, *Parasitology: The Biology of Animal Parasites*, Lea & Febiger, Philadelphia and London, 1989, p. 516.

Chapter 6 Design in Nature: Evidence for a Creator

1. For more details see W.S. Romoser & J.G. Stoffolano Jr, *The Science of Entomology*, 3rd edn, Wm. C. Brown Publishers, Dubuque, Iowa, 1994, pp. 125–163; R.J. Elzinga, *Fundamentals of Entomology*, 4th edn, Prentice Hall, Upper Saddle River, NJ, 1997, pp. 106–135.
2. For results see A.A. Roth, C.D. Clausen, P.Y. Yahiku, V.E. Clausen & W.W. Cox, 'Some effects of light on coral growth', *Pacific Science*, vol. 36 (1982), pp. 65–81.
3. J. Cousteau & P. Diolé, *Octopus and Squid: The Soft Intelligence*, trans. J.F. Bernard, Doubleday & Co., Garden City, NY, 1973.
4. M.J. Wells, *Brain and Behaviour in Cephalopods*, Stanford University Press, Stanford, 1962, p. 13.
5. R.T. Hanlon & J.B. Messenger, *Cephalopod Behaviour*, Cambridge University Press, Cambridge, 1996, pp. 18–19.

6. J.B. Messenger, 'Comparative physiology of vision in mollusks', in H. Autrum (ed.), *Comparative Physiology and Evolution of Vision in Invertebrates*, VII, 6, C: Invertebrate Visual Centers and Behavior II, Springer-Verlag, Berlin, Heidelberg, 1981, pp. 93–200.
7. K.N. Nesis, *Cephalopods of the World: Squids, Cuttlefish, Octopuses, and Allies*, trans. B.S. Levitov, T.F.H. Publications, Ascot, Berkshire, 1987, p. 72.
8. M.F. Land, 'Molluscs', in M.A. Ali (ed.), *Photoreception and Vision in Invertebrates*, Plenum Press, New York, London, 1984, pp. 699–725.
9. The classic detailed study on the muscles of the eyes of these kinds of animals was done on the cuttlefish *Sepia*. See J.S. Alexandrowicz, 'Contribution a l'étude des muscles, des nerfs et du mécanisme de l'accommodation de l'oeil des céphalopodes', *Archives de Zoologie Expérimentale et Générales*, vol. 66 (1927), pp. 71–134.
10. J.Z. Young, 'The statocysts of *Octopus vulgaris*', *Proceedings of the Royal Society of London*, 152 (1960), pp. 3–29.
11. For a good illustration of the arrangement and reversal by the chiasmata, see Figure 1 in J.Z. Young, 'The optic lobes of *Octopus vulgaris*', *Philosophical Transactions of the Royal Society of London (Series B)*, 245 (1962), pp. 19–58.
12. Nesis, pp. 65–68.
13. G.L.G. Miklos & G.M. Rubin, 'The role of the genome project in determining gene function: Insights from model organisms', Berkeley Drosophila Genome Project, 1996. <http://www.fruitfly.org/about/pubs/miklos96.html>, viewed September 29, 2004.
14. D.L. Gilbert et al. (eds), *Squid as Experimental Animals*, Plenum Press, New York, London, 1990; R.R. Llinás, *The Squid Giant*

Synapse: A Model for Chemical Transmission, Oxford University Press, New York, Oxford, 1999.
15. B.J. Schnapp et al., 'Single microtubles from squid axoplasm support bidirectional movement of organelles', *Cell*, 40 (1985), pp. 455–462.
16. For example, H. Lodish et al., *Molecular Cell Biology*, 4th edn, W.H. Freeman & Co., New York, 2000, pp. 811–815.
17. E.N.K. Clarkson & R. Levi-Setti, 'Trilobite eyes and the optics of Des Cartes and Huygens', *Nature*, 254 (1975), pp. 663–667; K.M. Towe, 'Trilobite eyes: calcified lenses in vivo', *Science*, 179 (1973), pp.1007–1009.
18. G.D. Pollak & J.H. Casseday, *The Neural Basis of Echolocation in Bats*, Springer-Verlag, Berlin, Heidelberg, 1989, p. 12.
19. I am indebted to several good reviews of this topic including: J.E. Hill & J.D. Smith, *Bats: A Natural History*, Rigby Publishers, Adelaide, Sydney, 1984, pp. 107–126; A.D. Grinnell, 'Hearing in bats: an overview', in A.N. Popper & R.R. Fay (eds), *Hearing in Bats*, Springer-Verlag, New York, Berlin, 1995, pp. 1–36; G. Neuweiler, *The Biology of Bats*, trans. Ellen Covey, Oxford University Press, 2000, pp. 140–209.
20. G. Neuweiler, ibid., p. 162.
21. For a discussion of the various topics considered in this section, related topics and many literature references, see the recent book by the writer: A.A. Roth, *Origins: Linking Science and Scripture*, Review and Herald Publishing Association, Hagerstown, MD, 1998, pp. 80–115, 130–144.
22. M.J. Behe, *Darwin's Black Box*, Freepress, New York, 1996.
23. Miklos & Rubin, op. cit.
24. For an example of this scenario see A.S. Romer, *Man and the Vertebrates*, The University of Chicago Press, Chicago, 1941, pp. 290–292; D.J. Futuyma, *Evolutionary Biology*, Sinauer

Associates, Sunderland, MA, 1998, pp. 146–151.
25. A.R. Møller, 'Neurophysiological basis of the acoustic middle-ear reflex: Basic principles and clinical applications', in S. Silman (ed.), *The Acoustic Reflex: Basic Principles and Clinical Applications*, Academic Press, Orlando, San Diego, 1984, pp. 1–34.
26. For further discussion see the chapter 'The search for an evolutionary mechanism' in Roth, *Origins: Linking Science and Scripture*, pp. 80–93.
27. Psalms 41:13; 90:2; 106:48; Isaiah 43:10; Revelation 1:8; 22:13.

Chapter 7 The Scientific Case for Creation
1. T. McKee & J.R. McKee, *Biochemistry: The Molecular Basis of Life*, 3rd edn, McGraw Hill Publishers, New York, 2003, p. 58.
2. M. Delbruck, *Mind from Matter?*, Blackwell Scientific Publications, Oxford, 1986, p. 33.

Chapter 8 A Question of Biology
1. C. Wieland, 'Goodbye, peppered moths: A classic evolutionary story comes unstuck', *Creation*, vol. 21, no. 3 (1999), p. 56.
2. J.A. Coyne, 'Not black and white', *Nature*, 396 (1998), pp. 35–36.
3. The offspring of eyed surface-dwelling banded tetra and their eyeless cave-dwelling cousins have eyes that are intermediate in size. <http://www.bmb.psu.edu/597a/stdnts98/ags107/coursework/rsrch.htm>, viewed May 29, 2000.
4. R. Borowsky & L. Espinasa, 'Antiquity and origins of troglobitic Mexican tetras, *Astyanax fasciatus*', *Proceedings of the 12th International Congress of Speleology*, vol. 3, 1997, pp. 359–361. Accessed from <http://www.nyu.edu/gsas/dept/bio/faculty/borowsky/cave_publications/asty.html>, viewed June 8, 2000.
5. C. Wieland, 'Blind fish, island immigrants and hairy babies', *Creation*, vol. 23, no. 1 (2001), pp. 46–51.

6. Genesis 1:11–12, 21, 24–25.
7. Genesis 1:31.
8. Genesis 3:19; Romans 5:12, 8:19–22; 1 Corinthians 15:21–22, 26.
9. A. Lamb, 'The mutant "feather-duster" budgie', *Creation*, vol. 24, no. 1 (2001), pp. 54–55.
10. C. Wieland, 'Beetle bloopers—even a defect can be an advantage sometimes', *Creation*, vol. 19, no. 3 (1997), p. 30.
11. There are designed (that is, created) mechanisms for generating limited new information under strict cellular control. See, for example, D. Batten, 'The adaptation of bacteria to feeding on nylon waste', *TJ*, vol. 17, no. 3 (2003), pp. 3–5 (also at <http://www.answersingenesis.org/home/area/magazines/tj/docs/v17n3_nylon.asp>); and R. Truman, 'The unsuitability of B-cell maturation as an analogy for neo-Darwinian theory', <http://www.trueorigin.org/b_cell_maturation.asp>.
12. V.A. McKusick, Online Mendelian Inheritance in Man, McKusick-Nathans Institute for Genetic Medicine, Johns Hopkins University (Baltimore, MD), and National Center for Biotechnology Information, National Library of Medicine (Bethesda, MD). <http://www.ncbi.nlm.nih.gov/entrez/query.fcgi?db=OMIM>, viewed October 7, 2004.
13. L. Spetner, *Not by Chance*, Judaica Press, New York, 1999.
14. C. Wieland, 'Superbugs: Not super after all', *Creation*, vol. 20, no. 1 (1998), pp. 10–13.
15. C. Wieland, 'Darwin's finches: Evidence supporting rapid post-Flood adaptation', *Creation*, vol. 14, no. 3 (1992), pp. 22–23.
16. P.R. Grant, 'Natural selection and Darwin's finches', *Scientific American*, 265, no. 4 (1991), pp. 60–65.
17. D. Catchpoole & C. Wieland, 'Speedy species surprise', *Creation*, vol. 23, no. 2 (2001), pp. 13–15.

18. A.A. Hoffman, R.J. Hallas, J.A. Dean & M. Schiffer, 'Low potential for climatic stress adaptation in a rainforest *Drosophila* species', *Science*, 301 (2003), pp. 100–102; D. Roff, 'Evolutionary danger for rainforest species', *Science*, 301 (2003), pp. 58–59.
19. D. Batten, 'Are look-alikes related?' *Creation*, vol. 19, no. 2 (1997), pp. 39–41.
20. Romans 1:18–23.
21. John 1:3, Colossians 1:16.
22. D. Batten, (ed.), *The Answers Book*, Answers in Genesis, Brisbane, 1999, Chapter 7.
23. J. Sarfati, *Refuting Evolution 2*, Answers in Genesis, Brisbane, 2002, Chapter 6.
24. J. Sarfati, *Refuting Evolution*, 3rd edn, Answers in Genesis, Brisbane, 2004.
25. R. Grigg, 'Ernst Haeckel: Evangelist for evolution and apostle of deceit', *Creation*, vol. 18, no. 2 (1996), pp. 33–36.
26. Stephen Jay Gould, 'Dr Down's syndrome', *Natural History*, vol. 89 (1980), p. 144, cited in H. Morris, *The Long War Against God*, Baker Book House, Grand Rapids, MI, 1989, p. 139.
27. K. Thompson, 'Ontogeny and phylogeny recapitulated', *American Scientist*, vol. 76 (1988), p. 273.
28. For example, P.H. Raven & G.B. Johnson, *Biology*, 3rd edn, Mosby-Year Book, St Louis, 1992, p. 396.
29. *World Book Encyclopedia 2000*, vol. 6, World Book, Inc., Chicago, IL, pp. 426–434.
30. M.K. Richardson et al., 'There is no highly conserved embryonic stage in the vertebrates: Implications for current theories of evolution and development', *Anatomy and Embryology*, vol. 196, no. 2 (1997), pp. 91–106; R. Grigg, 'Fraud rediscovered', *Creation*, vol. 20, no. 2 (1998), pp. 49–51. The photographs can be viewed at <http://www.answersingenesis.org/docs/1339.asp>.

31. E. Pennisi, 'Haeckel's embryos: Fraud rediscovered', *Science*, 277 (1997), p. 1435; Anonymous, 'Embryonic fraud lives on', *New Scientist*, vol. 155, no. 2098 (1997), p. 23.
32. J. Bergman & G. Howe, '"Vestigial organs" are fully functional', Creation Research Society Monograph No. 4, Creation Research Society Books, Terre Haute, Indiana, 1990.
33. K. Ham & C. Wieland, 'Your appendix—it's there for a reason', *Creation*, vol. 20, no. 1 (1997), pp. 41–43.

Chapter 9 The Geological Evidence for Creation

1. Genesis 7:8–9.
2. For further reading refer to the following:

    K.P. Wise, *Faith, Form and Time*, Broadman & Holman, Nashville, Tennessee, 2002.

    J.D. Morris, *The Young Earth*, Master Books, Colorado Springs, 1994.

    D. Batten (ed.), *The Answers Book*, Answers in Genesis, Acacia Ridge, Qld, 1999.

    L. Vardiman, A.A. Snelling & E.F. Chaffin (eds), *Radioisotopes and the Age of the Earth*, Institute for Creation Research, El Cajon, CA, and Creation Research Society, St Joseph, MO, 2000.

    S.A. Austin (ed.), *Grand Canyon: Monument to Catastrophe*, Institute for Creation Research, El Cajon, CA, 1994.

Chapter 10 Where Do Thoughts Come From?

1. M. Minksy, 'Decentralized mind', *Behavioral and Brain Sciences*, vol. 3 (1980), p. 439.
2. Paul Churchland, *The Engine of the Soul*, MIT Press, Cambridge, 1985, p. 19.
3. S. Stich, *From Folk Psychology to Cognitive Science: A Case Against Belief*, MIT Press, Cambridge, 1985.

4. See, for example, L.R. Baker, *Saving Belief: A Critique of Physicalism*, Princeton University Press, Princeton, 1987.
5. See, for example, G. Strawson, *Mental Reality*, MIT Press, Cambridge, 1994.
6. J. Kim, *Mind in a Physical World*, MIT Press, Cambridge, 1998, p. 2.
7. Churchland, p. 211.
8. See, for example, F. Jackson, 'What Mary didn't know', *Journal of Philosophy*, vol. LXXXIII, no.5 (1986).
9. See Strawson, op. cit.
10. See C. McGinn, *The Mysterious Mind*, Basic Books, New York, 1999.
11. See, for example, E. Lepore & R. Van Gulick, *John Searle and His Critics*, Blackwell, Oxford, 1991.
12. See Kim, *Mind in a Physical World*.
13. D.M. Armstrong, *A Materialist Theory of Mind*, Routledge, London, 1968, p. 30.
14. McGinn, p. 15.
15. R. Swinburne, *The Evolution of the Soul*, Clarendon Press, Oxford, 1986.
16. C. Taliaferro, *Consciousness and the Mind of God*, Cambridge University Press, Cambridge, 1994.

Chapter 11 The Question of Moral Values

1. Quote from the transcript of an interview with Jeffrey Dahmer on *Dateline NBC*, broadcast November 29, 1994.
2. ibid.
3. F.J. Beckwith & G. Koukl, *Relativism: Feet Firmly Planted in Mid-Air*, Baker Books, Grand Rapids, MI, 1998, p. 59.
4. J.L. Mackie, *The Miracle of Theism*, Clarendon Press, Oxford, 1982, p. 115.

5. R. Dawkins, 'God's utility function', *Scientific American*, 273 (1995), p. 85.
6. See Charles Darwin, *The Descent of Man*, in R.M. Hutchins (ed.), *The Great Books of the Western World*, vol. 49, Encyclopedia Britannica, Chicago, 1952, p. 320.
7. M.D. Linville, *Is Everything Permitted? Moral Values in a World Without God*, Ravi Zacharias International Ministries, Norcross, GA, 2001, p. 36 (emphasis his).
8. J. Rachels, *Created from Animals: The Moral Implications of Darwinism*, Oxford University Press, Oxford, 1990, p. 77.
9. M. Ruse, 'Evolutionary theory and Christian ethics', in *The Darwinian Paradigm*, Routledge, London, 1989, p. 269.
10. Shankara, *Vedanta Sutras*, 2.3.48.
11. See Genesis 1:26–27.
12. J.P. Moreland & K. Nielsen, *Does God Exist? The Great Debate*, Thomas Nelson, Nashville, TN, 1990, p. 119.
13. This way of stating the argument for moral relativism, as well as some of the following responses, can be found in J. Rachels, 'Morality is not relative', in L.P. Pojman (ed.), *Philosophy: The Quest for Truth*, 4th edn, Wadsworth, Belmont, CA, 1999, pp. 375–382.
14. P. Copan, 'Is the Naturalistic Fallacy a Case of Question-Begging? An Assessment of Contemporary Naturalistic Moral Realism.' Unpublished paper presented at the Evangelical Philosophical Society, Colorado Springs, Colorado, November, 2001.
15. Linville, *Is Everything Permitted?*, pp. 50–51 (emphasis his).

Chapter 12 The Problem of Evil
1. Jeremiah 31:3.
2. 1 John 4:8.
3. Genesis 1:11–12; 2:8–9.

4. Genesis 2:19–20.
5. Ecclesiastes 3:13; 5:19; James 1:17.
6. Genesis 1:31.
7. Genesis 1:26–28; 2:9, 16–17.
8. Isaiah 14:12–14; Ezekiel 28:13–15.
9. Revelation 12:7–9.
10. Genesis 3:1–7.
11. Genesis 3:8–24.
12. Genesis 6:5.
13. C.S. Lewis, *Mere Christianity*, MacMillan Paperbacks Edition, New York, 1960, pp. 54–55.
14. Matthew 2:1–25; Luke 2:1–20.
15. Acts 10:38; Matthew 8:1–17; John 4:46–54; Luke 7:11–17; John 11:1–44; John 2:1–11; Mark 6:30–44; John 6:1–15.
16. One liners: Matthew 7:12; Matthew 5:39; John 13:35. Unforgettable stories: Luke 10:25–37; Luke 15:11–32; Matthew 13:18–23. Memorable encounters: John 3:1–21; John 4:1–42; John 11:1–44.
17. Lewis, p. 56.
18. 1 Peter 2:21–24.
19. Isaiah 53:1–12.
20. Matthew 28: 2,5; Luke 24:4.
21. Romans 6:3–6.
22. D. Cauchon, 'For many on Sept. 11 survival was no accident', *USA Today*, December 19, 2001.
23. 'The problem with miracles', Editorial, *Adventist Review*, March 21, 2002.
24. Job 1:21 (NIV). Other quotations from Job are either NIV or author's own translation.
25. Job 38:1–42:6.
26. Acts 13:32–33.

Chapter 14 Will the Real God Please Stand Up?
1. Psalm 115:3; Isaiah 46:9–11; Isaiah 14:24–27.
2. Ephesians 1:11.
3. Romans 9:15–21 (NASB).
4. Ephesians 1:12.
5. Ephesians 1:3–6.
6. Ephesians 1–2 (NASB).
7. Leviticus 26:4–20.
8. Amos 3:6.
9. Proverbs 16:4; 1 Kings 22:23; Ezekiel 14:9–10.
10. Proverbs 21:1; Proverbs 16:9; Acts 4:27–28 (emphasis added); Revelation 17:16–17 (emphasis added).
11. For example, Genesis 17:1.
12. Genesis 18:14; Jeremiah 32:27; Jeremiah 32:17; Job 42:2; Matthew 19:26; Luke 1:37.
13. Hebrews 6:18; James 1:13; 2 Timothy 2:13 (NASB).
14. 1 John 3:20; Hebrews 4:13; Colossians 2:3.
15. Matthew 11:21–24.
16. Psalm 90:2; Isaiah 44:6; Revelation 1:8; 22:13.
17. Malachi 3:6; Psalm 102:27; Hebrews 13:8.
18. Colossians 1:16.
19. Genesis 1:31.
20. 1 Corinthians 14:33.
21. Genesis 1.
22. Genesis 8:22.
23. Romans 8:20–22.
24. Leviticus 26:4–20.
25. Amos 3:6.
26 James 1:13.
27. Jeremiah 17:9.
28. Romans 8:23; 1 Corinthians 15:54–57.

29. Habakkuk 1:13.
30. 2 Corinthians 6:17; 1 Peter 1:16.
31. Romans 2:14–15 (NASB).
32. Romans 1:18–19.
33. Romans 1:18, 24, 26, 28.
34. James 3:9.
35. Matthew 5:21–28.
36. Romans 6:17; Ephesians 2:1–3.
37. Romans 8:5–8.
38. Romans 7:7–11.
39. Ezekiel 18:23.
40. Genesis 6:7.
41. 2 Peter 2:5.
42. 1 John 4:8, 16.
43. Romans 5:6–10.
44. Romans 4:4–5.
45. Romans 8:1.
46. Romans 7:6.
47. 1 John 4:9–10.
48. 1 John 4:19.

Chapter 15 Can the Bible Be Relied On?

1. A. Negev (ed.), *Archaeological Encyclopedia of the Holy Land*, G.P. Putnam's Sons, New York, 1972, p. 89.
2. F.M. Cross & D.W. Parry, 'A preliminary edition of a fragment of 4QSam^b (4Q52)', *Bulletin of the American Schools of Oriental Research*, vol. 306 (1997), p. 66.
3. J.A. Sanders, 'Keep each tradition separate', *Bible Review*, vol. 16, no. 4 (2000), p. 47 (italics original).
4. R.S. Hendel & J.A. Sanders, 'The most original Bible text: How to get there', *Bible Review*, vol. 16, no. 4 (2000), p. 27.

5. L.H. Schiffman, 'Review of *The Dead Sea Scrolls Bible*', *Bible Review*, vol. 17, no. 4 (2001), p. 42.
6. E.M. Meyers, 'Unraveling the Qumran texts', *Archaeology*, vol. 51, no. 6 (1998), p. 83.
7. C.E. Pfeiffer, *The Dead Sea Scrolls and the Bible*, Baker Book House, Grand Rapids, MI, 1969, p. 105.
8. B.M. Metzger & M.D. Coogan (eds), *The Oxford Companion to the Bible*, Oxford University Press, New York and Oxford, 1993, p. 685.
9. ibid., p. 487.
10. W. Duckat, *Beggar to King: All the Occupations of the Bible*, Doubleday, Garden City, NY, 1968, pp. 215–216.
11. Metzger & Coogan, p. 488.
12. ibid.
13. ibid., p. 489.
14. The Diatessaron was a combination of all four Gospels, compiled around 170 AD.
15. M. Burrows, *What Mean These Stones?*, American School of Oriental Research, New Haven, CT, 1941, p. 42.
16. Genesis 6:8.
17. Genesis 9:20–24.
18. Genesis 17:4; 18:18.
19. Genesis 12:13–19.
20. Genesis 20:1–12.
21. Genesis 26:1–8; 26:9 (KJV).
22. Numbers 12:8 (KJV).
23. Numbers 20:1–11. In the Sinai it is occasionally possible to strike a rock surface hard enough so that underground water will trickle out. This is due to the fact that pockets of water sometimes form in Sinai's porous limestones (Harvey Arden, 'In search of Moses', *National Geographic*, vol. 149, no. 1 [1976],

p. 31). The first confirmed report of this occurred on October 31, 1917, when General Allenby's British army had finished driving the Turks out of the Negev, the southern part of Israel, during World War I. An Australian private named Stephen Kelly went up to a ledge that jutted up from the desert sand and thrust his bayonet between two loose rocks at the foot of the ledge. The crack between the rocks suddenly widened, and water came streaming out (E.G. Schurmacher, *Strange Unsolved Mysteries*, Paperback Library, New York, 1962, p. 60). A similar incident occurred in the Sinai itself. Major C.S. Jarvis, British Governor of the Sinai in the 1930s, was an eyewitness to such an event. He observed several men of the Sinai Camel Corps digging around the foot of a rock face in a dried-out riverbed. Jarvis saw their target: water trickling out of the limestone rock face. The top-ranking sergeant in the group took a spade and violently struck the rock face, and water gushed forth. The Sudanese Muslims who were part of the group shouted, 'Look at him! The prophet Moses!' (W. Keller, *The Bible as History*, 2nd edn, trans. W. Neil, William Morrow & Co., New York, 1981, p. 136). Moses' sin, therefore, lay in the fact that he publicly defied God's command to merely speak to the rock, and instead attempted to obtain water the natural way, by striking the limestone rock face. The credit for finding water for the Israelites would thus have been his, not God's.
24. Numbers 20:12.
25. 1 Samuel 13:14.
26. The story is told in 2 Samuel 11–12.
27. 2 Chronicles 1:1–12.
28. 1 Kings 4:29–34.
29. 1 Kings 11:7.
30. Matthew 8:26; Luke 9:51–56; Mark 10:35–41; Mark 8:33; Matthew 26:33–35, 69–75.

31. J.L. Sheler, 'Is the Bible true?' *US News and World Report*, October 25, 1999, p. 54.
32. ibid.
33. Exodus 32.
34. This apostasy has been confirmed by archaeology. See, for example, E. Stern, 'Pagan Yahwism: The folk religion of ancient Israel', *Biblical Archaeology Review*, vol. 27, no. 3 (2001), pp. 21–29.
35. Philippians 1:7, 17; 1 Peter 3:15.
36. H. Schlossberg, *Idols for Destruction*, Crossway Books, Wheaton, IL, 1990, p. 28.
37. Matthew 11:2–6.
38. John 20:24–30.
39. 2 Peter 1:16.
40. Acts 1:22.
41. 1 Corinthians 15:3–19.
42. R.P. Martin, *New Testament Foundations: A Guide for Christian Students (Vol. 1, The Four Gospels)*, William B. Eerdmans, Grand Rapids, MI, 1975, p. 42.
43. ibid., p. 43.
44. N. Anderson, *Jesus Christ: The Witness of History*, Inter-Varsity Press, Leicester, 1985, pp. 28–29.
45. Josephus, *Antiquities of the Jews*, book 18, chapter 3, paragraph 3.
46. Acts 5:30.
47. F.F. Bruce, *Jesus and Christian Origins Outside the New Testament*, William B. Eerdmans, Grand Rapids, MI, 1974, p. 56.
48. John 13:1; Matthew 9:34; Mark 14:50.
49. Bruce, p. 56.
50. Matthew 9:27.
51. Quoted in G. Cornfield (ed.), *The Historical Jesus*, MacMillan, New York, 1982, p. 162.

52. John 5:18; Mark 14:62; John 6:62.
53. E.A. Speiser, *The Anchor Bible: Genesis*, Doubleday & Co., Garden City, NY, 1964, p. 6; J.P. Free, *Archaeology and Bible History*, revised & expanded by H.F. Vos, Zondervan, Grand Rapids, MI, 1992, pp. 27–28.
54. Job 37:18 (KJV); 2 Samuel 22:43 (KJV).
55. Genesis 1:20 (KJV).
56. L. Thomas, *The Lives of a Cell*, Bantam, Toronto, 1974, p. 50.
57. Mark 15:25; John 19:14.
58. J. Finegan, *Handbook of Biblical Chronology*, Princeton University Press, Princeton, NJ, 1964, p. 291.
59. Mark 14:12–17; John 13:1; 18:28; 19:14.
60. Matthew 27:62; Mark 15:42; Luke 23:54; John 19:31.
61. Metzger & Coogan, p. 120. Nisan is a month in the Jewish calendar, covering the second half of March and the first half of April.
62. ibid. For further on this problem, see Finegan, pp. 290–291.

Chapter 16 Historical Evidence for the Biblical Flood

1. I wish to thank Clifford Lillo, Helen Fryman, Tim Wallace, John Woodmorappe and other reviewers for their comments on an earlier draft of this chapter.
2. P. Freund, *Myths of Creation*, Washington Square Press, New York, 1965, p. 6.
3. B. Sproul, *Primal Myths: Creation Myths around the World*, Harper, San Francisco, 1991; P. Colum, *Myths of the World* (Original title: *Orpheus*), The Universal Library, Grosset & Dunlap, New York, 1930.
4. B. Nelson, *The Deluge Story in Stone*, Augsburg, Minneapolis, MN, 1931.
5. H.A. Guerber, *Myths and Legends Series: Greece and Rome*,

Bracken Books, London, 1986. The Christian and Muslim creation stories are both based on the Hebrew and all three are substantially the same in most of the major details (see Sproul, *Primal Myths*).
6. S.L. Fahs & D.T. Spoerl, *Beginnings: Earth, Sky, Life, Death*, Beacon Hill, Boston, 1960, p. 53.
7. M.-L. Von Franz, *Creation Myths*, Shambhala, Boston, 1995, p. 3.
8. C. Long, *Alpha: The Myths of Creation*, Collier Books, New York, 1963.
9. D. Steindl-Rast, 'Views of the cosmos', *Parabola*, vol. 2, no. 3 (1977), p. 7.
10. D.A. Leeming & M.A. Leeming, *A Dictionary of Creation Myths*, Oxford University Press, New York, 1995, p. 59.
11. A. Christie, *Chinese Mythology*, Paul Hamlyn, London, 1968, p. 47.
12. V. Hamilton, *In the Beginning: Creation Stories from Around the World*, Harcourt Brace Jovanovich, San Diego, CA, 1988, p. 127.
13. For example, Jeremiah 18:1–9.
14. Leeming & Leeming, p. 58.
15. J. Levy, *In the Beginning: The Navajo Genesis*, University of California Press, Berkeley, CA, 1998.
16. Leeming & Leeming, pp. 126–128.
17. J. Mbiti, *African Religions and Philosophy*, Doubleday & Co., Garden City, NY, 1970, pp. 39–40.
18. G. Hasel, 'The polemic nature of the Genesis cosmology', *The Evangelical Quarterly*, vol. 46, no. 2 (1974), p. 87.
19. Leeming & Leeming, pp. 228.
20. Von Franz, pp. 224–232.
21. Christie, pp. 49, 53–54.
22. ibid.

23. E. Lerner, *The Big Bang Never Happened*, Random House, New York, 1991, pp. 11–57.
24. R. Van Over, *Sun Songs: Creation Myths from Around the World*, New American Library, New York, 1980, pp. 15–16.
25. 1 John 1:5.
26. Genesis 1:2–3.
27. Nelson, pp. 1–20.
28. J.C. Allen, *The Legend of Noah*, University of Illinois Press, Urbana, IL, 1963.
29. See Long, pp. 36–220.
30. For an example of the embellishments common to non-Genesis creation myths and how they developed, see Guerber, pp. 1–26.
31. J. Strickling, 'A statistical analysis of flood legends', *CRSQ*, vol. 9, no. 3 (1972), p. 152.
32. Leeming & Leeming, p. 204.
33. F. Warshofsky, 'Noah, the flood, the facts', *Reader's Digest*, no. 3 (1977), p. 129 (emphasis added).
34. J. Morris, 'Why does nearly every culture have a tradition of a global flood?', *Back to Genesis*, 154 (2001), p. 4.
35. ibid.
36. C.H. Gordon, *The Ancient Near East*, W.W. Norton, New York, 1965, p. 50; H. Morris, *The Genesis Record*, Baker Book House, Grand Rapids, MI, 1976, pp. 25–26.
37. Van Over, p. 10 (emphasis added).
38. C. Levi-Strauss, 'The structural study of myth', in R. Kearney & M. Rainwater (eds), *The Continental Philosophy Reader*, Routledge, New York, 1996, p. 308.
39. C. Kluckhohn, 'Recurrent themes in myths and mythmaking', in R. Chmann (ed.), *The Making of Myth*, G.P. Putnam's Sons, New York, 1962. (See also Levi-Strauss, ibid., and C. Kluckhohn, 'Myths and rituals: A general theory', in W. Lessa et al. (eds),

Reader in Comparative Religion, Harper & Row, New York, 1958.)
40. Van Over, p. 11.
41. Freund, pp. 21–23, 81.
42. Hasel, pp. 90–91.
43. Genesis 2:9; 3:22.
44. G. Woodcock, 'The lure of the primitive', *The American Scholar*, vol. 45, no. 3 (Summer 1976), pp. 387–402.
45. H.W. Ellis, 'Creationism discussion continues', *Physics Today*, vol. 35, no. 10 (1982), pp. 11, 13; T. Sebeok (ed.), *Myth: A Symposium*, Indiana University Press, Bloomington, IN, 1968.
46. S. Diamond (ed.), *Primitive Views of the World: Essays from Culture in History*, Columbia University Press, New York, 1964.
47. Steidl-Rast, p. 7.
48. ibid.
49. D. Jones, 'Burn, books, burn: The death of the Alexandria museum', *Intellectual Digest*, vol. 2, no. 8 (1972), p. 53. See also S.N. Kramer, *Sumerian Mythology: A Study of Spiritual and Literary Achievement in the Third Millennium BC*, Harper & Row, New York, 1961.
50. J. Harrison, *Myths of Greece and Rome*, Ernest Benn Ltd, London, 1927, pp. 9–11.
51. Levi-Strauss, p. 308.
52. See J. Carcopino, *Daily Life in Ancient Rome*, edited and annotated by H.T. Rowell, trans. E.O. Lorimer, Yale University Press, New Haven, 2003.
53. E. Chiera, *They Wrote on Clay*, University of Chicago Press, Chicago, 1938, p. 110.
54. G. Glotz, *Ancient Greece at Work*, W.W. Norton, New York, 1967, pp. 7–13.

55. R. Graves & R. Patai, *Hebrew Myths: The Book of Genesis*, Greenwich House, New York, 1983.
56. I. Asimov, *In the Beginning*, Crown Publishing Co., New York, 1981, p. 3.
57. John 1:1–3 (KJV); Acts 4:24; 14:15; 17:23–25; Revelation 4:11.
58. H. Morris, *The Genesis Flood: The Biblical Record and Its Scientific Implications*, Presbyterian & Reformed, Philadelphia, 1967.
59. See Chapter 15 of this book.
60. See C. Doria & H. Lenowitz, *Origins: Creation Texts from the Ancient Mediterranean*, Anchor Press/Doubleday, Garden City, NY, 1976.
61. Hasel, pp. 91, 102.
62. ibid., p. 91.
63. Long, p. 19.

Chapter 17 Archaeological Evidence for the Exodus

1. 1 Kings 6:1. Bible quotations in this chapter are from the New King James Version.
2. 'Is the Bible fact or fiction?', *Time*, December 18, 1995, pp. 54ff.
3. P. James, *Centuries of Darkness*, Pimlico, London, 1992, p. 318.
4. ibid, pp. xiv.
5. D.M. Rohl, *A Test of Time: The Bible from Myth to History*, Century Ltd (Random House), London, 1995, p. 143.
6. ibid., pp. 11, 38.
7. Professor Bryant Wood, who lectures at the University of Toronto in Canada, has now also joined the ranks of those who consider that the Exodus events should be found in the Twelfth Dynasty, though he arrives at this conclusion by moving the Bible dates rather than by revising the Egyptian dates. See B. Wood, 'Wood lecture', Institute of Archaeology, *Horn*

Archaeological Museum Newsletter, vol. 23, no. 4 (Fall 2002), pp. 2–3.
8. Genesis 41.
9. H. Brugsch-Bey, *Egypt under the Pharaohs*, Trafalgar Square Publishers, London, p. 158.
10. Genesis 41:48.
11. Brugsch-Bey, p. 162.
12. Genesis 41:42–43.
13. Exodus 1:7.
14. Exodus 1:8.
15. Josephus, *Antiquities of the Jews*, book 2, chapter 9, paragraph 1.
16. Exodus 2:5.
17. Exodus 2:11–15.
18. I.E.S. Edwards, C.J. Gadd, N.G.L. Hammond & E. Sollberger (eds), *The Cambridge Ancient History*, vol. 2, part 1, Cambridge University Press, Cambridge, 1973, p. 43.
19. ibid., p. 49.
20. *Encyclopaedia Britannica*, 1964, vol. 8, p. 35.
21. A. Gardiner, *Egypt of the Pharaohs: An Introduction*, Oxford University Press, London, 1966, p. 133.
22. Josephus, op. cit.
23. Exodus 5:6–7.
24. W.M.F. Petrie, *Ten Years Digging in Egypt*, Ares Publishers Inc, Chicago, 1976, pp. 112–113.
25. R. David, *The Pyramid Builders of Ancient Egypt*, Guild Publishing, London, 1986, pp. 175, 188, 190–191.
26. Genesis 46:5–7.
27. David, pp. 189–190.
28. Petrie, pp. 116–117.
29. Exodus 1:15–16.
30. Exodus 1:22.

31. Exodus 2:3.
32. D.A. Courville, *The Exodus Problem and its Ramifications*, Challenge Books, Loma Linda, CA, 1971.
33. David, pp. 195–199.
34. Exodus 12:41.
35. Exodus 14:28.
36. Exodus 7–12.
37. Ipuwer Papyrus, Leiden Museum. In A. Erman (ed.), *The Ancient Egyptians: A Source Book of Their Writings*, Harper Torchbooks, New York, 1996, pp. 94–101.
38. K.M. Kenyon, *Archaeology in the Holy Land*, Thomas Nelson, Nashville, TN, 1985, p. 134.
39. J.B. Pritchard, *Gibeon, Where the Sun Stood Still*, Princeton University Press, Princeton, 1962, p. 153.
40. Joshua 10:2.
41. Pritchard, pp. 157–158.
42. See Numbers 13–14.
43. R. Cohen, 'The mysterious MBI people', *Biblical Archaeology Review*, July/August 1983, pp. 16–29.

Chapter 18 The Historical Reliability of the Old Testament

1. G. Mendenhall, 'Mari', *Biblical Archaeologist*, vol. XI, no. 1 (1948), p. 15.
2. Ephesians 1:18. Bible quotations in this chapter are from the New Revised Standard Version.
3. The average archaeological site in Palestine is about 15 acres. A. Hoerth, *Archaeology and the Old Testament*, Baker, Grand Rapids, MI, 1998, p. 214.
4. A. Malamat, *Mari and the Early Israel Experience*, Oxford University Press, Oxford, 1989, pp. 55–62.
5. ibid., p. 56.

6. W.L. Moran, *The Amarna Letters*, Johns Hopkins Press, Baltimore, 1987, letters 148, 227, 228, 364.
7. Ahlstrom is a critic who almost completely denies the historicity of the Old Testament. G. Ahlstroem, *The History of Ancient Palestine*, Fortress Press, Minneapolis, 1993, p. 246.
8. Joshua 11:10.
9. Joshua 11:1ff.
10. Judges 4:2.
11. Malamat, p. 58.
12. W. Horowitz & A. Shaffer, 'A fragment of a letter from Hazor', *Israel Exploration Journal*, vol. 42 (1992), pp. 165–166.
13. Judges 4:1ff.
14. A. Ben-Tor & M.T. Rubiato, 'Did the Israelites destroy the Canaanite city?', *Biblical Archaeology Review*, vol. 25, no. 3 (1999), pp. 25, 30.
15. C. Krahmalkov, 'Exodus itinerary confirmed by Egyptian evidence', *Biblical Archaeology Review*, vol. 20, no. 5 (1994), p. 61.
16. Joshua 11:10–15.
17. Deuteronomy 7:5.
18. Ben-Tor & Rubiato, pp. 22, 35.
19. ibid., pp. 35–36.
20. ibid., cover, p. 38.
21. Alhstrom, p. 259.
22. B. Hesse, 'Pig lovers and pig haters: Patterns of Palestinian pork production', *Journal of Ethnology*, vol. 10, pp. 195–225; I. Finkelstein, 'Ethnicity and origin of Iron I Settlers in the highlands of Canaan: Can the real Israel stand up?', *Biblical Archaeologist*, vol. 59, no. 4 (1996), pp. 198–212.
23. J. Hoffmeier, *Israel in Egypt*, Oxford University Press, Oxford, 1997, pp. 38–43.

24. J.K. Hoffmeier, 'The structure of Joshua 1–11 and the Annals of Thutmose III', in A.R. Millard, J.K. Hoffmeier & D.W. Baker (eds), *Faith, Tradition and History*, Eisenbrauns, Winona Lake, IN, 1994, p. 176.
25. N.K. Gottwald, *The Hebrew Bible, A Socio-Literary Introduction*, Fortress Press, Philadelphia, 1985, p. 262.
26. Gottwald hastens to add that Hazor still being in the hands of King Jabin in Judges 4 is problematic (ibid., p. 266). However, it has already been observed that 'Jabin' was probably a dynastic name, and cities being lost by the Israelites to their enemies fits well the scenario of apostasy in Judges.
27. Y. Yadin, 'Hazor', in *The New Encyclopedia of Archaeological Excavation in the Holy Land*, E. Stern (ed.), Simon & Schuster, New York, 1993, pp. 599, 601.
28. As quoted in A. Mazar, *Archaeology of the Land of the Bible*, Doubleday, New York, 1990, pp. 384ff.
29. ibid., p. 400, note 16. See also 'Hazor', in A. Negev & S. Gibson (eds), *Archaeological Encyclopedia of the Holy Land*, Continuum, New York, 2001, p. 198.
30. G. Barkay, 'The Iron Age II-III', in A. Ben-Tor (ed.), *The Archaeology of Ancient Israel*, Yale University Press, New Haven, pp. 307–308.
31. Negev and Gibson, p. 198.
32. G. Davies, *Megiddo*, Lutterworth Press, Cambridge, 1986, pp. 83–92.
33. C. Meyers, 'Kinship and kingship', in M.D. Coogan (ed.), *The Oxford History of the Biblical World*, Oxford University Press, Oxford, 1998, pp. 187–188.
34. Davies, op. cit.
35. I. Finkelstein, 'The Archaeology of the United Monarchy: An Alternate View', *Levant* 28 (1996), pp. 177–187.

36. Television interview on the History Channel program 'Digging for Truth', October 14, 2002.
37. B. Halpern, *David's Secret Demons*, Eerdmans, Grand Rapids, MI, 2001, pp. 451–453.
38. Y. Shiloh, 'Megiddo', in Stern, *New Encyclopedia of Archaeological Excavation in the Holy Land*, pp. 1016–1019.
39. Revelation 16:16.
40. Joshua 17:11–12; Judges 1:27.
41. 1 Kings 4:12.
42. Mazar, pp. 355–356, 374.
43. Davies, pp. 78–85.
44. T.P. Harrison, 'The battleground—who destroyed Megiddo?', *Biblical Archaeology Review*, vol. 29, no. 6 (2003), pp. 28–35, 60–62.
45. Judges 1:19ff.
46. Harrison, pp. 34–35.
47. 1 Samuel 14:22.
48. Harrison, p. 62.
49. These are associated with a temple. See J. Gibson, *Phoenician Inscriptions*, Clarendon Press, Oxford, 1982, pp. 17–21.
50. 2 Kings 11:1ff.
51. H.L. Katzenstein, *History of Tyre*, Ben Gurion University Press, Beer Sheva, 1997, p. 69.
52. J.A. Wilson, 'Egyptian myths, tales and mortuary texts', in J. Pritchard (ed.), *Ancient Near Eastern Texts*, University of Princeton Press, Princeton, 1969, p. 27a.
53. A. Biran & J. Naveh, 'An Aramaic stele fragment from Tel Dan', *Israel Exploration Journal*, vol. 43, no. 2/3 (1993), pp. 81–85.
54. 2 Kings 13:1ff.
55. 2 Samuel 8:3–10.
56. G. Rendsburg, 'On writing BYTDWD in the Aramaic inscrip-

tion from Tel Dan', *Israel Exploration Journal*, vol. 45, no. 1 (1995), pp. 22–25.
57. R. Degen, *Altaramaeische Grammatik*, Deutsche Morgenlandische Gesellschaft, Wiesbaden, 1969, pp. 27–28.
58. D.N. Freedman & J. Geoghegan, 'House of David is there', *Biblical Archaeology Review*, March/April 1995, pp. 78–79.
59. A. Lemaire, '"House of David" restored in Moabite inscription', *Biblical Archaeology Review*, vol. 20, no. 3 (1994), pp. 30ff.
60. 2 Samuel 8:2.
61. S.R. Driver, *Introduction to the Literature of the Old Testament*, Meridian, New York, 1956 reprint, p. 324.
62. A. Feuillet, 'Les Sources du Livre de Jonas', *Revue Biblique*, vol. 54 (1947), p. 161.
63. For an analysis of attempts to find historical mistakes in Jonah, see P. Ferguson, 'Who is the King of Nineveh in Jonah 3:6?', *Tyndale Bulletin*, vol. 47, no. 2 (1996), pp. 301ff.
64. Jonah 1:2.
65. Jonah 3:4.
66. Jonah 3:5–8.
67. S. Parpola (ed.), *Letters from Assyrian and Babylonian Scholars*, volume X in State Archives of Assyria, University Press, Helsinki, 1993, #43, lines 7 to 14, p. 33.
68. ibid., pp. vii-xiv.
69. H. Frankfort, *Kingship and the Gods*, University of Chicago Press, Chicago, 1948, p. 254.
70. Parpola, p. 187.
71. ibid., pp. 215–216.
72. H. Ringgren, *Religions of the Ancient Near East*, Westminster Press, Philadelphia, 1973, p. 85; G. Van Driel, *The Cult of Ashur*, Van Gorcum, Assen, 1969, pp. 145, 149, 162–165.
73. Ringgren, pp. 83, 84; S.H. Hooke, *Babylonian and Assyrian Religion*,

University of Oklahoma, Norman, OK, 1963, pp. 110–111.
74. Frankfort, p. 333.
75. Parpola, #277, p. 217.
76. A.L. Oppenheim, *Letters from Mesopotamia*, University of Chicago Press, Chicago, 1967, pp. 166–167.
77. Frankfort, pp. 259–262.
78. S. Parpola, *Letters from Assyrian Scholars*, Kevalaer-Neukirchen/Vluyn, 1971, pp. 54–60, 87, 215–217; D.J. Wiseman, 'Jonah's Nineveh', *Tyndale Bulletin*, vol. 30 (1979), p. 47.
79. 2 Kings 14:25.
80. D.J. Wiseman, pp. 46–51.
81. A.T. Olmstead, *History of Assyria*, University of Chicago Press, Chicago, 1951, p. 174.
82. A. Millard, *Eponymns of Assyrian Empire 910–612*, Neo-Assyrian Text Corpus Project, Helsinki, 1994, p. 58; A.K. Grayson, *Assyrian Rulers of the Early First Millennium*, State Archives of Assyria Studies II, Toronto, 1991, pp. 200, 231–232, 239.
83. J. Reade & S. Parpola, *Astrological Reports to Assyrian Kings*, Vol. VIII in *State Archives of Assyria*, H. Hunger (ed.), University Press, Helsinki, 1992, #384, pp. 220–221.
84. ibid., pp. 192, 271.
85. D.J. Wiseman, 'Nineveh', *New Bible Dictionary*, J.D. Douglas (ed.), Tyndale Press, Wheaton, IL, 1982, pp. 836–837.
86. A.T. Olmstead, *History of Palestine and Syria*, Baker Book House, Grand Rapids, MI, 1965 reprint, p. 416.
87. F.J. Stephens, 'Sumero-Akkadian hymns and prayers' in Pritchard, p. 391.
88. C. Keller, *Jonas*, Delachaux et Nestle, Neuchatel, 1965, p. 268.
89. Wiseman, p. 51.
90. A. Heidel, *The Gilgamesh Epic*, University of Chicago Press, Chicago, 1949, pp. 135–136.

91. Wiseman, p. 51.
92. William Dever cites an article of this title that was a lead story in the July 2000 issue of the *New York Times*. See W. Dever, *Who Were the Early Israelites and Where Did They Come From?*, Eerdmans, Grand Rapids, MI, 2003, p. 3.
93. This is the general approach of D. Lazare, 'False testament', in *Harpers*, March 2002, pp. 39–47.
94. P. Bahn, *Bluff Your Way Through Archaeology*, Egmont House, London, 1989, p. 5.
95. Lecture by James Hoffmeier at St Olaf's College, Northfield, Minnesota, July 26, 2004.
96. I. Finkelstein & N. Silberman, *The Bible Unearthed*, Simon & Schuster, New York, 2002, pp. 81–82.
97. I. Finkelstein & D. Ussishkin, 'Back to Megiddo', *Biblical Archaeology Review*, January/February 1994, pp. 32–33.
98. Joshua 15:28; 19:2.
99. P. Fabian, 'Where art thou Beersheba?', *Eretz*, 92 (February–March 2004), p. 24.
100. A. Zertal, 'Israel enters Canaan', *Biblical Archaeology Review*, September/October 1991, pp. 330–331.
101. Lecture at University of Oklahoma, October 15, 2004.

Chapter 19 Archaeology and the Reliability of the New Testament

1. John 20:30–31. Bible quotations in this chapter are from the New Revised Standard Version.
2. Matthew 26:3, 57; Luke 3:2; John 11:49; 18:13–14, 24, 28.
3. Acts 4:6.
4. Matthew 2:1.
5. Matthew 27:2; Mark 15:1; Luke 23:1; John 18:29.
6. Romans 16:23.

7. Acts 18:12.
8. Acts 18:11–12.
9. Acts 18:2.
10. Acts 17:6.
11. Acts 19:31.
12. 2 Timothy 4:14.
13. Acts 21:27–40.
14. Luke 3:23; 4:16–30.
15. Matthew 8:14; Mark 2:1.
16. Mark 1:21; 3:1; John 6:59.
17. John 5:1ff.
18. John 9:7.
19. Matthew 4:5; Luke 4:9.
20. Matthew 28:2.
21. Acts 19:8–10.
22. Acts 18:12.
23. Acts 18:14–17.
24. Matthew 8:28–34; Mark 5:1–10; Luke 8:26–39.
25. Luke 18:35; Mark 10:46.
26. For further reading refer to the following:

   J.D. Currid, *Doing Archaeology in the Land of the Bible*, Baker Book House, Grand Rapids, MI, 1999. A brief introduction to the history and methods of archaeological excavation.

   T. Dowley (ed.), *Discovering the Bible*, William B. Eerdmans, Grand Rapids, MI, 1986. This book explores the results of archaeological finds before 1986 which illuminate the backgrounds to the Bible.

   J. Finegan, *The Archaeology of the New Testament*, Vol. 1, The Life of Jesus and the Beginning of the Early Church, Princeton University Press, Princeton, NJ, 1969; Vol. 2, The Mediterranean World of the Early Christian Apostles,

Westview Press, Boulder, 1981. These volumes constitute an authoritative, well-written overview of the archaeological data available at the time of their composition about the life of Jesus and the origin and spread of the church he founded. A more extensive account of the Mediterranean world, including the many centuries preceding and following the founding of Christianity, is Finegan's *Light from the Ancient Past*, Princeton University Press, Princeton, NJ, 1959.

J. McRay, *Archaeology and the New Testament*, Baker Book House, Grand Rapids, MI, 1991. A discussion of the archaeological sites in the Mediterranean world that relate to the New Testament.

—— *Paul: His Life and Teaching*, Baker Book House, Grand Rapids, MI, 2003. The first half of this book covers the journeys of Paul, exploring the archaeological sites of the cities he visited.

—— 'Archaeology', in J.D. Douglas (ed.), *New International Dictionary of the Bible*, Zondervan, Grand Rapids, MI, 1987.

—— 'The Bible and archaeology', in T. Dowley (ed.), *Discovering the Bible*, William B. Eerdmans, Grand Rapids, MI, 1986, pp. 7–26. Dowley's book is a brief well-illustrated introduction to excavation methodology and artefacts.

—— 'The contribution of archeology to the study of the Old Testament', in F. Kearly & E. Myers (eds), *Biblical Interpretation: Principles and Practice*, Baker Book House, Grand Rapids, MI, 1986, pp. 146–157.

—— 'The excavation of low-level settlement sites', in J. Drinkard et al. (eds), *Benchmarks in Time and Culture*, Festschrift for J. Callaway, Scholar's Press, Atlanta, GA, 1988. This article deals with excavating ancient sites that are not built on a 'tell' (that is, a hill created by the successive destruction and rebuilding of ancient cities on the same site).

—— 'Recent biblical archaeology', *Moody Monthly*, November 1984, pp. 102–111. Overview of archaeological discoveries until the 1980s.

E. Meyers (ed.), *Oxford Encyclopedia of Archeology in the Near East*, Oxford University Press, New York, 1997. This multi-volume work is an authoritative compilation of articles on the general field of archaeology.

R. Price, *The Stones Cry Out*, Harvest House Publishers, 1997. A discussion of what archaeology reveals about the truth of the Bible.

W.E. Rast, *Through the Ages in Palestinian Archeology*, Trinity Press International, Philadelphia, 1992. A semi-popular synthesis of the history and archaeology of ancient Palestine, comprehensive, authoritative and clearly written.

H. Shanks (ed.), *Recent Archaeology in the Land of Israel*, Biblical Archaeology Society, Washington, 1984. A collection of top-notch scholars reporting on all the major areas and periods of Israel's culture and history; a summary of important discoveries since the founding of the state of Israel in 1948.

—— *Archaeology and the Bible: The Best of BAR*. This well-illustrated series contains thoughtful articles published in the periodical *Biblical Archaeology Review*, from different periods of history. Volume 2, Archaeology in the World of Herod, Jesus and Paul (Biblical Archaeology Society, 1990) covers material related to the period of the New Testament. Subsequent volumes cover other discrete areas.

—— *Jerusalem: An Archaeological Biography*, Random House, New York, 1995. The story of Jerusalem is told in its archaeological setting with a wonderful collection of photos celebrating the city's tri-millennium as Israel's capital. This book makes Jerusalem as accessible to the layman as to the scholar.

Chapter 20 What about the Bible's Scientific Reliability?
1. Aristotle, *The History of Animals*, Book V, Part 15, trans. D.W. Thompson, available at <http://www.philosophy.ru/library/aristotle/history_anim_en/>.
2. K.R. Miller & J.L. Levine, *Biology*, Chapter 16, 'The origin of life', Prentice Hall, Englewood Cliffs, NJ, 1991, pp. 338–348.
3. Revelation 4:11.
4. Genesis 2 makes this very clear, stating that both humans and animals were formed 'out of the ground'. Interestingly, according to the Bible the plants were not *formed* out of the ground, but God caused them to *grow* out of it.
5. See Genesis 1:22.
6. Probably the best book detailing this myth and its origin is J.B. Russell, *Inventing the Flat Earth: Columbus and Modern Historians*, Praeger Publishers, Westport, CT, 1991, p. 117.

Chapter 21 Amazing Biblical Prophecies That Came to Pass
1. Isaiah 46:10.
2. Jeremiah 50:26, 39–40; 51:26, 29, 37.
3. Daniel 2:37–38.
4. Daniel 2:39.
5. Daniel 7:5.
6. Daniel 8:20.
7. Daniel 2:44–45.
8. Daniel 7:9–14, 26–27.
9. Nahum 2:5–10.
10. Zephaniah 2:13–15.

11. Daniel 8:20.
12. Daniel 8:2–8, 21.
13. Daniel 8:8, 22.
14. Josephus, *Antiquities of the Jews*, book 11, chapter 8, paragraphs 337–338.
15. Tables for the sabbatical years of the Jews can be found in B.Z. Wacholder, 'The calendar of sabbatical cycles during the second temple and the early rabbinic period', *Hebrew Union College Annual* (1973), pp. 153–196.
16. Matthew 24:15–20; Mark 13:14–18; Luke 21:20–21.
17. Josephus, *The Wars of the Jews*, book 2, chapter 19, paragraphs 515–520.
18. For further reading refer to the following:

   For the texts relating to Belshazzar and a synthesis of them, see R.P. Dougherty, *Nabonidus and Belshazzar*, Yale University Press, New Haven, 1929.

   For a very useful collection of cuneiform texts in translation, especially Assyrian sources related to topics in the chapter, see J.B. Pritchard (ed.), *Ancient Near Eastern Texts Relating to the Old Testament*, Princeton University Press, Princeton, NJ, 1955.

   For the Chronicles of the Babylonian kings Nabopolassar and Nebuchadnezzar, see D.J. Wiseman, *Chronicles of Chaldean Kings*, British Museum, London, 1956.

   For a useful review of Assyrian and Babylonian history, see H.W.F. Saggs, *The Greatness That Was Babylon*, Hawthorn Books, New York, 1962.

   For a brief review of the history, archaeology and exploration of Nineveh, see A. Parrot, *Nineveh and the Old Testament*, trans. B.E. Hooke, The Philosophical Library, New York, 1955.

   For a review of the archaeology of an extended list of sites in Israel and the Ancient Near East, see D.W. Thomas (ed.),

Archaeology and Old Testament Study, The Clarendon Press, Oxford, 1967.

For the history of the interpretation of the four world kingdoms in the prophecies of Daniel through the ages and the presence of Alexander the Great in Daniel 8, see L.E. Froom, *Prophetic Faith of Our Fathers*, vols. I–IV, Review and Herald, Washington, DC, 1950–1954.

Chapter 22 The Messianic Prophecies Fulfilled in Jesus

1. A.G. Fruchtenbaum, *Jesus Was a Jew*, Ariel Ministries, Tustin, CA, 2003.
2. The text quoted here is from the American Standard Version (1901). All other Bible quotations in the chapter, unless otherwise indicated, are from the New American Standard Bible.
3. Genesis 18:10, 14.

Chapter 23 The Evidence for Jesus' Resurrection

1. Moroni 10:3–5.
2. The Qur'an, sura 2:23.
3. For a critique of Mormonism, see M. Licona, *Behold, I Stand at the Door and Knock*, TruthQuest Publishers, Virginia Beach, 1998. (For more about this book, go to <http://www.risenjesus.com>). Also see F.J. Beckwith, C. Mosser & P. Owen (eds.), *The New Mormon Challenge*, Zondervan, Grand Rapids, MI, 2002.
4. Although the Gospel of John is filled with such claims, critics hold that it was written nearly seventy years after Jesus' death and reflects a theology of Jesus that had developed by that time. Many of these critics, however, also acknowledge that Jesus made these claims in the earlier gospels of Matthew, Mark and Luke. See Matthew 11:27 where Jesus claims to have a special relationship with the Father; Mark 12:1–12 where Jesus as Son to the

vineyard owner (God) is of a status above the prophets; Mark 13:32 where the Son is above all people and the divine angels.
5. Matthew 11:27; John 3:36.
6. Matthew 12:38–40; 16:1–4; John 2:18–21. Cf. Mark 14:58; Luke 11:29–30.
7. For a good discussion on the trustworthiness of the gospels, see C. Blomberg, *The Historical Reliability of the Gospels*, InterVarsity Press, Downers Grove, IL, 1987.
8. In a chapter of this brevity, it is impossible to provide adequate documentation. The data that will be presented are a partial reflection of a massive study completed recently by Gary R. Habermas, who has specialised in the resurrection of Jesus for more than a quarter of a century. The scope of his study was to discover where scholars from 1975–2003 stand on more than 100 issues related to Jesus' resurrection. The full results of this study have not yet been published. However, some of the results have been published in G.R. Habermas, *The Risen Jesus and Future Hope* (Rowman & Littlefield, Lanham, 2003). In addition, a much more thorough defence of Jesus' resurrection has been presented in G.R. Habermas & M.R. Licona, *The Case for the Resurrection of Jesus* (Kregel, Grand Rapids, MI, 2004).
9. The Greek word translated by the English 'hand' can mean anything from the finger tips to the elbow.
10. 1 Clement 42:3–4.
11. There is some dispute regarding this passage. See further the discussion in the notes in chapter 3 of Habermas & Licona, op. cit.
12. For a detailed analysis of these sources, see Habermas & Licona, chapter 3.
13. 1 Corinthians 15:8–10; Galatians 1:12–16; Philippians 3:6–7.
14. Galatians 1:22–23.

15. Acts 7:52–58; 9:1–2.
16. Mark 6:4; John 7:5.
17. John 19:26–27.
18. Acts 15:12–21; Galatians 1:19. 1 Corinthians 9:5 and Acts 1:14 indicate that more than one of Jesus' brothers became believers.
19. Galatians 2:12–13.
20. *Antiquities of the Jews*, book 20, chapter 9, paragraph 1. James' martyrdom is likewise reported by Hegesippus and Clement of Alexandria.
21. Mark 2:24; 3:1–6.
22. The works of Hegesippus have been lost. Eusebius in *Ecclesiastical History* 2:23 (c. 325 AD) devotes a chapter to the martyrdom of James and says that the most accurate account of the event was reported by Hegesippus whom he proceeds to quote on the matter.
23. 1 Corinthians 15:7.
24. Tacitus speaks of an 'immense multitude' in Rome being convicted as Christians (*Annals* 15:44).
25. Matthew 28:13.
26. *Antiquities* book 4, chapter 8, paragraph 15 (author's translation).
27. J Sotah 19a; Kiddushin 82b; Rosh Hashannah 1:8; Gaius Suetonius, *The Twelve Caesars*, Augustus 44.
28. That critics capitalised on ridiculing the reports since women were witnesses is clear in Origen, *Contra Celsum*, Book 2, chapter 59.
29. R. Brown, 'The resurrection and biblical criticism,' *Commonweal*, November 24, 1967, especially p. 233.
30. For a comprehensive treatment of alternative theories to Jesus' resurrection that offer natural explanations, see Habermas & Licona, Part 3.

31. See ibid., chapter 8.
32. David Hume, *An Inquiry Concerning Human Understanding*, in R. Cohen (ed.), *Essential Works of David Hume*, Bantam, New York, 1965, p. 129.
33. Thomas Paine, *Age of Reason*, Part I, 63–64.
34. It would not do to claim that the religious context was later added. Historical Jesus research, which is generally quite hesitant to grant historicity unless good reasons exist, has verified the high likelihood that Jesus made many of the claims attributed to him and that he was known as a miracle worker prior to his death.

Chapter 24 The Absurdity of Life Without God

1. P. Fernandes, *No Other Gods: A Defense of Biblical Christianity*, Xulon Press, Fairfax, 2002, pp. 81–82. See also P. Fernandes, *The God Who Sits Enthroned: Evidence for God's Existence*, Xulon Press, Fairfax, 2002, pp. 95–101.
2. Augustine, *On Free Will*, 2:1–15.
3. Fernandes, *No Other Gods*, pp. 79–80.
4. Blaise Pascal, *Pensees*, trans. A.J. Krailsheimer, Penguin Books, New York, 1966, p. 165.
5. ibid., p. 66.
6. Fernandes, *The God Who Sits Enthroned*, pp. 19–21.
7. F. Schaeffer, *The Complete Works of Francis A. Schaeffer*, vol. 1, Crossway Books, Westchester, 1982, p. 229; G.R. Lewis, *Testing Christianity's Truth Claims*, University Press of America, Lanham, 1990, pp. 231–236, 253; L.J. Crabb Jr., *Effective Biblical Counseling*, Zondervan, Grand Rapids, MI, 1977, p. 61.
8. Fernandes, *The God Who Sits Enthroned*, pp. 23–27.
9. B. Russell, *Why I Am Not a Christian*, Touchstone Books, New York, 1957, pp. 106–107.

10. N. Geisler & W. Corduan, *Philosophy of Religion*, Baker Book House, Grand Rapids, MI, 1988, p. 69.
11. ibid., pp. 69–72.
12. Friedrich Nietzsche, *The Portable Nietzsche*, W. Kaufmann (ed.), Penguin Books, New York, 1982, p. 441.
13. Geisler & Corduan, pp. 72–73.
14. Sigmund Freud, *The Future of an Illusion*, trans. W.D. Robson-Scott, Doubleday, New York, 1964, p. 20.
15. N. Smart, *The Religious Experience of Mankind*, Charles Scribner's Sons, New York, 1976, p. 40.
16. R.C. Sproul, *If There's a God, Why Are There Atheists?* Tyndale House Publishers, Wheaton, 1978, p. 49.
17. ibid., pp. 12, 58, 101.
18. J.P. Moreland, *Scaling the Secular City*, Baker Book House, Grand Rapids, 1987, p. 229. Cf. P.C. Vitz, *Faith of the Fatherless: The Psychology of Atheism*, Spence Publishing, Dallas, 1999, p. 13.
19. Vitz, pp. 17–57.
20. Moreland, p. 229.
21. Geisler & Corduan, pp. 74–75.
22. Matthew 4:4.
23. John 6:35.
24. Jeremiah 2:13.
25. Augustine, *Confessions*, trans. R.S. Pine-Coffin, Penguin Books, London, 1961, p. 21.
26. Moreland, p. 231.
27. ibid., pp. 232–234.
28. ibid., p. 234.
29. Moreland, pp. 159–183. See also W.L. Craig, *The Son Rises*, Wipf and Stock Publishers, Eugene, OR, 1981; G.R. Habermas, *The Resurrection of Jesus*, University Press of America, Lanham, 1984; S.T. Davis, *Risen Indeed*, William B. Eerdmans, Grand

Rapids, MI, 1993; N. Geisler, *Christian Apologetics*, Baker Book House, Grand Rapids, MI, 1976, pp. 346–351; N.T. Wright, *The Resurrection of the Son of God*, Fortress Press, Minneapolis, 2003.
30. Jeremiah 29:13; Psalm 145:18–19; Acts 17:27.
31. Acts 17:28.